Big Data and Mobility as a Service

Big Data and Mobility as a Service

Edited by

Haoran Zhang
Center for Spatial Information Science, The University of Tokyo, Kashiwa, Chiba, Japan

Xuan Song
Department of Computer Science and Engineering, South University of Science and Technology, Nanshan, China
Center for Spatial Information Science, The University of Tokyo, Kashiwa, Chiba, Japan

Ryosuke Shibasaki
Center for Spatial Information Science, The University of Tokyo, Kashiwa, Chiba, Japan

ELSEVIER

Elsevier
Radarweg 29, PO Box 211, 1000 AE Amsterdam, Netherlands
The Boulevard, Langford Lane, Kidlington, Oxford OX5 1GB, United Kingdom
50 Hampshire Street, 5th Floor, Cambridge, MA 02139, United States

Library of Congress Cataloging-in-Publication Data
A catalog record for this book is available from the Library of Congress

British Library Cataloguing-in-Publication Data
A catalogue record for this book is available from the British Library

ISBN: 978-0-323-90169-7

For information on all Elsevier publications
visit our website at https://www.elsevier.com/books-and-journals

Publisher: Joe Hayton
Editorial Project Manager: Howi M. De Ramos
Production Project Manager: Punithavathy Govindaradjane
Cover Designer: Miles Hitchen

Typeset by STRAIVE, India

Contents

Contributors

Numbers in parenthesis indicate the pages on which the authors' contributions begin.

Shreyas Bharule (xiii), Location Mind Inc., Tokyo, Japan

Jinyu Chen (1, 25), Center for Spatial Information Science, The University of Tokyo, Kashiwa, Chiba, Japan

Rong Guo (265), Key Laboratory of Cold Region Urban and Rural Human Settlement Environment Science and Technology, Ministry of Industry and Information Technology, School of Architecture, Harbin Institute of Technology, Harbin, China

Renhe Jiang (245), Center for Spatial Information Science, The University of Tokyo, Kashiwa, Chiba, Japan

Wenxiao Jiang (113), Center for Spatial Information Science, The University of Tokyo, Kashiwa, Chiba, Japan

Weifeng Li (117), Key Laboratory of Road and Traffic Engineering of the Ministry of Education, Tongji University, Shanghai, China

Wenjing Li (1, 25), Center for Spatial Information Science, The University of Tokyo, Kashiwa, Chiba, Japan

Wen-Long Shang (203), Beijing Key Laboratory of Traffic Engineering, College of Metropolitan Transportation, Beijing University of Technology, Beijing, China

Fengjing Shao (113), Institute of Smart City and Big Data Technology, Qingdao, China

Ryosuke Shibasaki (xiii, 1, 25, 77, 245), Center for Spatial Information Science, The University of Tokyo, Kashiwa, Chiba, Japan

Xiaoya Song (265), Key Laboratory of Cold Region Urban and Rural Human Settlement Environment Science and Technology, Ministry of Industry and Information Technology, School of Architecture, Harbin Institute of Technology, Harbin, China

Xuan Song (229), Department of Computer Science and Engineering, South University of Science and Technology, Nanshan, China; Center for Spatial Information Science, The University of Tokyo, Kashiwa, Chiba, Japan

Yi Sui (113, 203), College of Computer Science and Technology, Qingdao University, Qingdao, China; Center for Spatial Information Science, The University of Tokyo, Kashiwa, Chiba, Japan; Institute of Smart City and Big Data Technology, Qingdao, China

Rencheng Sun (113), College of Computer Science and Technology, Qingdao University, Qingdao, China; Center for Spatial Information Science, The University of Tokyo, Kashiwa, Chiba, Japan; Institute of Smart City and Big Data Technology, Qingdao, China

Haoning Xi (143), School of Civil and Environmental Engineering, University of New South Wales, Sydney; Data61, CSIRO, Canberra, Australia

Hongbin Xie (229), Department of Computer Science and Engineering, South University of Science and Technology, Nanshan, China

Chuang Yang (245), Center for Spatial Information Science, The University of Tokyo, Kashiwa, Chiba, Japan

Dongyuan Yang (177), Key Laboratory of Road and Traffic Engineering of the Ministry of Education, Tongji University, Shanghai, China

Yuhao Yao (77), Center for Spatial Information Science, The University of Tokyo, Kashiwa, Chiba, Japan

Qing Yu (177), Key Laboratory of Road and Traffic Engineering of the Ministry of Education, Tongji University, Shanghai, China

Haoran Zhang (xiii, 1, 25, 77, 113, 203, 229, 265), Center for Spatial Information Science, The University of Tokyo, Kashiwa, Chiba, Japan

Introduction

Shreyas Bharule[a], Haoran Zhang[b], and Ryosuke Shibasaki[b]
[a]LocationMind Inc., Tokyo, Japan, [b]Center for Spatial Information Science, The University of Tokyo, Kashiwa, Chiba, Japan

1 Background

Data was originally manual. Transportation data for the longest time were collected through clickers at junctions. However, in the last decade, we have seen a rapid transformation of the mobility landscape. The convergence of technologies such as smartphones, sensors, the Internet of Things, and apps within smartphones have revolutionized how we look at every form of transportation, particularly to interpret and predict the movement of goods and people in real time. In recent years, mobility services such as Uber for taxi service, DHL for logistics, and others have used various technologies that collectively impact how we move in long- and short-haul transportation.

Also, the number of sensors also used to collect data, both stationary and mobile, has increased multifold. They are generating and collecting relational data to examine issues and resolve challenges. Such a convergence of technology and the resultant emerging services are changing how cities respond to newer types of goods and people movements through Mobility as a Service(MaaS). The stated advantages of using these services are convenience, real-time information, and connectivity to go wherever the traveler wants to go. Though such possibilities are marketed as limitless, there are technical limitations.

Big data-driven MaaS development is an emerging area both in academic and industrial aspects. Though several research studies and technical reports are available, a clear link to understand big data in MaaS appears vague and fragmented. In this book, we aim to fill this gap by systematically summarizing the knowledge in this field—the chapters in the book outline the areas to streamline the understanding of big data analytics for MaaS.

2 Big data: Definition, history, today

Big data is made up of larger, complex datasets that originate from increasingly new sources of data. These datasets are so massive that the abilities of traditional data processing software are insufficient to manage them. Nevertheless, these massive volumes of data can be used to address business problems that would not have been tackled before.

A simple definition of big data organized around big data's three Vs could be large *volumes* of data that contain immense *variety* and are generated at increasing *velocity*. Recently, six other Vs and one C have been added to define the truthfulness and meaningfulness of data as *veracity*, infrequency as *validity*, extracting *value* from the collected data, replicability in the form of *visualizations*, processability in a *virtual* cloud platform, data *variability*, and *complexity*, purely in computation form.

Though the concept of big data is still emerging, the origins of large datasets go back to the 1970s when the world of organized data in the first data centers was emerging. In addition, the development of the relational database further consolidated the concept. Around 2005, researchers identified just how much data users generated through SNS, video streaming services, and other online services. In parallel to these developments, data organization and storage systems were developed as well. Hadoop, an open-source framework created specifically to store and analyze large datasets, was developed around the same year. This development of open-source frameworks such as Hadoop, Spark, and others was essential for the growth of big data because they make data storage cheaper and enhance the ease of engaging with big datasets.

Nevertheless, the volume of big data has been at an all-time high, as is the dependency on it for decision making. Users of the system are still generating significant amounts of data. However, it is not just humans that are engaging with systems. Evolution in smartphone and sensor technology now allows devices to communicate through the Internet of Things (IoT) network. Besides, advancements in devices and communication network systems have eased data gathering for various indicators and drawing performance insights. Moreover, the emergence of artificial intelligence and machine learning has churned out diversity and cross-operability across various platforms.

Furthermore, cloud computing has expanded big data possibilities in truly elastic scalability. Such scalability is of critical importance in analyzing and supporting demands in services. We focus on this aspect of big data to solve challenges that hinder future development, particularly in understanding the emerging area of *MaaS*.

3 MaaS: Definition, history, today

MaaS is a new concept in the transport sector; it provides a new way of thinking about how the delivery and consumption of transport (or mobility) are managed. In order to have a common starting point within the book project group, it was important to have a common definition for MaaS. Although MaaS is regarded as a major paradigm shift in transportation toward more environmentally friendly and efficiently used transport modes, describing all important MaaS facets is relevant for determining the main research scope.

Though there is relatively little literature on the planning and concepts of MaaS systems, MaaS as a concept was first described in 1996 as an "intelligent information assistant" for travel needs. We describe MaaS as a shift away from privately owned modes of transportation and toward mobility solutions that are consumed as a service.

The one significant application of MaaS is the emergence of shared transportation systems (STS). STSs are enabled by combining transportation services from person-to-person transportation providers through a unified gateway that creates and manages the trip. The key concept behind STS is to offer travelers mobility solutions based on their travel needs. For local authorities and policymakers, the potential to use STS to source data on travel movements could open the door to new transport management tools, planning for sustainable growth, and more efficient use of capacity.

UK's Transport Systems Catapult predicted that by the end of 2020, more than 50 billion connected devices would collect more than 2.3 ZB of data globally. In the coming decade, these numbers are expected to increase each year. STS leverages this opportunity and is an example of a business model supported by the growth in smartphone use. STS is a digital, data-driven service that uses several technological capabilities associated with intelligent mobility and innovation. The system relies on building an ecosystem of stakeholders that agrees to manage the supply and demand of the services that travelers demand.

4 Big data X MaaS

Big data-driven MaaS development is an emerging topic both in academic and industrial aspects. Therefore, currently all studies about big data in MaaS are fragmented. No work has summarized the systemic knowledge in this field. This book is designed to fill this gap.

The book organizes cornerstone technologies in the sphere of big data and MaaS, focusing on introducing how to screen and process the potential value from the "deluge" of unverified, noisy, and sometimes incomplete information for MaaS development. Furthermore, chapters are written in the form of summaries of frontier technologies applicable in MaaS, including MaaS system development and APPs, spatiotemporal data preprocessing technologies, travel similarity estimation and clustering, data fusion technologies for MaaS, data-driven optimization technologies for MaaS, data-driven estimation for urban travel shareability, MaaS system data mining technologies, IoT technologies for MaaS, MaaS system visualization, and MaaS for urban sustainable development (Fig. 1).

Within the ambit of MaaS, the book offers demonstrated answers to several questions, such as:

a) How to efficiently extract and effectively utilize key feature information of high-dimensional city people flow data?

b) How to efficiently express the massive and large-scale people movement and predict mobility sharing potential at an urban scale?

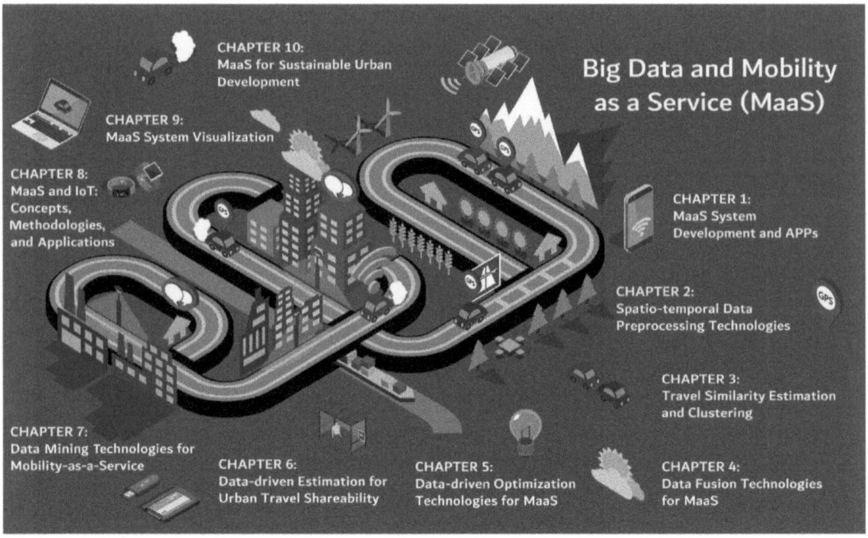

FIG. 1 Chapter overviews.

c) What solutions can facilitate ethical, secure, and controlled mobility sharing in different transportation types?
d) How to build shared transportation services beneficial to citizens, businesses, and society safely and responsibly?
e) How to apply novel technologies to shared transportation management and MaaS development?

Furthermore, chapters in the book are organized to help audiences understand:

a) A systematic outline to define and reinvent data-driven mobility models by studying urban dynamics, urban mobility, transportation behavior, and sharing potential.
b) Within a global urban mobility framework, how we can characterize the nature of data-enabled MaaS.
c) Understanding the existing positive and successful MaaS can be identified in the studied domain and it can be determined how best to apply it for practical success.

Collectively, the knowledge in this book is of immense significance for stakeholders in MaaS and those planning to enter the industry, such as researchers, engineers, operators, company administrators, and policymakers in related fields, to comprehensively understand current technology infrastructure knowledge structures and limitations.

5 Summary

Each chapter deals with one component of big data concerning MaaS. However, the overall takeaways for the readers of this book are several. First, the book outlines methods to assess the urban mobility characteristics as well as the potential to adopt shared transportation and match their performance as conditions and constraints in a transport system. The underlying assumption is that transport systems are undergoing automation and are highly dependent on software, navigation systems, and connectivity. Second, the book introduces how to design MaaS platforms for STS that adapt to the evolving mobility environment, new types of transportation, and users based on integrated solutions that utilize the sensing and communication capabilities to tackle the major challenges the MaaS industry faces. Finally, the book aims to demonstrate:

a) In addition to policy and market factors, the most significant development bottleneck is technical limitation. This book systematically summarizes the current fundamental technologies of MaaS to help promote the development of the MaaS industry. In particular, crowd data's multimode urban and inflow-outflow is of immense value for urban traffic administration and shared transportation management to monitor more diverse aspects of mobility and associated information for policy and market formulation.
b) The introduction of technologies from a technical standpoint can help people understand the potential of MaaS in the future city more intuitively than the conceptual overview. Therefore, in the context of ubiquitous data analytics-enabled MaaS, when there are effective enabling structures and processes for stakeholder participation and engagement, the governance and decision-making process of shared transportation services will be more effective and sustainable.
c) Artificial intelligence (AI), machine learning, and big data are very hot and popular but lesser-understood topics. Ubiquitous data-based services based on more disruptive technologies such as AI and big data are more likely to succeed and be sustained than shared transportation services based on frugal technologies and inflexible methods. Thus, this book is written to help the audience understand MaaS technologies from the perspectives of AI researchers and data scientists.

We organize the chapters in the book to create an easy-to-interpret handbook that introduces the fundamental technologies for utilizing human mobility data to develop the MaaS system and contribute to the literature and policy dialogue around big data analytics to provide MaaS.

Chapter 1

MaaS system development and APPs

Wenjing Li, Ryosuke Shibasaki, Haoran Zhang, and Jinyu Chen
Center for Spatial Information Science, The University of Tokyo, Kashiwa, Chiba, Japan

1 The development history of MaaS

1.1 The conception

Although the term "Mobility as a Service (MaaS)" has only attracted widespread attention in recent years, the concept of MaaS had appeared a long time ago.

A similar concept of MaaS dates back to 1996, the ENTER Conference in Innsbruck, Austria [1], with the idea of integrating travel services into an intelligent platform. At that time, Nico Tschanz and Hans-Dieter Zimmermannen visioned an "intelligent information assistant" platform on which people could search and book trip ways and do many other travel-related things like booking hotels and buying tickets. Considering that the World Wide Web just developed in 1990 and Microsoft just made its first web browser, Internet Explorer, to access the internet in 1995, one year before 1996, Nico Tschanz and Hans-Dieter Zimmermann's idea was way ahead of their time. Their idea would eventually formulate into the MaaS system that we are moving toward implementing today.

1.2 The early application

In the first 15 years of the 2000s, some discussions about MaaS have emerged and some forward-looking companies have already started their exploration in the MaaS business.

In 2006, the startup BlaBlaCar was set up in France. It developed a long-distance ride-sharing website platform that connects drivers and passengers who will share the cost of long car journeys. In 2008, the second version of BlaBlaCar website was launched. It included a community module allowing users to show their profiles and biographies, as well as recommend their preferences. In 2012, an online reservation service was added. Although the word "MaaS"

Big Data and Mobility as a Service. https://doi.org/10.1016/B978-0-323-90169-7.00002-6

had not appeared yet, BlaBlaCar has shown the MaaS characteristic that integrated the mobility services.

Another early explorer is Zimride (predecessor of Lyft), founded in 2007. It started with the idea that matches drivers and passengers who want to share rides through social network services to eliminate the anxiety that they do not know each other before. The early users of Zimride were mainly university students and company workers. Zimride only connected people who go to the same campus or work at the same company.

Uber, now one of the largest firms of MaaS economy, is founded in 2009. It is developed for reducing transportation fees by sharing rides. The beta version was launched in 2010. Originally, the application only allowed users to call for luxury vehicles that the price was higher than that of a taxi. Until 2012, Uber allowed users to request regular taxis and personal vehicles with background checks.

In June 2012, Agrion (the "Energy & Sustainable Development" brand of the EBG, a leading think-tank on digital innovation in France) sponsored a half-day conference in San Francisco, US named "E-Mobility as a Service." The discussion topics included how to link the private automobile and the publicly funded and operated transit system, the potential of the integration between the smartphone and shared vehicles, the impact of shared-use vehicles, and so forth. The "E-Mobility as a Service," which is very similar to the concept of MaaS, highlights a digitally connected seamless multi-modal transportation network. Through real-time connectivity from a smartphone, mobility services could be ubiquitous and seamless. At that year that the smartphones began to spread and replace feature phones, this idea gave people a huge imagination of the commercial prospects.

From 2013 to 2014, an integrated mobility service experiment named UbiGo launched in a commercial pilot of Gothenburg, Sweden. It is the first-ever development of what today is called MaaS. This project aimed to explore how new business models can reduce the use of private cars and how the seamlessness and multimodality use of information technology can promote sustainable travel. The UbiGo achieved this by combining public transport, bike-sharing, car-sharing, car rental service, taxi all into one application and one service on a smartphone. Under the prepaid monthly subscription, the users can access all these travel services. The public trial operated last for 6 months and involved 195 users. The service was welcomed by citizens. However, it was discontinued due to the lack of the government level support for third-party on-selling of public transport tickets.

Though a little tortuous, these early business explorations and applications helped to develop the MaaS concept into a feasible future.

1.3 MaaS alliance

In Intelligent Transportation System (ITS) Congress held in 2014 in Helsinki, Finland, the word "Mobility as a Service" was formally proposed. In 2015, ITS

Congress in Bordeaux, France, "Mobility as a Service" became the popular topic of discussion. At that congress, more than 20 European organizations jointly formed the world's first MaaS Alliance. Its members include transportation service providers, public transportation operators, MaaS service operators, IT system providers, user groups, government organizations, etc. The foundation of MaaS Alliance is an incredibly significant event that the MaaS concept is widely concerned. Among people, skepticisms also began to arise, raising questions such as "how could it work?" and whether or not it was a "complete revolution or simply some changes to integrate into the current travel business". In 2016, ITS Congress, there were six subforums dedicated to discussing the technology, business model, and engineering application issues faced by MaaS.

1.4 Development

Since 2016, explosive developments of smartphone technology have lifted MaaS to a higher integration level.

Whim was released in 2016, marking the beginning of a prototype system with a higher integration level. This platform integrated public transit, bikes, e-scooters, ferry, taxis, and affordable rental cars in one app. The users could search and choose the preferred mode of transport on their smartphone. Trips could be paid on the app per time or by seasonal ticket subscription.

In 2017, DIDI Chuxing, the MaaS giant in China, expanded their business of taxi-hailing and ride-sharing to bike-sharing services and e-bike-sharing services. Car maintenance, refueling, and recharging services were also provided to driver-partners and independent vehicle owners on the platform.

2018 ITS Congress in Copenhagen, Denmark provided participants with the use of an experimental MaaS prototype system MinRejseplan. It was a similar project but built on a case from Denmark. It offered users various travel options to get from A to B—train, metro ferries, sharing services, and even a self-driving shuttle bus. Electronic tickets with all public modes in including special discounts for sharing services and taxis would show on the app. In addition to the regular public transport services all over Denmark, Rejseplanen also integrated several demand-responsive and shared mobility services such as Flextrafik, taxis as well as carpooling.

In September 2019, Berlin's public transport authority Berliner Verkehrsbetriebe (BVG) launched a city-owned MaaS project named "Jelbi", together with a Lithuanian mobility startup Traffic. Jelbi integrated bus, metro, bike, motor scooter, ride-sharing, taxi in one app. All the available ways to get to the destination were showed and compared clearly by duration and price so that users could book just what they need for the occasion, weather conditions, price, or the mood.

1.5 Revolution and innovation

Meantime, with the development of MaaS and the explosion in the amount of information, MaaS is more closely integrated with big data. Developments such as IoT Technologies, artificial intelligence also promote this trend. Massive data was collected from platforms such as smartphones, desktops, and other digital devices to essentially reconstruct MaaS services. Many data-driven high-tech solutions are coming forth in MaaS systems.

From 2016, Uber started to establish Uber AI Labs, a research arm dedicated to cutting-edge research of MaaS business in artificial intelligence and machine learning. Based on massive ride records, Uber uses AI for matching drivers and riders, route optimization, risk assessment, safety processes, and so forth.

In 2017, Moovit launched a suite of MaaS solutions, powered by AI and big data, covering simulation, real-time operations, and optimization. It shows positive significance in growing ridership, increasing operational efficiency, and reducing urban congestion.

In 2017, Didi Chuxing officially launched AI Labs to expand its business on AI-driven innovation. In 2018， Didi Chuxing launched an integrated solution for smart city traffic management named "DiDi Smart Transportation Brain". Leveraging cloud computing and AI-based technologies, it combines video cameras, GPS data, and other sensor data from Didi's cars with data from government and other partners. This smart brain can help improve the efficiency of driving schedules. For example, utilizing real-time mobility data and predictive algorithms, it anticipates traffic jams and routes its own riders and drivers around existing traffic to alleviate traffic congestion. Ride-hailing supply and demand in different regions, weather dates and time can also be forecasted. The smart brain can also promote a range of transportation infrastructure improvements. For example, by analyzing data from billions of rides of Didi's drivers, the system could predict when and where the traffic jam is likely to form and adjust traffic lights signal timing. Congestion during rush hours dropped after applying this solution.

In 2018, Hacon of Siemens demonstrated the MyMobility plugin for HAFAS application. This plugin enables learning the users' behavior and habits from historical records data by machine learning techniques to deliver personalized and proactive mobility recommendations.

In 2020, an intelligent backend system powered by PTV Visum was integrated into the operations of the MaaS service of the traffic control center in Hamburg, Germany. This software is now used for simulating and calculating the impact of traffic disruptions. The frontend system of ROADS, a software for the **coordination** of construction measures and analysis of changes in traffic flow, will also be integrated into the system. This system is now used for congestion forecasting for its long-term and short-term traffic planning based on big data.

Also in 2020, Toyota Motor Corporation announced to set up a new business alliance with NTT DATA, a leading IT services provider in Japan. This alliance aims to accelerate Toyota's MaaS initiatives by operating its cloud data center for big data collected from connected cars.

2 The category of MaaS system

In different studies, MaaS system can be categorized according to different ways. For example, according to **(1) transportation type**, MaaS can be categorized into bike-hailing, ride-sharing, bus-sharing, train services, micromobility, etc.; according to **(2) services type**, MaaS can be categorized into journey management, journey planning, booking, navigation, flexible transactions, personalized customization, etc.; according to **(3) solution type**, MaaS can be categorized into technology platforms, navigation solutions, payment engines, ticketing solutions, telecom connectivity providers, etc.; according to **(4) transportation type**, MaaS can be categorized into public or private; according to **(5) business model**, MaaS can be categorized into business-to-business, business-to-government, business-to-consumer, and peer-to-peer; according to **(6) operating system**, MaaS can be categorized into Android, iOS, and PC.

Sochor, Arby [2] characterized MaaS system from Levels 0 to 4 as characterized by different levels of services integration, that is no integration, integration of information, integration of booking and payment, integration of the service offer, and integration of societal goals (Fig. 1). Service integration is the precondition of MaaS. Based on this foundation, the integration of information, data, platform, etc. are realized. The category way of Sochor, Arby [2] is representative and has been widely recognized by MaaS-related research.

FIG. 1 Summary of MaaS integration level.

2.1 Level 0: No integration

At this level, the services provider provides separate services of transportation mode. The services system is single. It can provide mature service for certain travel modes but lacks integration with other travel modes. There is no data interaction between different transportation modes. Hellobike (bike-sharing), Hertz (car rental), Nippon Rent-a-Car (car rental), and Sunfleet (car-sharing) are examples of services providers of this level.

2.2 Level 1: Information integration

The MaaS service of this level has a centralized information platform that can provide information comparison and travel recommendations of multiple transportation modes. It helps users to find the best trip. It typically focuses on a single trip rather than single customers. Representative services include Google and AutoNavi. The standardized information is typically for free and every user can use its service. Some information such as public transport, catering information, and reservation information would be integrated into the services platform of this level. However, the platform operator only acts as an information collector and a connector between the user and the chosen provider, rather than the service directly operator. The application of this level does not reflect the core concept of MaaS and has no direct effects in pushing travelers to abandon private cars to choose public transport.

2.3 Level 2: Integration of booking and payment

Application of this level also focuses on a single trip. Based on providing travel planning services, they provide travelers with search, reservation, and payment services of public transit, taxi, bike, or any other type of travel mode. Representative services include Moovit, HANNOVER mobile, and Xiecheng. The additional value of Level 2 is that it enables services from multiple operators easier to be accessed to. Users can find, compare, book, and pay with the same app.

The application of this level is mainly responsible for ticketing, booking, and purchase but they typically are not directly responsible for the passenger and cargo transportation process. The application of this level does not reflect the core concept of MaaS too. They just provide the integration of payment but have no direct effects in pushing travelers to abandon private cars to choose public transport.

2.4 Level 3: Integration of the service offering

Representative services of this level include UbiGo and Whim. Application of this level focuses on the user's total mobility needs rather than only single trip from one place to another place. The MaaS operator takes two-way responsibility to both end-user and supplier. The MaaS operator buy services from different transport services providers, then reorganize, integrate, and sale them to the end-user.

There is two-side data integration at this level. On the one hand, as the MaaS operator takes responsibility for the service delivered to its customers, the information of the user side, such as mobility needs, mobility preferences, and mobility patterns, should be carefully integrated. It is not only for a single user but probably for the whole family or the whole community. On the other hand, MaaS operators should build a good connection with the transport services supplier. For example, for taxi-hailing, the MaaS operation should analyze the booking data to help better vehicle scheduling.

However, this does not mean that the data integration level is higher than level 2. Because the services of this level are typically local-based, MaaS operator needs to find the best supplier of each mode to develop the service with. Due to fewer suppliers and less interaction, the complexity of the technical integration can be lower for a Level 3 service than for a Level 2 service. Besides, Level 3 service usually needs more cooperation with the regional or local public transport authorities to find politically acceptable contract models.

2.5 Level 4: Integration of societal goals

The MaaS service of this level integrates local, regional, national policies and goals. For example, public transportation is integrated into the MaaS customizable packages with commercial services. The MaaS services operator can provide incentives for desired travel mode to influence the travel behavior of the user. Besides, traditional public transport service is a one-size-fits-all service with non-flexible business models, while an attractive MaaS service needs to be unified and flexible.

Integration of societal goals requires the high integration of data. It is not only integrating data from the end-user side or transport services supplier side but also from public infrastructure, public space, and public transportation. The MaaS service of this level can help reduced private car ownership, improve the transportation efficiency and promote a more sustainable and livable city.

3 Study case

The development of MaaS has brought new opportunities and challenges to transportation in various countries. At present, various countries and regions are actively involving the implementation and application of MaaS. Dozens of App services have been launched worldwide. Jittrapirom, Caiati [3] summarized the projects that have been run or piloted in the first few years of MaaS development. On their basis, we have updated these projects and added new application results in recent years. Table 1 shows the summary of some of the integrated MaaS systems around the world.

TABLE 1 Summary of some of the integrated MaaS systems around the world (till 2021 May).

Name	Area	Launch year	Current status	Platform	Transport modes													Available function							Tariff option				Integration level
					Public transit	Bike-sharing[a]	Bike rental[b]	Car rental[d]	Car-sharing[c]	Car hailing[e]/ride sharing[f]	Taxi	Train	Scooter	Ferry	Mopeds/motorcycle	Plane	Parking	Trip planning	Booking	Navigation	Payment	Real-time information	Refueling/battery charging	Freight	Pay-per-use	Weekly/Monthly/seasonal subscribe	Package according to number of tickets	Unlimited	
Uber	over 900 metropolitan areas worldwide	2009	Operational	App / web	●	●			●	●	●		●					●	●	●	●	●		●	●				2
Sked Go/Trip Go	Australia, US	2009	Operational	App	●	●			●	●	●	●	●		●		●	●	●	●	●	●			●				2
Citymapper	More than 70 metropolitan areas worldwide	2011	Operational	App/card	●	●			●	●	●	●	●					●	●	●	●				●	●		●	3
Cabify	Spain, Portugal and Latin America	2011	Operational	App		●				●	●							●		●	●				●				1
Moovit (formerly Tranzmate)	across 112 countries	2012	Operational	App / web	●	●				●		●	●	●				●		●	●	●			●				2
Didi Chuxing	China	2012	Operational	App / web		●				●	●							●	●	●	●	●	●		●				2
Lyft	US	2012	Operational	App	●	●				●			●					●		●	●	●			●				0
Grab	Southeast Asia	2012	Operational	App						●	●							●		●	●			●	●				0
UbiGo	Sweden	2013	Cease operations	App	●		●	●	●		●								●		●					●			3
Chariot	US	2014	Cease operations	App	●				●									●	●		●				●	●		●	2
Beeline	Singapore, Westchester	2015	cease operations	App	●						●							●		●	●				●				1
Whim	Helsinki, Turku, Veinna, Antwemp and West Miland	2016	Operational	App	●	●				●	●		●					●	●		●				●	●	●	●	3

REACH NOW (formerly Moovel)	Germany	2016	Operational	App	●	●			●		●		●					●	●	●	●	●			●				2
Cooltra	Barcelona, Madrid, Valencia, Lisbon, Rome and Milan	2016	Operational	App		●							●		●				●		●				●	●			1
My Cicero	Italy	2016	Operational	App	●							●					●	●			●	●			●	●			2
Lime	Asia, Europe, Middle East, North America, Oceania, South America	2017	Operational	App		●			●				●		●				●		●				●				1
Qixxit	German	2017	Cease operations	App	●	●			●		●	●		●		●		●	●	●		●			●				2
Wienmobil (formerly SMILE)	Vienna	2017	Operational	App	●	●			●		●						●	●	●	●	●				●	●	●		2
Beamrz	Nederland	2017	Operational	App	●	●			●	●	●	●	●					●	●	●	●				●				2
Mobility Mixx	Nederland	2018	Operational	App/card	●		●	●			●				●		●	●	●		●		●		●				2
Mobilleo	UK	2018	Operational	App						●	●	●				●		●	●	●	●				●				2
My route	Japan	2018	Operational	App	●	●		●	●		●	●					●	●	●	●	●	●			●	●	●	●	3
Rejseplanen (formerly MinRejseplan)	Copenhagen, Danmark	2018	Operational	App	●	●			●		●	●		●				●	●	●	●	●			●	●			3
Jelbi	Berlin	2019	Operational	App	●	●		●	●	●	●		●					●	●	●	●	●	●		●	●			3

Note:

a. Bike-sharing refers to the service that bicycles are available for shared use to individuals on a short-term basis. Users can pick up a bike from one dock and return it to another dock belonging to the same system. It is typically for short-term use, from a few minutes to a few hours.

b. Bike rental refers to the service that people rental the bike for long-term use. Bike can only be accessed to the corresponding user during the rental period, from 1 day to several months.

c. Car-sharing refers to the service that cars are available for shared use to individuals on a short-term basis. Users can pick up a car from one garage and return it to another garage belonging to the same system. It is typically for short-term use, from a few minutes to a few hours.

d. Car rental refers to the service that people rental the car for long-term use. The car can only be accessed by the corresponding user during the rental period, from 1 day to several months.

e. Car-hailing refers to the service that people can call for a car to pick them up and take them directly to their destinations.

f. Ride-sharing refers to the service that enables passengers who have a similar route to share the same car.

In the following part, we will introduce some representative applications: UbiGo, whim, Moovit, and Whim. UbiGo and Whim are considered to be the app closest to the ideal form of highly integrated MaaS. Similar apps include Jelbi, MyRoute, and MinRejseplan. Moovit is representative of navigation services and the promotion of public transport use. Simialr app is Beeline. Uber is representative of the sharing economy of MaaS and big data applications. Similar apps are Didi, Lyft Kabbee, and Grab.

3.1 UbiGo

3.1.1 *Introduction*

From the development time, UbiGo was the first-ever development of what today is called MaaS.

UbiGo started with the Go:Smart project in 2013 in Gothenburg, Sweden. Go:Smart project was an experiment to explore and evaluate if the new mobility service concept for more sustainable urban transport. Transport service providers, service users, ICT service developers, research institutes, society organizations, and the Gothenburg government were widely involved in this project. UbiGo is one of the subproject of Go:Smart. It offered customized transport services with an app on the smartphone to fit the individual's travel needs. UbiGo united already existing transport services, including public transport, taxi, car-sharing, bike-sharing as well as rental cars, and offered them in a package to customers through the flexible subscription service. It bridged the gap between private and public transit by taking on the role of a commercial actor.

The intended audience for the UbiGo service was urban households. Customers, in the form of households (comprised of any number of individuals including both adults and children but typically a family), paid a monthly subscription according to their transport needs. Different travel services included different personalized combinations, different access ranges of travel services, and different amounts of credit. The subscription could be modified monthly.

The beta version of UbiGo was tested in Gothenburg, lasting for 6 months from 2013 to 2014. The service was well received by citizens. However, it was discontinued due to the lack of support at the government level for third-party on-selling of public transport tickets. In 2019, the formal version of UbiGo launch in Stockholm. In 2021, UbiGo had to cease operations due to the loss caused by the Covid-19 epidemic.

3.1.2 *Services*

The formal version of UbiGo (from 2019) was based on a monthly subscription and classified by types of transportation modes. Each subscription was only corresponding to one type of transportation mode. The services include:

(1) Station-based car-sharing

Paying 330 SEK per month, the users could select cars that are available in the nearby area, and book via the app. Then they could pick up the car at the designated location with phones, get in the car and drive to wherever they want. After driving, they needed to leave the car where they picked it up and lock it via the app.

(2) Bike subscribing

Users could subscribe to a bike with 299 per month through the app. Repair fees and insurance fees were included. They could choose delivery or pick-up to get the bike. Then they had the access to use this bike all the subscription period.

(3) Public transportation

UbiGo offered a flexible subscription of day tickets at a competitive price (525 SEK per month) compared to using the pay-as-you-go service or the monthly pass. With UbiGo tickets, users could travel by public transport as many times as they want during the day.

(4) Car rental

Offered weekly rentals at competitive prices. User could book via the app and choose the station that suits them best for pick-up and drop-off.

(5) Taxi

Users could book a taxi via the app. Prices were predetermined and could be paid via the app.

3.1.3 Characteristics

(1) Integration

The UbiGo service offered its users one-stop access to a wide range of travel services through one application on smartphones. Under monthly subscription, Users could customize those travel packages by only downloading one app. Users could change the subscription according to their needs at any time. The account was shared by all members of a household. Every end-user could search and book routes via the app.

(2) Pricing

The fee of using a vehicle through UbiGo was cheaper than the fee that the user directly pays for. It was one of the most important ways to increase the stickiness of the application and keep users around, especially for people who travel a lot every day. UbiGo bought the transport services from transport service providers, restructured them, and sold them at a lower price than the market's price to the users. The income of UbiGo came from the difference between the purchase price and the package sale price. For each type of travel mode, UbiGo only had partnerships with one transport service provider. For example, the public transport service is cooperated with SL (a bus company), the car rental service cooperates with Hertz, and taxi services cooperate with Cab online. By this method, UbiGo could provide the best transport services to the users.

(3) Sustainability

From the very beginning, UbiGo was committed to promoting sustainable travel in the city. It provided a green travel reward system that compares the emissions generated by users using sustainable transportation and using private cars. UbiGo rewarded sustainable transport choices by giving points for each transport km and the points could be used to pay for other goods and services. This system reflects the advantages of MaaS services in advocating sustainable travel.

(4) Big data technologies

UbiGo conducted a wide range of surveys before, during, and after the development process to improve user experience and also optimize its services. For example, from 2013 to 2014, the development group conducted questionnaires and personal interviews around the UbiGo customers [4]. The participants were asked to fill out three web-based questionnaires; ex-ante, in-itinere, and ex-post the six-month trial period. The ex-ante questionnaire is concerned with expectations, motivations to the services, and self-reported travel behavior. The in-itinere questionnaire included the measurement of participants' current behavior and their experience of all aspects of the service. The ex-ante questionnaire concerned the changes in participants' travel behavior, attitudes, and willingness to continue using the service. 151 adults finally completed the completed all three questionnaires. Those data help the improvement of the released version of UbiGo in 2019. As a connected and multi-modal passenger transport network, UbiGo of course involved in processing a large amount of data for users' mobility. However, to the best of our knowledge, we could not find any materials, public or private, about their big data processing techniques in their MaaS algorithm.

3.2 Whim

3.2.1 Introduction

Whim is a service of MaaS Global Ltd., a company based in Helsinki, Finland. Whim first debuted in Helsinki in 2016, followed by a full commercial launch in November 2017. In late 2018, the service had over 70,000 registered users. Now it has expanded to Turku, Veinna, Antwernp and West Miland.

Whim combines transport options from different providers, including public transit, taxis, car rental, E-scooters, city bikes, and shared bikes, into one application. Users can access all modes of transport at every moment. Whim handles route planning, booking, and payment for the user and the service itself is free. Users only need to pay for their travels. Beyond having all transport at the user's disposal on-demand, Whim bundles all transport into different pricing plans to make the users' travel more convenient. Users can purchase individual tickets, tickets for the entire day, monthly, season, or select from some of the series tickets that come with various Whim benefits.

3.2.2 Services

There are minor differences in available services and prices in different cities. The services are classified by types of subscription. Taking Helsinki as an example, Whim offers five chooses of service (till May 2021):

(1) Pay-as-you-go

This option is for those who want to try Whim first before committing to a seasonal ticket with a recurring fee or those that simply don't travel that much. This option does not have a recurring fee. Users can pay for every type of transportation per ride on standard pricing without commitment or surcharges.

(2) Whim unlimited

For the price of owning a car, users can get unlimited access to public transport, a taxi, or a car according to their needs. This subscription renews automatically every 30 days according to the renewal cycle. The available services include Unlimited single public transport tickets, maximum 80 rides per month free taxi rides within a 5-km radius of users' location, unlimited amount of 30-min bike rides, unlimited access to basic rental vehicles for up to 30 days at a time, access to book and pay TIER and VOI e-scooters, one free ride within 30 min every month with JURO bikes and one free day pass to a coworking space.

(3) 10-ticket

The 10-ticket is a package for people who do not travel that often or do not want to purchase a season ticket. The package includes 10 adult public transport single tickets, a fixed price which is lower than the standard price for taxi trips for 3 km or 10 min rides, discount for car rental, access to book and pay TIER and VOI e-scooters with a standard price, one free ride within 30 min every month with JURO bikes and one free day pass to a coworking space.

(4) City bike seasonal pass

City Bikes Pass is available from April to October for city bike-sharing. The packages include an unlimited number of maximum 30 min rides of city bikes. Other benefits for the taxi, car rental, e-scooters, JURO bikes, and coworking place are the same as those of 10-ticket.

(5) 30-day season ticket

There are regular traveler's version and student's version. Users can access public transit as they want within 30days within prechosen municipality. Other benefits for the taxi, car rental, e-scooters, JURO bikes, and coworking place are the same as those of 10-ticket.

3.2.3 Characteristics

(1) Integration

Similar to UbiGo, Whim unites a wide range of travel services into one app. Whim negotiates with individual mobility providers before placing them in the app and takes a small commission when trips are booked. Users access those travel packages with a subscription. The services of Whim can only be for the individual but not for a household. And uses can access all kinds of transportation modes with one subscription. For taxi services, users can choose their preferred taxi company.

(2) Price

Whim may not save transportation costs. In some cases, the user will pay approximately the standard prices for public transport, taxi rides, and car rentals when they are subscribed to Whim. There is no difference if they choose to pay directly to the service provider concerned or via Whim. If the users are to utilize regularly different methods of public transports, a subscription to the needed services via Whim will be a cheaper choice. Besides, Whim provides fixed-priced taxi rides of 10€ within the 5-km radius, which is cheaper than paying it directly to taxi companies. It can also help save from rapidly increasing taxi ride fees during public transport strikes in Helsinki Region.

(3) Sustainability

Whim also encourages users to increase the use of public transport which is nice for the trending global issue such as global warming. With fixed-priced plans, a Whim user who needs to utilize regularly different kinds of public transports will get the best value of his/her money with a Whim plan.

However, there will also be some potential adverse effects to sustainability. Research from Hartikainen, Pitkänen [5] shows that Whim users travel by taxi 2.1 times more often than the typical Helsinki resident. Taxi services and public transit are all available for use within one subscription. Whim can help the user more easily to access taxi services. If a Whim user receives unlimited or discounted taxi service, the incentive for this user to use public transportation may fade, because the taxi is more convenient to get to the destination. If taxi trips increase, public transportation ridership would decrease, and vehicle miles and congestion would increase.

(4) Big data technologies

In 2019, Ramboll (engineering, design, and consultancy company) conducted a study about the first year's travel data (2018) of Whim [5]. The dataset spans from 1 November 2017 to 31 December 2018. The data contain information about trip origin zones, trip ticket types, ticket purchase times, and mode types. The anonymized data were provided by MaaS Global. This report analyzed spatially distribution and trip-combinations preferences of Whim users, the potential commonalities or differences in travel behavior between Whim users and the typical citizen, and geographical characteristics of MaaS users. Besides this report, to the best of our knowledge, we could not find any materials, public or private, about their big data technologies applied in their MaaS algorithm.

3.3 Moovit

3.3.1 Introduction

Moovit is a MaaS provider and journey planner app. Formerly called Tranzmate, Moovit was founded as a startup in Israel in 2012 and it is acquired by a subsidiary of Intel, Mobileye, since 2020.

Moovit offers real-time journey planning services and GPS navigation through mobile app across transit modes, including buses, metros, ferries, trains, trams, trolleybuses, ride-hailing, bike-sharing, car-sharing, and scooters. Users can access a live map, search nearby stops and stations based on their current GPS location, plan trips across transportation modes based on real-time data, as well as pay the travel fee.

Over the years Moovit has grown to now serve over 950 M riders in 3400 cities across 112 countries, in 45 languages.

3.3.2 Services

Moovit is an application based on the navigation function that helps users get the easier and more efficient way from point A to B. All the other MaaS services of Moovit are all built on the navigation function. The services of Moovit includes:

(1) Real-time arrivals times

View real-time arrival information which is directly taken from GPS devices positioned on public transits. It helps to avoid guessing train or bus arrival times and saving waiting time.

(2) Digital payment

Purchase tickets and validate the digital pass with the mobility wallet.

(3) Real-time alerts

Know about real-time states of the transportation ahead of time by receiving service alerts such as emergency, congestion, unexpected disruptions, delays, and more so that users can plan if bus time or train time changed.

(4) Live navigation

Get step-by-step navigation from A to B, for example, know how long it will take to walk to the station, view the arrival time of the line, receive get-off alerts at the destination.

(5) Stops' visualization

Use augmented reality to locate the bus or train stop.

(6) User reports

Report issues found with stations, line service, and schedules. It can help Moovit to inform all nearby riders about what's going on in their area.

(7) Bike routes

Get bike routes in addition to the bus, subway, train, or metro trip plans. Bike docking stations are updated in real-time.

3.3.3 Characteristics

(1) Integration

Moovit has a high integration level for public transport. The application integrates official public transit data collected from urban operators with real-time data collected from users via crowdsourcing. In addition to passively sharing data, users can actively send reports, including delays, overcrowding, unexpected disruptions, satisfaction with their bus driver, and Wi-Fi availability. It helps quick response to emergencies in transportation. But the services such as private car rental are not integrated into Moovit.

(2) Price

Users pay for every type of transportation per ride via the application on the smartphone. The pricing method was created by the Ministry of Transport and the Ministry of Finance. Typically, the user should pay the standard price. The one who takes rides that cost more than the daily limit, will not have to pay extra and will actually get a discount. Some groups such as students, minors, old people, and the disabled are entitled to travel on public transportation at discounted prices or even for free. Passengers pay for their rides only at the end of the month, depending on how many rides they've taken in the whole month.

(3) Sustainability

Moovit provides people a convenient and efficient way to access sustainable transport, including bus, metro, e-scooters, bikes, and car-sharing, that saves them time and reduce the need for a car for first and last-mile options. Moovit also cooperates with car manufacturing companies such as BMV to promote the use of electric vehicles [6]. Those all help reduce the use of private and promote a more sustainable city.

(4) Big data technologies

Every day, there are about 6B anonymous data points added to Moovit People's Movement Data repository. Such massive amount of data enable Moovit to provide leading MaaS solution that help improve the city transport systems. For example, Moovit developed an Urban Mobility Analytics tool to analyze the effectiveness of your transportation infrastructure using historical ridership data, simulate scenarios in potential service areas to identify the optimal framework for operations, verify the usability of on-demand service and tweak parameters to create the best service before launch day. Moovit developed On-Demand Operation Modules that can help optimize the transport services provider's service based on the real-time mobility data. An On-Demand Service Model is developed to provide a flexible demand response for smart shuttles.

3.4 Uber

3.4.1 Introduction

Found in 2009, Uber started with ride-hailing services, later spanned its business into the food delivery services, micro-mobility system (with bikes and scooters), public transport, freight, and peer-to-peer ride system. Users just enter where they want to go, and a nearby driver will help them get there reliably. The payment can be finished via the app. Uber currently operates in over 900 metropolitan areas worldwide.

3.4.2 Services

Uber provides multiple choices and finds the perfect ride to fit users' needs. The services include:

(1) Carpooling

Service for share rides. The Uber app will look for riders along the user's route to fill empty seats in the car. Maximum two seats per request. Transit fees can be saved but it may take a little longer to get to the destination, for Uber Vehicles will detour to find copassengers.

(2) Car-hailing

Uber dynamically matches drivers' supply and customers' demand, allow customers to hire a driver exactly when they want to be picked up and taken them directly to their destinations. The driver is an independent contractor. Drivers provide a vehicle themselves. The vehicles can be owned, rented, or leased. Drivers must meet requirements for age, health, car age, and type, have a driver's license and a smartphone, and are required to pass a background check. The users can choose different drivers according to affordability, car style (such as economical, comfort, green electric vehicles, and luxury), the number of passengers.

(3) Taxi-hailing

Users can call for a taxi anytime and anywhere. Unlike car-hailing, the driver is hired by a taxi company. The payment also is finished on the app. Uber will show the fare estimate before booking. User can always check their receipt for the fare breakdown in the Uber app.

(4) Bike-sharing and scooter sharing

Only available in some of the cities. Users can find an electric bike/scooter using the Uber app, reserve a nearby one, scan the QR code on the bike/scooters to unlock and start their rides. After the trip, they need to park the bike/scooters in the correct area shown on the app and lock it using the cable lock on the back wheel. The app will automatically charge.

(5) Public transit

Users can search nearby public transit stations and real-time departure information with the Uber app. Users can also enter the destination to compare the estimated price and arrival times between taking public transit and taking a car. Uber also provides step-by-step navigation to the destination through public transit. In some cities such as Denver, users are allowed to buy a ticket for public transit via the Uber app.

3.4.3 Characteristics

(1) Integration

Uber integrates the information from the user side, driver side, and public transport side. Users can easily access peer-to-peer ride services through one app. Through Uber API, ride-hailing services can be integrated into other businesses. For example, Uber is integrated into ticket order apps. Users can call Uber cars when they place orders for tickets and want to go to concerts or sporting events. Uber is also be integrated into a hotel booking app that they can immediately from their account order call for an Uber vehicle to the hotel in one click. The same service is integrated into healthcare services providers that allow patients not only to book a doctor visit but instantly order a taxi to the hospital. Besides, Uber is integrated into food and grocery apps. It offers providers one more food delivery method with Uber vehicles.

(2) Price

After the trip is finished, Users pay the fee for this ride via the application on the smartphone. Uber fare prices vary according to region and car service types. Uber has an upfront pricing model to determine the price. Transportation fee is first calculated on four main criteria: (1) Initial fee—A flat fee charged at the beginning of every ride; (2) Cost per minute—How much the users are charged for each minute inside the ride; (3) Cost per mile—How much the users are charged for each mile of the ride; (4) Booking fee– A flat fee to cover Uber's operating costs. Basically, Uber fee = Base Fee + (Cost per minute * time in ride) + (Cost per mile * ride distance) + Booking Fee. Besides, more luxury services like UberBlack or UberSUV will have an additional charge. Passengers also need to pay for any tunnel, bridge, or toll charges that are incurred during the ride. Besides, Uber has a dynamic pricing system. During a surge period when the demand for rides cannot be met by the number of current drivers on the road, prices will increase. Sometimes it may be two or three times the standard price. This can encourage more drivers to hit the road to guarantee the reliability and availability of cars. Users can obtain discounts according to their total distance of rides. In general, for car-sharing and economic car-hailing services, the transportation fee can be saved, but sometimes the user will be charged more compared to the standard fee. The advantage of Uber is not mainly the competitive price but provides the user an efficient way to access peer-to-peer ride services.

(3) Sustainability

It is controversial that whether Uber can promote sustainable urban transportation. On the one hand, Uber services can reduce the void cruising distance for drivers and help them find suitable passengers as soon as possible [7]. It improves the transportation efficiency and reduces the emission [8–10]. Car-sharing services can reduce vehicle emissions per capita by sharing rides. Besides, for people can easily travel from one point to another point via Uber services, their willingness to own a private car will decrease. On the one hand, some scholars questioned ride-hailing emission reduction performance. Uber increases the users' willingness to use a car, for its convenient to take them from one point to another point. It is reported that private-ride transportation companies' services including Uber have increased 180% more traffic to urban road networks and added billions of vehicle miles traveled in the nation's largest metro areas [11, 12]. This is also supported by the work of Henao and Marshall [13]. According to their estimation, ride-hailing leads to approximately 83.5% more vehicle miles travel than that if ride-hailing not existed. Another recent study also concluded that transportation service companies including Uber are the biggest contributor to the growth of traffic congestion in San Francisco [14]. Qian, Lei [15] also made a similar conclusion in their research. As ride-hailing services from MaaS services including Uber is an unstoppable trend, for they do bring people with convenient and efficient mobility services, Uber should make efforts in operation policies toward more sustainable urban transportation. For example, promoting the use of clean energy vehicles in its ride-hailing services [16].

(4) Big data technologies

Uber has revolutionized mobility services by powering billions of rides and deliveries connecting millions of riders, businesses, restaurants, drivers, and couriers. Petabytes of data are collected and processed per day. Uber relies on big data to make business-critical decisions such as better pricing and matching, fake rides detection, lowering estimation time of arrival, visualization, and experimentation.

In terms of data storage, before 2014, as the amount of Uber data was small, Uber used traditional online transaction processing (OLTP) databases (MySQL and PostgreSQL) to store data. Data was stored in scattered OLTP databases. From 2014 to 2015, Uber used Vertica as their data warehouse software and developed multiple tools to aggregate all Uber data in one place. Standardized SQL interface allowed Uber engineers to access data and built online queries. From 2015 to 2016, Uber rearchitected its big data platform around the Hadoop ecosystem. This second-generation big data platform enables horizontal scaling. Incorporating technologies such as Parquet, Spark, and Hive, tens of petabytes of data were ingested, schematized, stored, and served in an efficient and stable way. Since 2017, to achieve incremental data ingestion, as well as more efficient storage and serving of data, Uber built Hadoop Upserts and Incremental (Hudi), the third generation big data platform. Besides, this more extensible data transfer platform allowed Uber to easily aggregate all data pipelines in a standard way under one service as well as support connectivity between any data source and data sink.

Uber has its own machine learning platform called Michelangelo which is used to create different models for Uber's various services based on its massive data. Michelangelo is an internal ML-as-a-service platform that democratizes machine learning and makes AI meet the needs of Uber business as easy as possible. It covers the whole machine learning workflow: data management, model training, performance evaluation, models deployment, and monitor predictions.

According to massive historical data, Uber uses a variety of spatiotemporal forecasting models to predict ride demand and driver availability at various places and times in the future. External factors, include market conditions, global news events, user historical data, weather, holidays, time, traffic, etc., are considered in the predicted model. Based on predicted imbalances between supply and demand, Uber can optimize the dispatch system to encourage drivers ahead of time to go where there will be the greatest opportunity for rides.

Besides, a robust and reliable dynamic pricing system can also be established to adjust fares according to current traffic conditions and demand.

Uber also uses big data and machine learning to estimate times of arrival (ETA) for drivers and the whole trip. When users open the app and request a ride or Uber Eats, the first thing they see is ETA. Accurate ETAs are important for a positive user experience. But ETAs are most difficult to get right since it involves many factors like traffic, time of the day, weather, etc. UberMap Services Team developed a sophisticated segment-based routing system that can estimate base ETA values. These base ETAs have consistent patterns of errors. The machine learning model can be utilized to predict these errors based on historical data and then use the predicted error to make a correction.

In terms of data visualization, Uber developed Kepler.gl, an open-source platform that helps visualize large-scale spatial/temporal datasets. It can visualize a large amount of location data, playback geo-temporal trends over time in a city or region, explore, filter, and deeply engage with location data.

It can be said that big data and data science is the heart of Uber's mobility services platform. It is the insights derived from millions of users that enable Uber to build all these excellent products and deliver extreme customer service.

4 Future development trend of MaaS system

In the first two decades of the 2000s, from the development history of MaaS services, we have seen the great impacts that the MaaS has on the way we live and the changes in what MaaS stands for. However, we are only at the early stages of MaaS development. We would expect to see more changes appear within the next few years as we take the steps toward revolutionizing travel and integrating MaaS into our daily lives. The future trend of the MaaS system includes:

4.1 Data-integrated

There is no doubt that the MaaS system will be integrated with big data more and more closely, and the integration level will be higher and higher. It puts forward higher requirements of the cross-functional collaboration between big-data platforms, data foundation, data streaming, real-time platform, and big data developer. Multiple sources of data will be integrated into MaaS system, including but not limited to:

- **Movement data**: such as user's mobility data from mobile phone network data and GPS, vehicle data from connected vehicles, including passenger cars and goods vehicles.
- **Transportation sensor**: such as traditional traffic count sites, Wi-Fi/Bluetooth vehicle counters, cameras capable of registration plate or facial recognition.

- **Infrastructure data**: such as signal locations, road geometry, and layout, roadworks and road closures, bridge health, EV charge point status, road ride quality.
- **Public transit data**: public transport timetables, real-time vehicle locations, location of the public transit station.
- **Transportation policy**: form Transportation department, such as parking regulations, traffic regulations orders.
- **Environmental data**: such as air quality and carbon level to measured pollution levels, emissions data from emissions and dispersion modeling tools, Weather data such as temperature, precipitation, wind, flooding, etc.
- **Socioeconomic data**: such as mosaic geodemographic data and behavioral analysis, social activity data from smartphone apps, and social media content.
- **Transaction data**: from credit cards and ticketing, including public transport, taxi, and shared micro-mobility.
- **Future developments modeling**: Future road layout and supporting infrastructure, projected traffic flow , transit occupancy, travel demand, etc.

Independently, these data sources describe part of the picture of how a transport network is operating. When these data sets are combined, they create a rich tapestry of MaaS services and public infrastructure, where interrelationships can be easily identified. Combining and analyzing this data enable MaaS service operators to create effective transport networks and provide efficient MaaS services. In turn, the data created from the MaaS services will help transportation infrastructure improvements, promote an open, efficient, collaborative, and sustainable transportation ecosystem at a local, regional, and national level.

4.2 Future-oriented

With the development in transportation technologies, new travel modes will appear. They may change the service way, business models, and regulations of MaaS services.

For example, auto-driving cars, delivery robots, and shuttles are already driving out of labs and onto streets to help deliver food, goods, and medical supplies. Chinese tech giants Baidu, one of the leaders in autonomous vehicle technology, used its driverless vehicles for frontline work during the Covid-19 pandemic, such as cleaning, disinfecting, logistics and transportation. In the US, autonomous vehicles are being used to transport medical supplies and Covid-19 tests at the Mayo Clinic in Florida. In 2020, Moovit joined as part of Intel, aiming at advancing the company's MaaS strategy and the global adoption of autonomous transportation.

In the future, fleets of networked cars and autonomous self-driving cars will become commonplace in the years ahead. Other autonomous vehicles, such as flying drones may join the transportation network. Those technologies will be integrated into the MaaS ecosystem as a whole. A flexible and open MaaS system will help better prepare to embrace this possible future revolution.

4.3 Sustainable

The potential for MaaS to be a real urban mobility solution becomes clear. If online ride-hailing platforms are integrated seamlessly into the public transportation system, it will be an important enabler in reducing today's share of private car traffic, contributing to the reduction of both carbon emissions and pollution [8–10].

For example, Uber is committing to becoming a fully electric, zero-emission platform by 2040, with 100% of rides taking place in zero-emission vehicles, on public transit, or with micro-mobility. They will do this by offering riders more ways to ride green, helping drivers go electric, and partnering with NGOs and the private sector to help expedite a clean and just energy transition.

MaaS will have a more and more important role to play in the sustainable cities of the future. It could address mobility's pain points in cities, while revolutionizing urban mobility, making it more affordable, efficient, user-friendly, environment friendly, and available to everyone.

References

[1] Tschanz N, Zimmermann H-D. The electronic mall bodensee as platform for the development of travel services. In: Information and communication technologies in tourism. Springer; 1996. p. 200–10.

[2] Sochor J, Arby H, Karlsson IM, Sarasini S. A topological approach to mobility as a service: a proposed tool for understanding requirements and effects, and for aiding the integration of societal goals. Res Transp Bus Manag 2018;27:3–14.

[3] Jittrapirom P, Caiati V, Feneri A-M, Ebrahimigharehbaghi S, Alonso González MJ, Narayan J. Mobility as a service: A critical review of definitions, assessments of schemes, and key challenges; 2017.

[4] Karlsson IM, Sochor J, Strömberg H. Developing the 'Service'in mobility as a service: experiences from a field trial of an innovative travel brokerage. Transp Res Proc 2016;14:3265–73.

[5] Hartikainen A, Pitkänen J, Riihelä A, Räsänen J, Sacs I, Sirkiä A, et al. WHIMPACT: Insights from the world's first mobility-as-a-service (MaaS) system. Rambøll Consulting: Helsinki; 2019.

[6] Sekai M. Industrial digitalisation in Europe: the growth of start-ups and the transformation of established firms. In: Mitsui & Co Global Strategic Studies Institute monthly report; 2017.

[7] Sui Y, Zhang H, Song X, Shao F, Yu X, Shibasaki R, et al. GPS data in urban online ride-hailing: a comparative analysis on fuel consumption and emissions. J Clean Prod 2019;227:495–505.

[8] Martin E, Shaheen SA, Lidicker J. Impact of carsharing on household vehicle holdings: results from North American shared-use vehicle survey. Transp Res Rec 2010;2143:150–8.

[9] Martin EW, Shaheen SA. Greenhouse gas emission impacts of carsharing in North America. IEEE Trans Intell Transp Syst 2011;12:1074–86.

[10] Shaheen SA, Cohen AP. Carsharing and personal vehicle services: worldwide market developments and emerging trends. Int J Sustain Transp 2013;7:5–34.

[11] Schaller B. The new automobility: Lyft, Uber and the future of American cities; 2018.

[12] Schaller B. Can sharing a ride make for less traffic? Evidence from Uber and Lyft and implications for cities. Transport Policy 2021;102:1–10.

[13] Henao A, Marshall WE. The impact of ride-hailing on vehicle miles traveled. Transportation 2019;46:2173–94.

[14] Erhardt GD, Roy S, Cooper D, Sana B, Chen M, Castiglione J. Do transportation network companies decrease or increase congestion? Sci Adv 2019;5, eaau2670.

[15] Qian X, Lei T, Xue J, Lei Z, Ukkusuri SV. Impact of transportation network companies on urban congestion: evidence from large-scale trajectory data. Sustain Cities Soc 2020;55:102053.

[16] Jenn A. Emissions benefits of electric vehicles in Uber and Lyft ride-hailing services. Nat Energy 2020;5:520–5.

Chapter 2

Spatio-temporal data preprocessing technologies

Jinyu Chen, Haoran Zhang, Wenjing Li, and Ryosuke Shibasaki
Center for Spatial Information Science, The University of Tokyo, Kashiwa, Chiba, Japan

1 Introduction

The preprocessing of data plays a fundamental and essential role in the overall operation system. Every algorithm or system has its own desired input format. Normally speaking, the mission of data preprocessing is to convert raw data to the desired one of algorithm or system. Further, we can understand and use processed data to mine deeper knowledge. In the field of MaaS, almost all of the spatio-temporal data is recorded by global positioning system (GPS). In current studies, many scholars used trajectory data from certain transportation methods for analysis. Hong et al. [1] adopted commuting vehicle trajectory data to analyze the potential of sharing. Christian et al. [2] used bike-sharing trajectory data to estimate the influencing factors. Cai et al. [3] used taxi trip data to compute the environmental benefit that could be brought by taxi-sharing. Trajectory dataset with hybrid travel mode can help us better understand the urban mobility pattern and operate various research [4, 5]. Such data can provide many upper-level insights into sharing mobility because it is easy to obtain an overview on comprehensive urban-level mobility. However, raw GPS data from mobile phones in daily life are not always marked with travel mode and they are not suitable to be input into the algorithms. Therefore, preprocessing raw GPS data is essential.

To introduce how the preprocessing works, in this chapter, we will first introduce the raw GPS data and workflow of raw data preprocessing and the function of every part in Section 2. Then, in Section 3, we will further focus on the details and techniques of each step. In Section 4, we will provide some samples of former research as a case study. Section 5 will give a summary of this chapter.

Big Data and Mobility as a Service. https://doi.org/10.1016/B978-0-323-90169-7.00003-8

2 Raw GPS data and workflow of data preprocessing

The raw GPS record is usually in the form of a tuple:

$$(x, y, t) \tag{1}$$

where x refers to the longitude; y refers to the latitude; and t refers to the corresponding timestamp.

Of course, there may be other variants like an additional identity number that indicates the individuals. Usually, the big dataset of GPS trajectory has the scale of thousands, millions, or even billions of this kind of tuple. It is impossible to compute with the raw dataset of this size. High-dimensional information should be extracted to help us better understand the mobility pattern. In MaaS, many studies that consider the mobility data-driven problems [6–8] are based on trajectory data with higher dimensional features. Fig. 1 gives a sense of the technologies and corresponding possible applications. It includes outlier removement, stay location detection, travel segmentation, travel mode detection, and map matching. For their applications, we give some simple samples like stay location detection can be used for the life pattern analysis; travel segmentation and mode detection for potential analysis; map matching for navigation. Notice that each step of technology is not independent. They are also connected. We give a general workflow of technologies in data preprocessing shown in Fig. 1.

The raw GPS dataset on the first hand will be dealt with outlier removement. This tries to eliminate the not properly recorded GPS points. Next, we will divide the GPS data into several travel segmentations. This is a necessary front step for travel mode detection. Then, a method to detect the travel mode of each

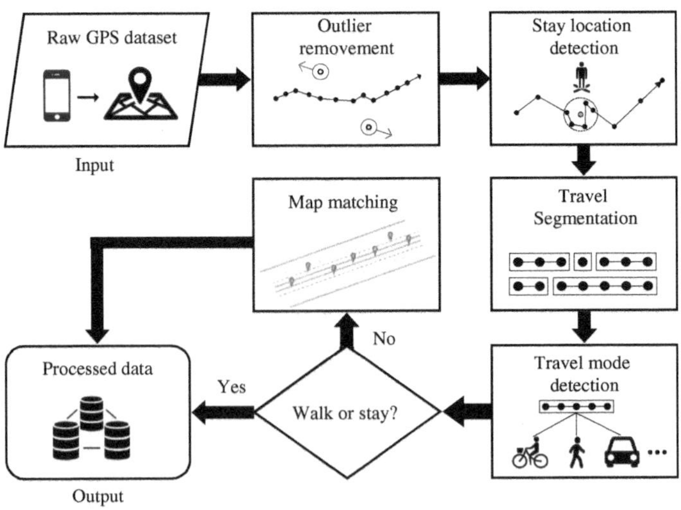

FIG. 1 General workflow of data preprocessing.

travel segmentation will be operated. Each travel segment can be marked with the label of bicycle, walk, stay, vehicle, bus, or railway. After that, judging by the marked travel mode, we will decide whether to operate the map matching method. The final processed data will be suitable for the data mining or analysis algorithms. In the next section, we will give details on the technical route of each step.

3 Key technologies and corresponding application

The components that formulate this workflow may function as a unity; besides, they are independent and have respective purposes. Raw GPS dataset can be processed by this series of technologies to become processed data that is suitable for mining algorithms. The output of each step also has its utility. In this section, we will focus on the technical details as well as the application.

3.1 Outlier removement

Due to the bad signals, computation error, or system error, there are always outliers in the GPS trajectory data. Fig. 2 gives a simple example of trajectory with outliers. Apparently, two dots leave away from the original trajectory. In some extreme cases when the GPS signal is unavailable, the GPS coordinate can be set default to $(0, 0, t)$ or remain at the last place where the signal is fine. They are harmful to the further analysis and application of data because it probably brings wrong knowledge. Thus, it is necessary to detect and remove the outliers.

In this chapter, we will introduce one typical method to achieve this purpose, which is mainly based on velocity. Suppose we have a sequence of GPS points:

$$\{p_0, p_1, p_2, p_3, \ldots\ldots, p_n\} \tag{2}$$

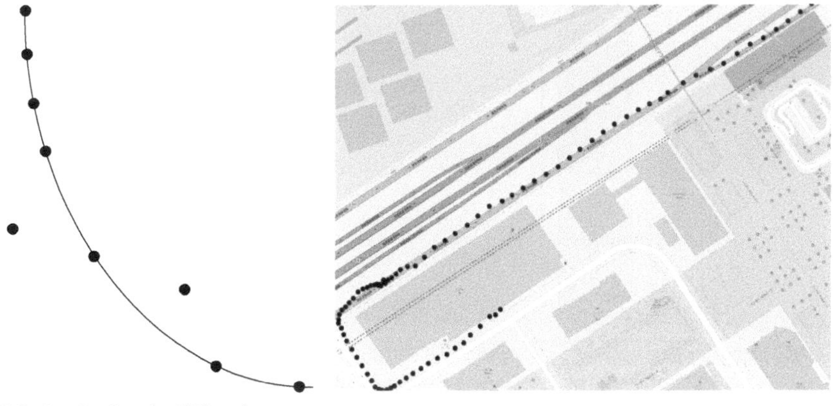

FIG. 2 Outliers in GPS trajectory.

Then, we can compute the discrete velocity in this sequence:

$$v_i = \frac{dis(p_i, p_{i+1})}{t_{i+1} - t_i}, i \in \{0, 1, 2, 3......, n-1\} \tag{3}$$

where t is the timestamp of the corresponding GPS record; *dis* refers to the function to compute the distance between two points. Notice that the distance can vary depending on the cases. Usually, it can be haversine distance, Manhattan distance, or Euclidean distance.

From the equation, we can get a velocity sequence. Next, we can apply statistical methods to see the distribution of v_i. Fig. 3 provides an example in the form of a box plot for better visualization.

The two points far from the plot on the right side are the outliers. By referring to Eq. (1), we can know that if v_i is the outlier in velocity sequence, p_{i+1} should be removed in the GPS trajectory.

Other than plot analysis, we can use a Gaussian distribution to represent the GPS points, with the mean μ and standard deviation σ being calculated as follows:

$$\sigma = \sqrt{\frac{1}{N}\sum_{1}^{N}(p_i - \mu)^2}, \mu = \frac{1}{N}\sum_{1}^{N}p_i \tag{4}$$

where p_i is the ith GPS point and N is the total number of points. If the distances from an inner point to its neighbors were larger than $(p_i - \mu)/\sigma > 2.6$, the point will be labeled as an outlier and removed from the trajectory.

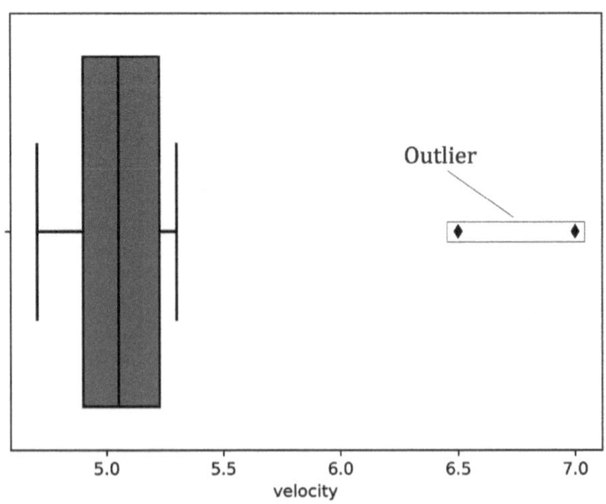

FIG. 3 Box plot of velocity.

3.2 Stay location detection

Stay location detection is to detect when and where the user stays in the trajectory. In daily life, individual mobility consists of multiple segments. For example, people probably first stay at home, next go out and travel with one travel mode like walk or vehicle, then, stay at one place for some time. Notice that staying at a place doesn't only mean that the GPS trajectory records are stuck or locked in the same coordinate for some time. Long time of wandering around a certain place should also be treated as staying. Technically speaking, a stay segment within a trajectory corresponds to a group of consecutive points representing a user who is stopped at a location. Consequently, the recognition of the segment of staying is more than recognizing the continuous points whose coordinates are unchanging. What is more, even if we have operated the outlier removement, there are probably still many points with little error in the trajectory. Those points also have a large offset from the true location. We should notice that it was possible for the distance from the noise points to the neighboring points to be larger than the threshold as if they corresponded to a stay status. A classical phenomenon is that even though the position of the object of GPS record is standing still, due to various factors, the recorded GPS points are leaping around. But this kind of error is not very significant. Therefore, we adopt the following algorithm to detect whether the segment is staying or traveling. So, the stay segments were extracted based on thresholds for distance and time.

Algorithm 1: Distinguishment of segment of staying and traveling.

Input: The starting point for detection.
Output: The detected segment of staying and traveling.
$i \leftarrow$ The serial number of the starting point of detection.
Suppose p_i is the first point in the segment of staying.
Append p_i to the segment.
While the last point in the segment is detected:
Compute average velocity from p_i to p_{i+1}.
If computed average velocity $<=$ threshold:
p_i to p_{i+1} is the segment of staying and $p_i$is not the last point in the segment.
Append p_{i+1} to the segment.
$i + 1$
Else:
p_i is the last point in the segment of staying.
Break from while
End while
If the quantity of points in the segment is larger than one:
The segment is the staying one.
Update p_i as the first point in the segment traveling.
Else:
p_i is the first point in the segment of traveling.
Do the while part to detect the last point in the segment of traveling.
End algorithm

The introduction of the algorithm is in the form of fake code. The real code can be very different depending on different languages of programming. The threshold of velocity also varies. If you are familiar with an error of GPS device, the threshold value should be determined by it. If not, it is recommended that the author either operate systematic analysis on an obvious cluster of staying points or refer to other studies and choose a conservative value. By repeating it, we can get several segments of staying and traveling which consist of the corresponding GPS points.

Next, we also need to further process with the segment of staying. As explained before, despite the segment of staying, GPS points could leap around due to the error. It is necessary to choose one coordinate as the most probable place of staying. Normally speaking, the coordinate that takes the majority of a segment can be assumed as the most probable one. There are also many other options. Former studies also developed many methods in spatial clustering to detect the stay point. Since GPS trajectory can be objected to a spatial Cartesian coordinate system, it is possible to operate clustering. Gathering points can well indicate the stay location. Former studies have proposed many classical clustering methods as well as their variations, such as DBSCAN [9], ST-DBSCAN [10], DB-SMoT [11], and OPTICS [12]. These methods have long been proven to be efficient and effective in spatial clustering. In addition, they are easy to apply as they are highly integrated. If one is interested in this part, he can read other references regarding this part. We will not give further discussion.

3.3 Travel segmentation

The travel segmentation mainly helps to judge whether the study target moving and more importantly when they change the travel mode to another one.

The distinguishment of different travel segments in the traveling procedure is according to the change points, at which the users changed transportation mode. The change points were detected by using two features: the velocity change rate (VCR) and points on train lines. The VCR was defined as the average speed of the current segment compared with the current observed speed, such as

$$VCR = \left|S.Speed_{average} - P.Speed\right| / S.Speed_{average} \tag{5}$$

where $S.\ Speed_{average}$ is the average speed of the segment and $P.\ Speed$ is the speed of one point in the segment. The use of VCR index has been proven to be more effective and more stable than the acceleration index in the case of sparse GPS data.

The role of travel segmentation is mainly to serve travel mode detection.

3.4 Travel mode detection

Then, we will introduce how to detect the travel mode of segment of traveling. Travel mode detection is currently a hot topic for a long time [13, 14]. Former studies have showed numerous methods. Xiao et al. developed a Bayesian network consisting of a K2 algorithm and conditional probability tables [15]. The model can distinguish walk, bike, bus, and car travel with an accuracy of 86%. Nguyen et al. [16] adopted a random forest model for travel mode detection. The accuracy can reach 89.1% on the classification of bus, motorcycle, car, walk, and bike. Zhou et al. [17] utilized random forest but different features to accomplish the same goal. In their case, the accuracy reaches 94.1%. Besides, there are also other algorithms or methods like deep learning [18], Support Vector Machine [19], or a hybrid combination of deep learning and optimization [20]. Whatever kind of method, a common thing is that they all need groundtruth trajectory data labeled with real travel mode as the basis. Generally, groundtruth data is not open source. The author probably needs to collect data by employing volunteers. The groundtruth data should follow the principle of (1) collection from as many people as possible to eliminate the difference in travel behavior among people. (2) covering various functional areas like commercial, residential and rural areas. (3) covering long time period to ensure the variety of road conditions and weather. (4) covering all of the travel modes that want to be detected. The GPS device can either be smartphone [21] or professional GPS data logger [22].

As introduced before, former studies have developed enormous methodologies to detect travel mode. In addition to the method itself, the input feature is also an essential factor. Despite the features found by former studies, it is encouraged that the author use regression model to find more novel features and methodologies. A good feature should show high correlation with variable of travel mode. In Table 1, we provide a partial selection of typical studies considering the problem of travel mode detection.

From Table 1, it can be informed that there are a large variety of candidate methods. Each method has its own character and significance. Here, we will give a simple introduction to classical technologies.

Random Forest: The group of random forest is an ensemble learning method for classification, regression and other tasks that operates by constructing a multitude of decision trees at training time and outputting the class that is the mode of the classes or classification or mean/average prediction or regression of the individual trees. Generally, it is an assembly of decision trees. Each single decision tree is trained to decide the classification of the target, so random forest employees have many trees to vote for the output. To avoid the same output produced by all the trees, each decision tree only learns a part of the features of total dataset. In the long term of application, it has been proved to outperform the decision tree. Random forest is a supervised learning method. Therefore, it relies on the quality of data.

TABLE 1 Table of studies.

Study	Feature	Methodology
Stenneth et al. [23]	1. Distance to all buses 2. Distance to selected buses 3. Distance to nearest railway line 4. Number of GPS records that are closer to bus station than a threshold 5. Accuracy of GPS records 6. Average speed 7. Average change rate of direction 8. Average acceleration	(1) Naive Bayes (2) Bayesian Network (3) Decision Trees (4) Random Forest
Widhalm et al. [24]	Totally 77 features regarding: 1. Velocity 2. Acceleration 3. Deceleration 4. Direction change 5. Signal 6. Power Spectrum of device	Decision trees followed by a Discrete hidden Markov Model
Xiao et al. [25]	1. Average speed 2. Speed of 95th percentile 3. Average absolute acceleration 4. Travel distance 5. Proportion of records of low speed 6. Average change rate of direction	Continuous hidden Markov Model
Zong et al. [26]	1. Average speed 2. Maximal speed 3. Speed of 75th percentile 4. Acceleration 5. Acceleration of 75th percentile 6. Travel time 7. Travel distance 8. Standard Deviation of speed	Nested logit model and its variance
Rezaie et al. [27]	1. Travel time 2. Average speed 3. Travel distance 4. Distance between transit station and origin as well as destination	Random Forest model and label propagation

Bayesian Network: It is a probabilistic graphical model that represents a set of variables and their conditional dependencies via a directed acyclic graph. Bayesian network is ideal for taking an event that occurred and predicting the likelihood that any one of several possible known causes was the contributing factor. For example, a Bayesian network could represent the probabilistic relationships between diseases and symptoms. Given symptoms, the network can be used to compute the probabilities of the presence of various diseases. An effective and widely used algorithm to solve the Bayesian network is the Markov chain Monte Carlo algorithm. The strength of Bayesian network is that it can model cause and effect with uncertainty. Hence, it is powerful in solving uncertain problems. What's more, the Bayesian network can integrate and absorb various inputs and information effectively.

Hidden Markov Model: Hidden Markov Model is a statistical Markov model in which the system being modeled is assumed to be a Markov process with unobservable states. Different from normal Markov Model, the state of hidden Markov model is directly unobservable. The state should be inferred by the observable parameters affected by state. The states are usually expressed by a probability distribution. The model is usually used to deal with serial data or problems like natural language processing and trajectory computing. Currently, there are three kinds of algorithms to solve Hidden Markov Model: forward propagation; Viterbi algorithm, and Baum-Welch algorithm.

There is no obvious advantage or disadvantage. On the other hand, the selections of features in all of the studies are very similar. Almost all of them are time, distance, velocity-related. In addition, we can also see that in addition to the information extracted from trajectory data, much GIS information is also integrated into the feature, which is very helpful in judging the trip of public transportation. Trajectory of public transportation is especially important in the analysis of MaaS-related research as the relationship between traditional public transportation and MaaS is one of the essential topics [28–30].

3.5 Map matching

In the workflow of the whole preprocessing system, the map matching method aims at the travel segment with a mode of bike, car, railway, or other modes that ought to be in the designated area just like a car should travel on roads in urban areas. In most cases, the recorded GPS trajectories are not accurate. Due to the error of sensors or other factors, there are always vacancies and biases in the GPS records. When there was a vacancy, the continuity and completeness of trajectory would be lost. To deal with the problem, we can choose to cut the trajectories into subsets if there exist long time intervals. On the other hand, biases in the GPS sensors could cause the error in the true location of vehicles and the record of GPS location could usually be beyond the road grid. Thus, we need to match the GPS points on the road grid. Here, we introduce a sample technical routine. First, we can download shapefile of road grid of Tokyo area from OSM (Open Street Map). The original data from OSM isn't complete. So, we need to use PostgreSQL to do topology on these roads and each piece of road segmentation will be given a unique ID and logical connection. The logical connection is used to indicate the geographical connection among each road segmentation. The meaning of logical connection is that after the GPS trajectory is matched to the road network model, the driving distance between the neighboring points can be computed. After topology, in a large city, we have thousands of road segmentations with unique IDs.

Then, we can adopt the traditional map matching algorithm to find the match point of GPS trajectory point on the road grid. The principle of algorithm is to draw perpendiculars to road lines and match the point to foot of perpendicular with the least length for it is the most possible position on the road grid, which is shown in Fig. 4.

$$p_i' = argmin(verticaldis(p_i, R_i)), R_i \in (0, n) \tag{6}$$

where, p_i is the raw GPS point; p_i' is the matched point on-road grid; R_i is the road segmentation identity; *verticaldis*is the function that computes the vertical distance between the raw GPS point and road segmentation.

Obviously, there is flaw in the traditional map matching method. It cannot ensure finding the correct position of original GPS. The territory of error can be treated as a circle with the radius of error value. The true location can be any point in this circle. In a more complex road network, which is very common in metropolis, points that should be on the same road probably will be matched to

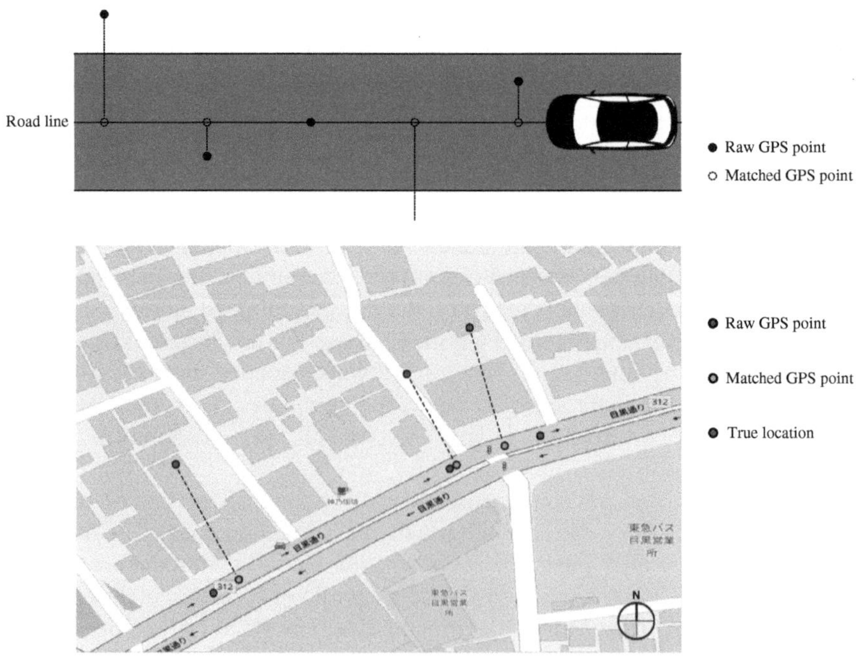

FIG. 4 Traditional map matching.

different ones. To overcome this issue, Singh et al. [31] developed a genetic-based map matching algorithm; Jagadeesh et al. [32] adopted classical Markov model in the map matching algorithm; for GPS trajectory with long time interval, Hsueh et al. [33] developed the method that considers the spatio-temporal relationship among GPS records.

3.6 Summary

Here is the introduction and overview of the key technologies in the process of data preprocessing. The typical output of the process is the travel segments marked with the corresponding travel mode. Data in this form is general for various algorithms of different purposes [34–36]. In the case study, we will give an introduction on some examples of our former works regarding the application of each key technology in the data preprocessing.

4 Case study

4.1 Stay location detection: Life pattern analysis

4.1.1 Introduction

The concerned field of Urban Computing integrates computational solutions to study how humans travel and conduct daily activities, making cities more humanized, more livable, and more sustainable [37, 38]. Life pattern, which describes the individual's activity regularity and general travel style, has gained significant attention from city and transportation planners and urban policymakers. However, individual daily activities are very diverse [39]. Abstracting the groups' characteristics of life patterns from diverse individual cases is essential for urban infrastructure improvement and policy making towards multiple types of citizens. Life pattern clustering is the core of abstracting the groups' characteristics of mobility patterns. The work is published in arxiv [40].

4.1.2 Problem and methodology

To cluster life patterns and mine the group characteristics of life patterns on a large-scale from unlabeled mobility data, we need to overcome three major challenges for clustering individual life patterns:

Cluster the life pattern data in a high dimension: from the aspect of one person 1 day, life pattern information, which couples temporal and spatial information, is high-dimensional. As the number of points required to represent any distribution exponentially increases with the number of dimensions, the data sets in high-dimensional tend to be sparse. In such a high-dimensional space, clustering becomes difficult due to the increasing sparsity of data. Besides, distinguishing distances between data points in high-dimensional space poses difficulties [41]. The metric to properly measure the similarities between groups should be carefully introduced.

Handling the uncertainty in the life pattern of an individual: if regarding the life pattern of one person in multiple periods, due to the uncertainty of human movements, the individual life pattern records extracted from GPS data contain massive uncertainties. Even one person can make different options for daily activity types, travel modes, time, and periods.

Represent the diverse life pattern between individuals in a suitable format: Achieving a suitable format for life pattern clustering is a challenging task. First, the individual daily life pattern is diverse. if regarding the life pattern of multiple persons in multiple periods, life pattern varies between different individuals, regions, and time. When regarding on a larger scale, the condition and situation will be more complex and uncertain. Thus, the information of time, activity semantics, temporal order, and frequency should be reflected in the presentation format of life patterns. The original features of individual life pattern data should be retained as much as possible. Besides, the metric to measure the similarities of life patterns should be computable through the representations.

To address three issues, we used the framework consisting of stay point detection shown in Fig. 5.

Significant places detection: The detection process includes three steps: stay location extraction, spatial clustering of stay location, and significant places identification.

(1) **Stay location extraction:** The algorithm is the same one introduced in former section. By implementing this process, we determined the travel state of each user during each period and obtained the stay location which allows life pattern detection in the next step. Here, we used $S_u = \{S_1, S_2, S_3, \ldots, S_i\}$ to represent the user u's stay location history. Each stay point Si represents a stay location. It retains the arrival time ($S.ar$) and the leaving time ($S.levT$) which, respectively, equals to the timestamp of the first and last GPS point constructing this stay point (Fig. 6).

(2) **Spatial clustering of stay location:** DBSCAN (Density-Based Spatial Clustering of Applications with Noise) was utilized for spatial clustering. Two parameters are requiredinDBSCANmethod: the minimum number of points that can form a cluster (*minpts*) and the maximum distance within which two points belong to a cluster (*eps*). Points with a neighborhood *eps* more than *minpts* are categorized as core points, and points reachable by a core point are border points of the cluster. Both core points and border points are considered as the clustered points. Points not within the *eps* search radius of any core point are treated as noise [42].

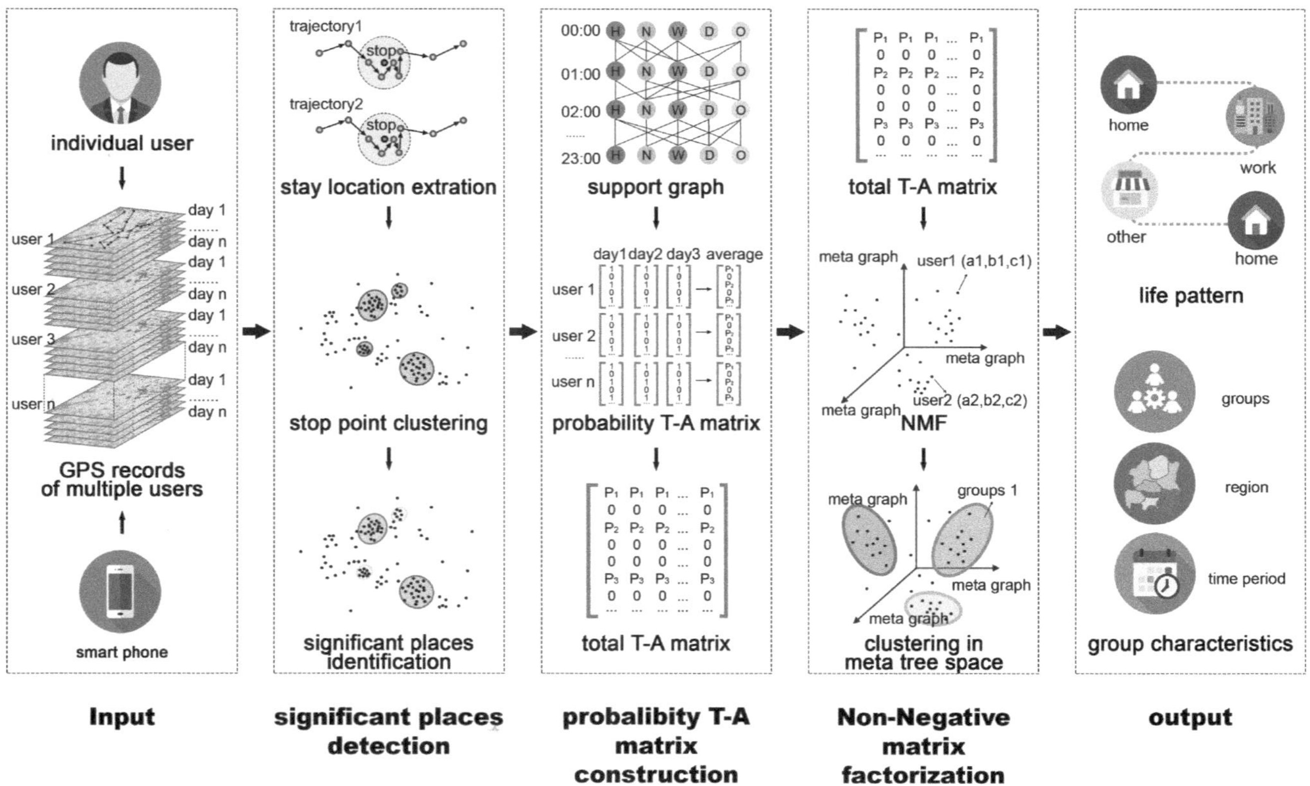

FIG. 5 Framework of metagraph-based life pattern clustering.

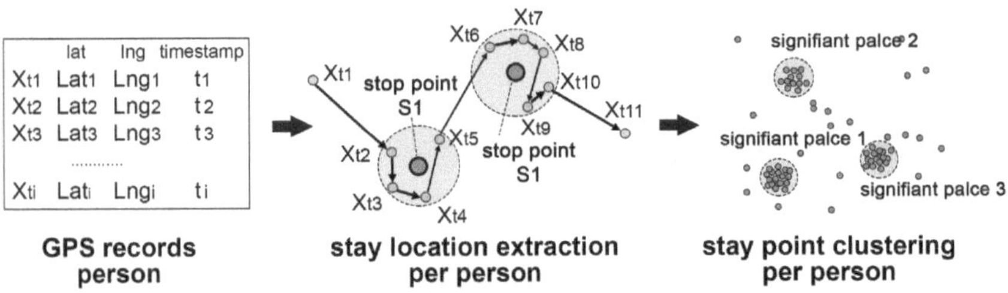

FIG. 6 The illustration of the stay location

(3) Significant places identification: We classify the significant places into five groups—home (H), other frequently visit nighttime spots (N), workplaces (W), other frequently visit daytime spots (D), and other significant places (O). Within each cluster, we extracted the stay points which meet the following conditions as candidate points of nighttime spots or daytime spots. Then, we compared the day of stop and the total stop duration for daytime spot candidates and nighttime spot candidates within each cluster to determine the semantic meaning. The cluster whose stop days and stops duration of nighttime spot candidates are highest will be identified as home, and the cluster whose stop days and stops duration of daytime spot candidates are highest will be identified as workplaces. Other clusters that contain daytime spot candidates or nighttime spot candidates will be identified as N or D. The cluster without daytime spot candidates or nighttime spot candidates that meet the conditions will be identified as other significant places. After obtaining the estimation of home, workplace, and other places, we can present the user u's stay location at day j as

$$S_{u,j} \in S_u,\ S_{u,j} = \left\{S_{(1,j)}, S_{(2,j)}, home_{(k\,j)}, S_{(3,j)}, \ldots, work_{(k,j)}, \ldots, Other_{(k,j)}, \ldots., S_{(i,j)}\right\},$$
$$Home_k \in H_u = \{home_1, home_2, home_3, \ldots, home_k\},$$
$$work_k \in W_u = \{work_1, work_2, work_3, \ldots, work_k\},$$
$$other_k \in O_u = \{other_1, other_2, other_3, \ldots, other_k\}.$$

Support graph and topology-attribute matrix construction: We present the life pattern of all users by using a support graph and topology-attribute matrix (T-A matrix) in this section. For each user, we first build an individual graph that represents the location of each hour within 1 day. Fig. 7A illustrates the construction of the individual graph, support graph, and edge indexing strategy based on a population of four individuals. Then, we use T-A matrix to represent the graphs. Let n denotes the number of edges in the support graph, a T-A matrix of one user 1 day will be generated with its row number being equal to the total index number of the support graph edges n. Fig. 7B is an illustration of the T-A matrix generation based on user 2.

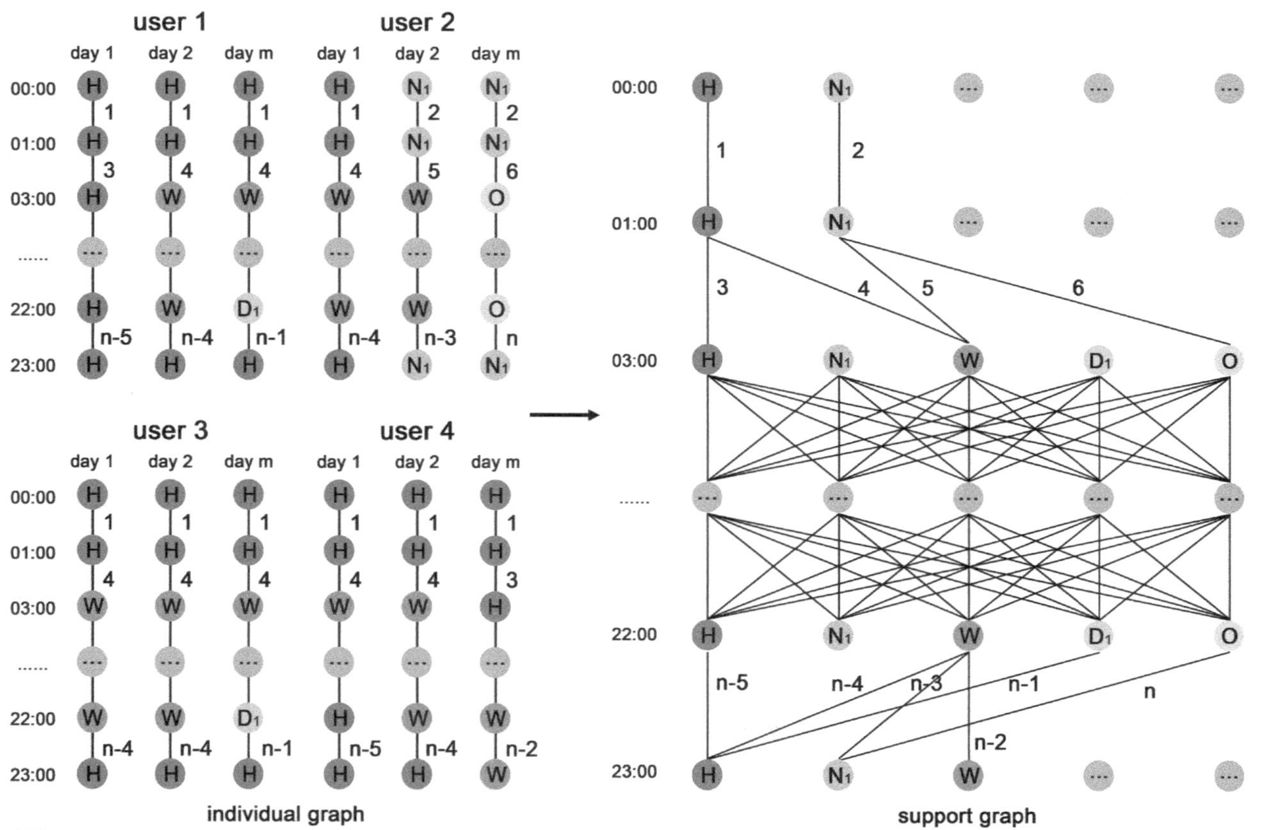

(A) Support graph construction and the branch indexing strategy based on a population of four individuals

FIG. 7 Individual graph, support graph, and topology-attribute matrix construction.

(Continued)

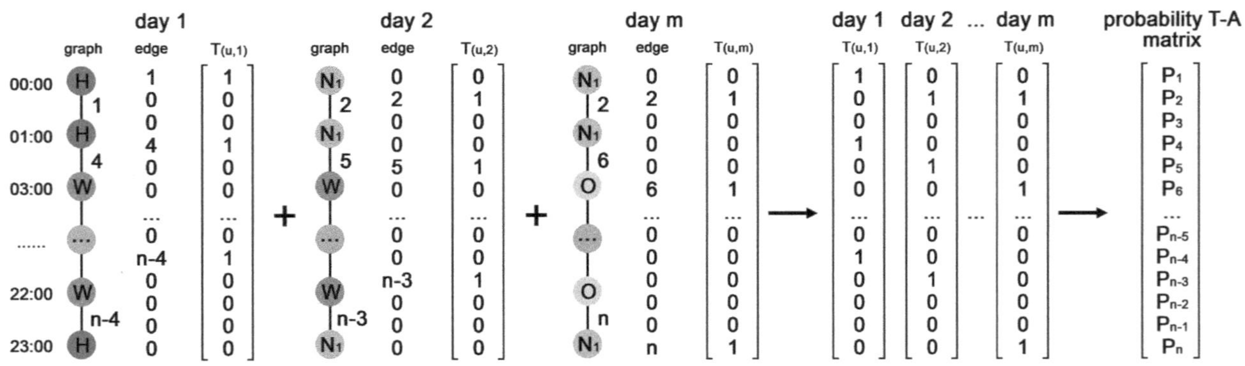

(B)

[T-A matrix] generation based on user 2

FIG. 7, cont'd

Structure constrained NMF and metagraph space: For a population of average probability T-A matrix $T=\{T_{average\ 1}, T_{average\ 2},\ldots,T_{average\ u}\}$, n denote the support graph with a total of n edges, u denote the total number of users. A total T-A matrix $T_{n\times \mathrm{u}}$ can be constructed for the total graph. With this representation, we can conduct graph clustering in the matrix space M. Once we identify the bases of M, each $T_{average\ u}$ can be decomposed into a linear combination of these bases. Then, the similarity between a pair of graphs can be defined based on the coefficients of the linear combination. The value of individual graph edges and the value of probability T-A matrix are all nonnegative, we use nonnegative matrix factorization (NMF) [43] to identify the bases of M.

4.1.3 Result illustration and analysis

Study area: The Great Tokyo area was chosen as the study area (Fig. 8). It includes Tokyo Metropolis and the surrounding area of prefecture Chiba, Ibaraki, Kanagawa, Saitama, Gunma, Yamanashi, and Tochigi. Both metropolitan areas and rural areas are included in the study area.

Significant places detection: There are a total of 74,693 users who can be clearly detected with the home location. There are a total of 74,693 home places, 17,033 other frequently visited nighttime spots, 50,125 workplaces, 63,183 other frequently visited daytime spots, and 266,895 other significant

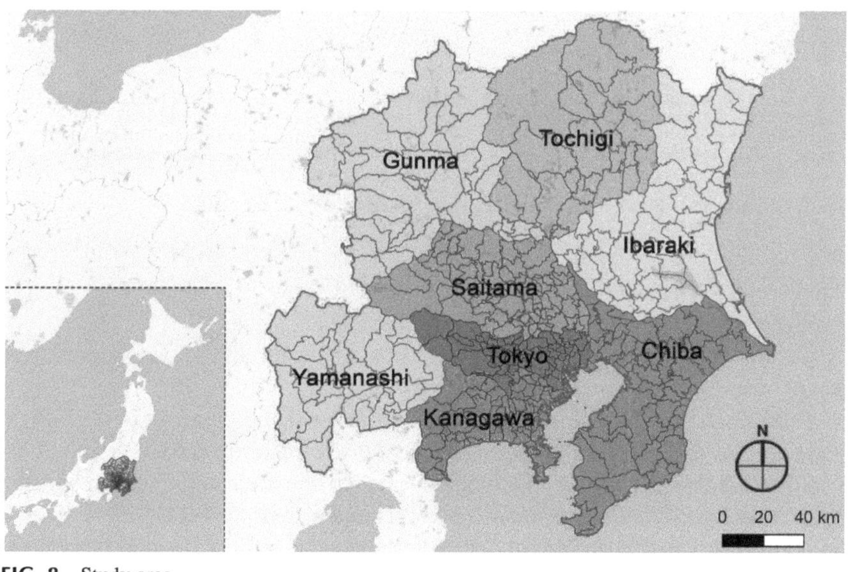

FIG. 8 Study area.

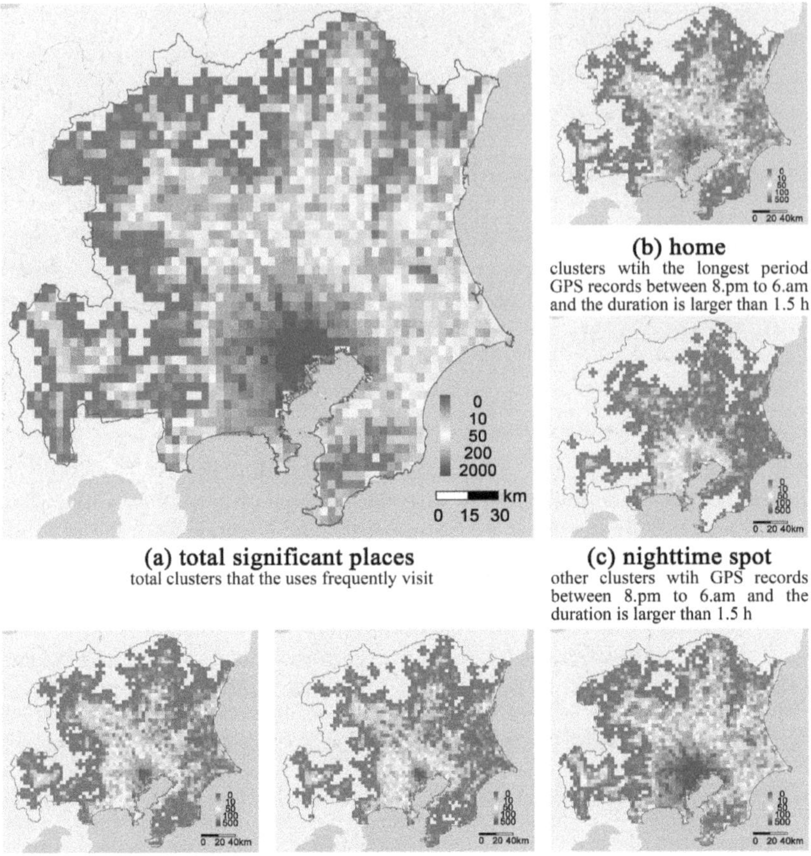

(a) total significant places
total clusters that the uses frequently visit

(b) home
clusters wtih the longest period GPS records between 8.pm to 6.am and the duration is larger than 1.5 h

(c) nighttime spot
other clusters wtih GPS records between 8.pm to 6.am and the duration is larger than 1.5 h

(d) work
clusters wtih the longest period GPS records between 9.am to 7.pm in the workday and the duration is larger than 1.5 h

(e) daytime spot
other clusters wtih GPS records between 9.am to 7.pm in the workday and the duration is larger than 1.5 h

(f) other significant places
other clusters that the uses frequently visit

FIG. 9 Heatmap and individual frequency distribution of significant places.

places detected for the raw GPS records. Fig. 9 shows the spatial distribution and individual frequency distribution of significant places.

Group characteristics of life pattern: We classify all the users into seven groups, which is determined by elbow method (Fig. 10).

Table 2 shows the life pattern characteristics of the central point of each group and gives a general description of the life pattern of each group. The X-axis refers to 24 h of a day. Each block refers to one node in the support graph, that is, one significant place of each hour. The height of the block in the Y-axis direction refers to the probability of which the users stay in the corresponding significant place.

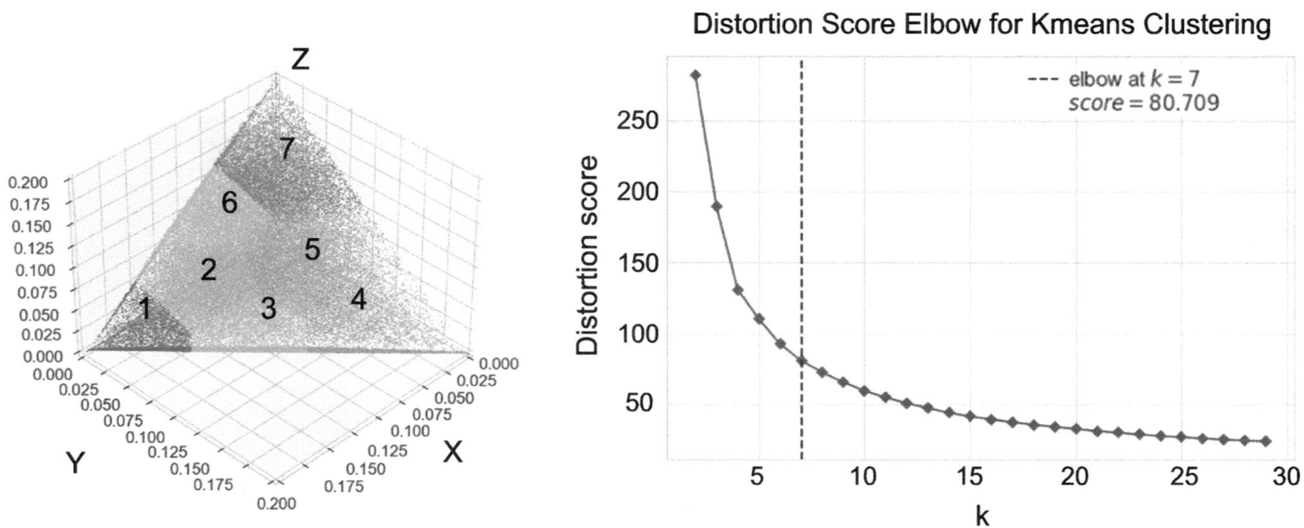

FIG. 10 Determine the number of cluster in Kmeans by elbow methods.

TABLE 2 Life pattern characteristics of each group.

Group characteristics	Description
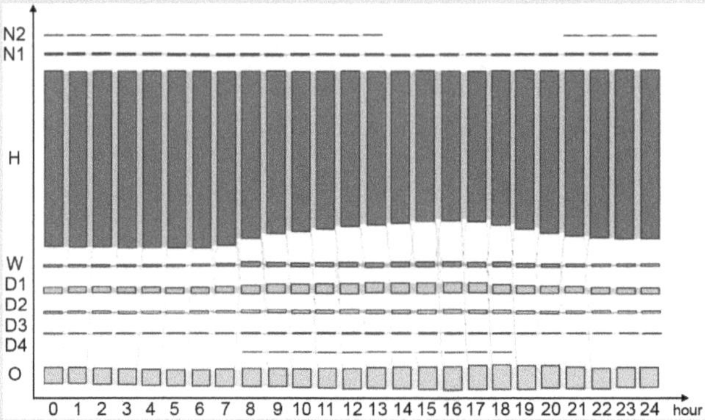	**Group 1. Home stayer: 85% Stay-at-home** The users of this group spent most of their time at home. They almost have no workplace and seldom go to other places. This group may refer to the old people or housewives who stay at home all day
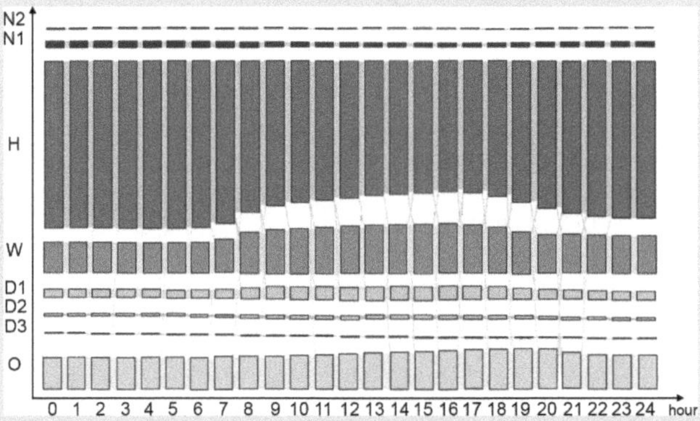	**Group 2. Home stayer: 60%-Stay-at-home** The users of this group have clear workplaces but they don't go to the workplaces often. This group may refer to the people who telework at home and go to the company sometimes
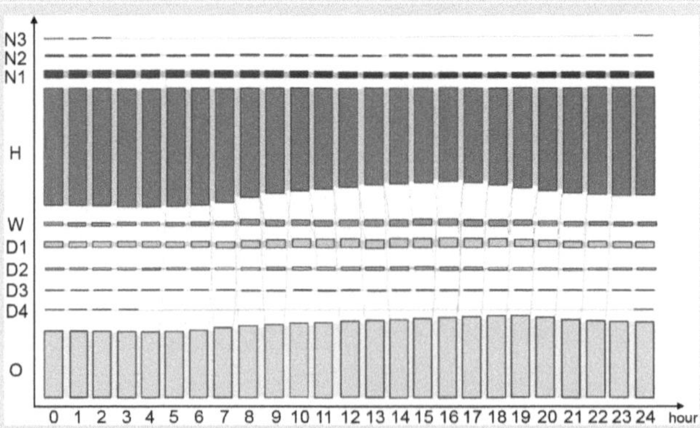	**Group 3. Home stayer: 50%-stay-at-home and 40%-stay at-other** The users of this group spend most of their time in the home or other places. They almost have no workplace

TABLE 2 Life pattern characteristics of each group—cont'd

Group characteristics	Description
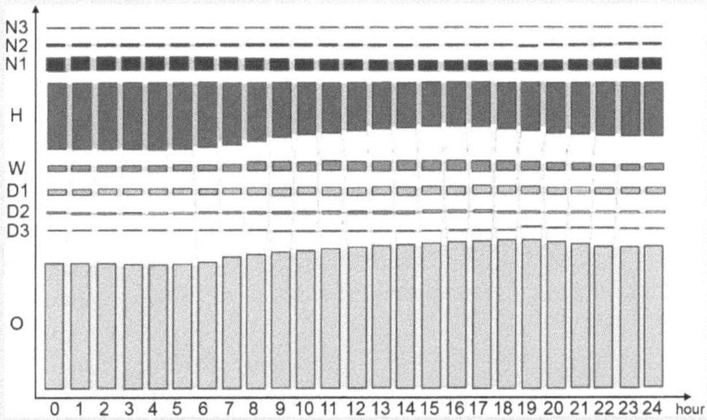	**Group 4. Traveler: 70%-stay-at-other** The users of this group spend most of their time in multiple other significant places
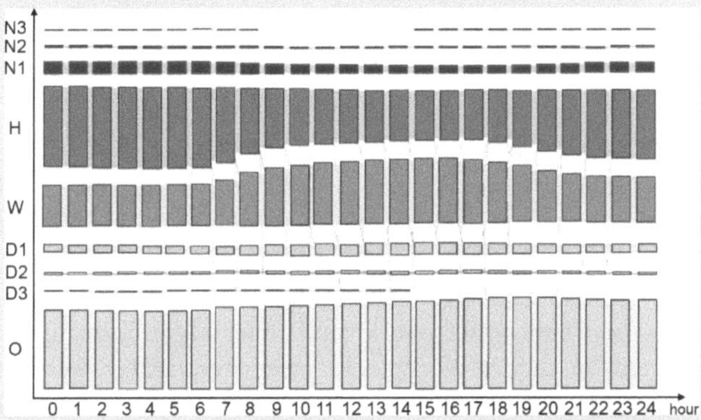	**Group 5. Work-life balance** The users of this group spend roughly equal time at home, work, and other places. Their working times are relatively short and they have much leisure time to go to other places
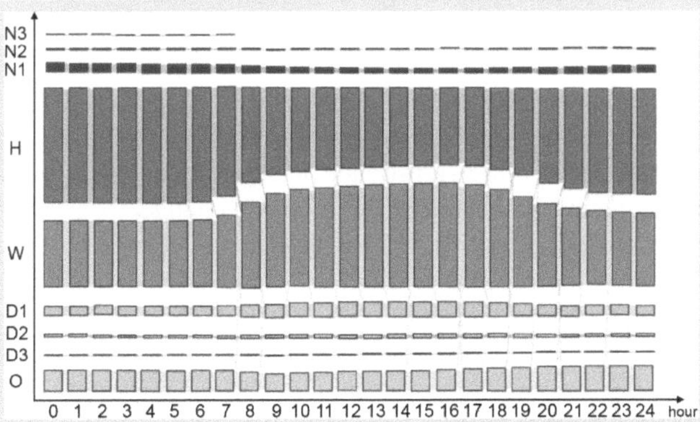	**Group 6. Regular office workers** The users of this group usually stay at home at nighttime and go to work in the daytime. They are typical office workers and have regular schedules. Little leisure time is spent in other places

Continued

TABLE 2 Life pattern characteristics of each group—cont'd

Group characteristics	Description
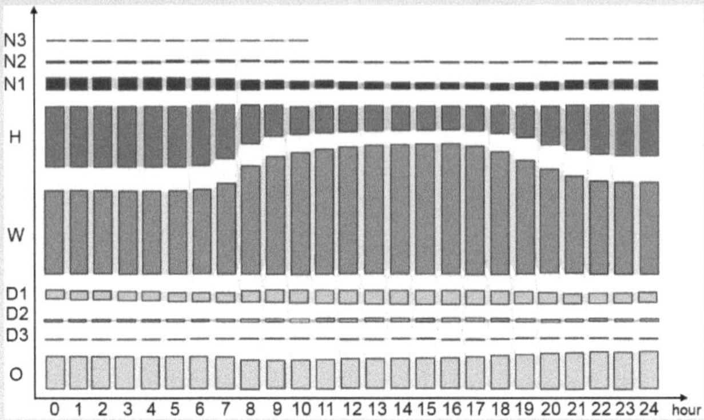	**Group 7. Workaholics** The users of this group have high-intensity work and spend most of their time in the workplace. The time spent in the home and other places is rare

Temporal changes of life pattern: We conducted statistics on how people change their life patterns groups on weekdays and weekends (Fig. 11). Group1 and Group 2 (Fig. 11) tend to stay at home more frequently; Group3 and Group 4 (Fig. 11) tend to stay at other significant places more frequently; Group 5 (Fig. 11) have balances between home, work, and other significant places; Group 6 and group 7 trend to stay at workplaces more frequently. From the aspect of people who stay at workplaces frequently on weekdays, from weekdays to weekends, 43.48% of group 6 change to group 1, 20.2% of group 6 change to group 2. In contrast to this, 7.00% of group 6 change to group 3, 0.93% of group 6 change to group 4. Similarly, from weekdays to weekends, 21.38% of group7 shift to group 1, 12.32% group 7 shift to group 2, while 12.9% of group 7 shift to group 3, 11.64% of group 7 shift to group 4. From the aspect of people who stay at other significant places frequently on weekdays, from weekdays to weekends, 24.62% of group 3 change to group 1, 2.70% of group 3 to group 2. In contrast, 0.05% of group 3 change to group 6, 0.03% of group 3 change to group7. Similarly, 7.32% of group 4 change to group1, 0.62% of group 4 change to group 2, while 0.01% of group 4 change to group 6, 0.069% of group 4 change to group 7. For people who stay at other significant places frequently on weekdays, the people who transform to stay at home on weekends are also more than those who transform to go to workplaces on weekends.

FIG. 11 Transformation of life patterns groups between weekdays and weekends

4.1.4 Conclusion

The finding has positive significance in urban planning. First, it provides an effective measurement for life pattern similarity. It is convenient for urban planners to directly compare the differences and diversification of life patterns in region scale, which helps in policy making for urban problems such as traffic congestion, home-work separation, public facilities mismatch, and regional unbalanced. Secondly, it reveals the representative life pattern groups. The representative activities and travel demand of these groups need to be considered in infrastructure planning and services improvement to provide them with better urban systematic support. Thirdly, as the complex life patterns including significant places semantics, time-sequential properties, and frequency are integrated into one vector in the metagraph space, it provides a simplified indicator the urban modeling and quantitative evaluation.

Stay location detection plays a foundational role in life pattern analysis. The stay locations represent the places that a user spends a considerable amount of time, for example, home, workplace, restaurant, market, gym, park, etc. Based on the stay location detection result that we can determine the significant place (e.g., home, workplace, other) which represents the individual's typical activity [44].

4.2 Travel segmentation and mode detection: Ride-sharing potential analysis

4.2.1 Introduction

Shared transportation has been widely argued as a key effective way to solve modern urban transportation problems, such as high energy consumption [45], road emission [46], and traffic congestion [47]. Ride-sharing, as one significant component of shared transportation, is an emerging transportation mode under a new concept of Mobility as a Service (MaaS) [48]. According to the existing cases, ride-sharing is generally introduced and operated by some emerging companies as a commercial project, rather than governmental public welfare [49]. Hence, the estimated production value of ride-sharing sector is the determinant for whether the investment would be determined and the preestimation for the potential of ride-sharing is necessary [50]. The work is published in Applied Energy [50].

4.2.2 Problem and methodology

ride-sharing potential analysis is an extremely complex issue [19]. There are certain knowledge gaps that need to be further solved:

Demand assessment (the scale of the user): Generally, the previous studies for the regional potential of ride-sharing are mainly based on the urban population distribution, which is rough and imprecise [20].

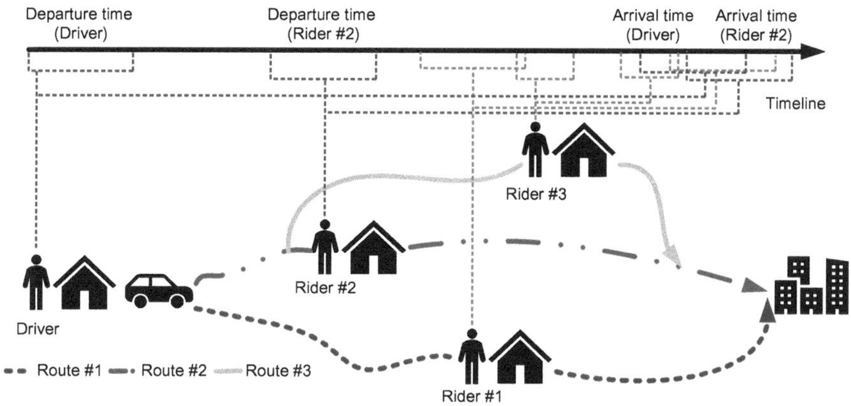

FIG. 12 Uncertainty of mobile trajectory and fuzziness of matching requirement.

Scenario analysis: The scenario analysis on the different extent of participation in ride-sharing from public transit riders.

Simulation of matching feasibility estimation: With the uncertainty of mobility trajectory and fuzziness of matching requirements, it is hard to predict the matching feasibility for each driver and rider [21]. As shown in Fig. 12, the commuting route for a driver is not unique, especially on an urban scale, there will be multiple routes for the same origin–destination (OD). Meanwhile, the possibility of choosing which route for a driver will depend on the current traffic condition, driver's habit, and even governmental traffic control scheme. For a real-world case, it is hard to observe that information from the individual level. The determinant for whether a driver and a rider would be matched or not is not only the similarity of their trip's OD but also that of their expected time scheme (departure and arrival time). Here, the asynchronism caused by driving time also should be taken into consideration.

Computational complexity: The existing optimization methods cannot support the big data-based potential and emission reduction analyses due to the computational complexity. A data-driven and high-effective algorithm is needed.

To address three issues, we used the framework consisting of stay point detection shown in Fig. 13.

Human travel mode detection model: The process contains four steps: extraction of staying and moving segments, splitting moving segments, traffic mode classification, and segment merging.

The stay segment of one participant's raw human trajectory is a group of consecutive points that represent the participant stopping at a location, while the moving segment represents the participant moving. In this research, we

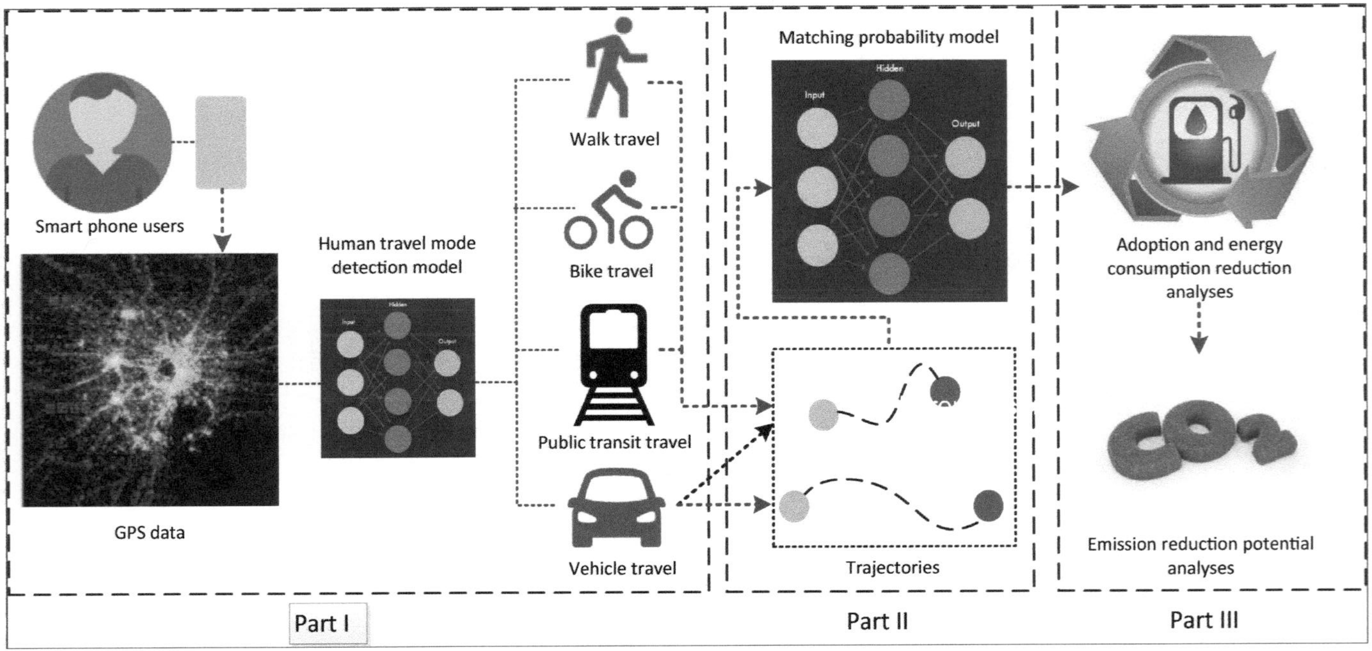

FIG. 13 Method framework.

extracted the stay segment based on the distance and time spamΔt of neighboring points which are less than the threshold. The ones which are greater than the threshold are identified as moving segments. We also removed the noise points based on the mean and standard deviation of Gaussian distribution [51].

The moving segments are split by the following procedures. Firstly, the segments were split into a walk and nonwalk trajectories by the speed threshold of 100 m/min. Then, we defined the velocity change rate as the average speed of the current segment compared with the current observed speed. The velocity change rate was utilized for extracting the change points which split the nonwalk trajectories. Moreover, the points in rail lines were further extracted to represent the trajectories of railway passengers.

Then, the traffic mode of each split GPS trajectory was detected and classified by the random forest method. In this research, the input features of each trajectory were as follows: total distance, time duration, percentage of points in the transportation network, and speed features including minimum speed, maximum speed, average speed, maximum acceleration, velocity change rate. The output labels of the model were four transportation modes including walk, bike, car, and public transit. We validated our model by test data sets of different transportations modes to check the performance of our model over these modes.

Once the classified travel segments had been obtained, we conducted segment merging to reduce trivial and uncertain segments. Consecutive segments with the same traffic mode were first extracted. Then, segments with the traffic modes that were uncertain were merged into neighboring segments. For other complex cases, classifiers with training data were used to merge the segments.

By implementing this travel mode detection model, we obtained the travel mode segments which determined the travel state of each participant during each period.

Matching feasibility estimation model: The model we used for matching feasibility estimation is the deep learning method, the structure of model training is shown in Fig. 14.

We used the GPS trajectory and did the training data preparation process as the base of the deep learning model. It includes the process of marking the matching feasibility of a pair of ODs. To make the deep learning model well trained, a comprehensive training dataset with different matching feasibility values is needed. However, in the original trajectory set, there is no direct mark of feasibility. Thus, a reliable strategy to generate training data is necessary. The detail of the generation and training process will be elaborated in the next paragraph. Finally, the generated data will be fed to the deep learning model to finish the training.

The training data is one of the essential factors that influence the performance of the deep learning model. In this case, using the deep learning model

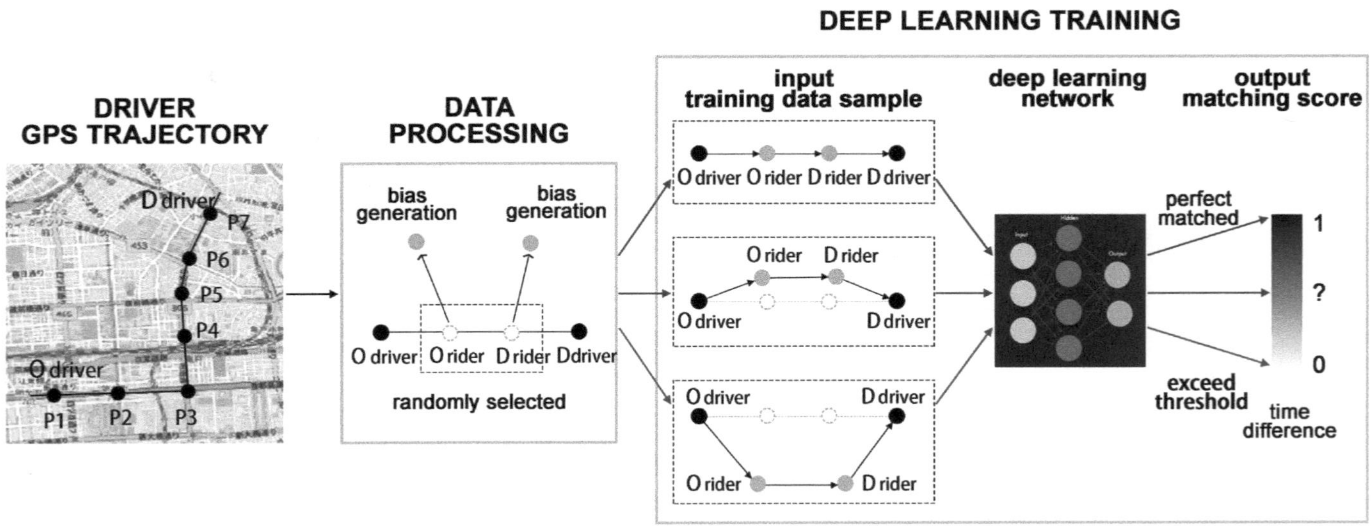

FIG. 14 Model training structure.

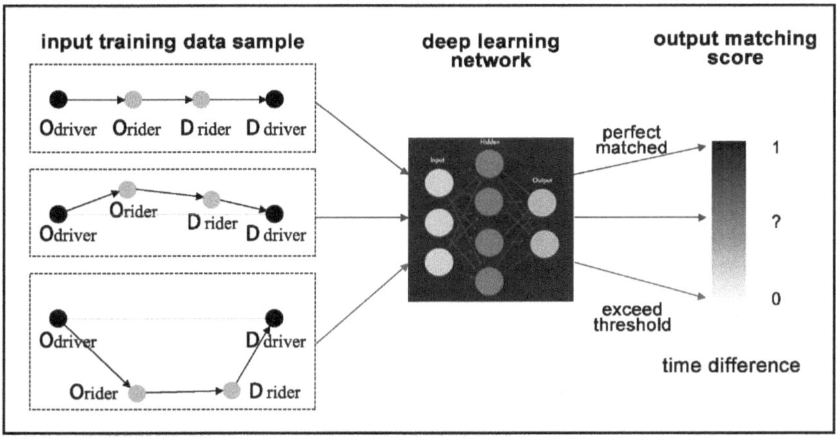

FIG. 15 Generation and training process of deep learning.

to compute the matching feasibility of two trajectories is a very complicated case. So, many factors should be considered. The first one is that the training cases should be comprehensive and cover various situations, which is the basic rule of training set.

The training data preparation process is shown in Fig. 15. Based on Definition 4 and 5 (in Section 3), we generate the training samples. We extracted a piece of complete and continuous trajectory record from the dataset, randomly chose the first and last two records in this trajectory as the OD of the driver. Then we randomly chose two other points in the interval of OD as the perfect matched rider (Rider*). Taking the OD information of the original trajectory (driver) $(ODx_{i,k}, ODt_{i,k})$, and perfectly matched OD information (rider) $(ODx, ODt)^*_{i,k}$ as the input, the corresponding feasibility output is $P_{(i,k),(i',k')}=1$. Then, we generated some noises z into the $(ODx, ODt)^*_{i,k}$ both from the aspect of location and time as the "not perfectly matched samples". In these samples, the feasibility output is a value between 0 and 1 based on the noise. We generated this kind of training data because comprehensive training data assures the accuracy of the deep learning model. When the noises are greater than a threshold, these samples can be as the "bad matched sample" $P_{(i,k),(i',k')}=0$. This process can be described as data augmentation [52]. The meaning of this step is to give the deep learning model a comprehensive understanding of the matching issue. Purely training on perfect matching will make the model tend to give the results of perfect. After that, we would put these training samples which contain the input (OD information of driver and rider) and output (feasibility value between 0 and 1) into the deep learning model. Finally, when finishing the training, this model could be used to operate feasibility estimation for real-world samples.

Deep learning architecture: Deep Learning is a branch of machine learning, the main idea of which is to simulate the mechanics of human brains. Therefore, deep learning applies layer-based construction to propagate information just like the electricity passing among brain neurons. Deep learning algorithms process data and imitate the thinking process of understanding the hidden key factors of output. Therefore, it is widely applied in complex data computation, image processing, natural language processing, and other fields. The detail of deep learning can be found in [53].

In this study, we adopted the input layer x_1 as: $(O_x^d, O_y^d, \triangle t_O, O_x^r, O_y^r, D_x^d, D_y^d, D_x^r, D_y^r, G(O_x^d), G(O_y^d), G(O_x^r), G(O_y^r), G(D_x^d), G(D_y^d), G(D_x^r), G(D_y^r))$ and the output layer y as $P_{(i,k),(i',k')}$. Where, O_x and O_y refer to the longitude and latitude of origin; D_x and D_y refer to the longitude and latitude of destination; d and r refer to driver and rider respectively; $\triangle t_O$ refers to the time difference between departure time; $G()$ function is used to compute the grid coordinate. We add the gridded coordinate of O and D to give the deep learning model a hierarchical understanding of the spatial distribution information and improve the computation efficiency [54]. The loss function we use for gradient descent is mean squared error and the optimizer is adam [55]. As elaborated in knowledge gap 2 and Fig. 2, we need to find if the OD of a rider is on or near to the trajectory of drivers to estimate the feasibility to match the driver and rider. Since in the process of training data preparation, we used the stochastically selected two GPS points in the trajectory of the driver (except origin and destination) as the base, in implementation, the deep learning will try to discover the similarity of passenger's trajectory with the driver.

In this model, given a set of $Rid_{i', k'}$ and $Dri_{i, k}$, the matching feasibility estimation model will construct a model $P_\theta\left(\widetilde{P}_{(i, k), (i', k')} | \left(Dri_{i,k}, Rid_{i',k'}\right)\right)$, where $\widetilde{P}_{(i, k), (i', k')}$ is the predicted matching feasibility. It will be built as a regression model, and its parameters can be obtained by minimizing the prediction error $L\left(\widetilde{P}_{(i, k), (i', k')}, P_{(i, k), (i', k')}\right)$ with generated training samples:

$$\text{Min}\left(\text{L}(\theta) = \frac{1}{n}\left(\sum_{i=1}^{n}(y - h_\theta)\right)\right) \tag{7}$$

In the Result and Discussion Section, we will illustrate and elaborate on the performance of the trained model.

4.2.3 *Result illustration and analysis*

Study Area: For the issue of ride-sharing and emission reduction potential analyses, the Tokyo area, Japan is a suitable study example. The reason is, after the accident of the Fukushima Daiichi nuclear power plant, Japan immediately shut down all 50 nuclear power plants in the country and aggrandize fossil fuel consumption, which brought on a significant increase in emissions. Therefore, emission reduction turned into a key mission for the Japanese government. Besides, Tokyo, as the capital and biggest city in Japan, is one of the most populated cities in the world. The production value of MaaS is considerable, but this industry is just getting started. Ride-sharing is an efficient and low-cost method for the Tokyo area to reduce fossil fuel consumption and air pollution and relieve traffic jam.

Result of matching and spatial analysis: we use our deep learning model to find the best match for all the trips respectively and compute the emission and travel distance that could be potentially reduced by shared transportation. Each individual trip will be matched with the best candidate trip that is selected by the deep learning model in the dataset despite himself. There are about 1.05×10^6pieces of individual trips in the study case. From the matching result, nearly 81.29% of trajectories, that is said to be about 8.5×10^5 pieces can be matched to another proper trajectory. The rest cannot find a candidate sharer, which means their shared trip cannot save travel distance. Averagely, each matching pair can reduce the travel distance by 26.97% compared to their original travel distance. It tells us that there are relatively high similarities in the travel pattern of people in Tokyo. The potential of ride-sharing in Tokyo area is considerable. In this case, totally over 1.35×10^8 kilometers of travel distance can be reduced. Here are two samples from the results of matching to illustrate how ride-sharing saves travel distance.

In Fig. 16, the probable trajectories if they are matched by ride-sharing and their original trajectories are shown in b and d. There are some similarities between their trajectories. In both cases, one can just detour a little distance or change his or her initial routine to pick the other up and complete both travels simultaneously. According to the computation, if this ride-sharing matching is adopted, the travel distance can be reduced by 26.95% and 26.94% respectively.

To illustrate the whole case study, we plot a heat map of all the origins of matchable trajectories in the study area, as shown in Fig. 17. We found that most of the cases were distributed in the center of Tokyo and Tokyo station, where many government offices and financial facilities are located around it, as well as the subcenters, typically the area of Ikebukuro, Shinjuku, Shibuya, and Kawazaki. They are the central business districts distributed around the city center. Commonly in Japan, more mobility flows are appearing in these places because of activities, such as commuting and daily shopping, thus, causing more travel demands around these areas. Here is a detailed explanation. According to the investigation conducted by the Ministry of Land, Infrastructure, Transport, and Tourism [56], in the Tokyo area, a considerable part of people travel every day with the railway. The stations with extremely high traffic volumes include the Ikebukuro, Shinjuku, Shibuya, Tokyo, and Kawasaki station. So, these areas become a large distribution center of origins of travels. Therefore, it is obvious

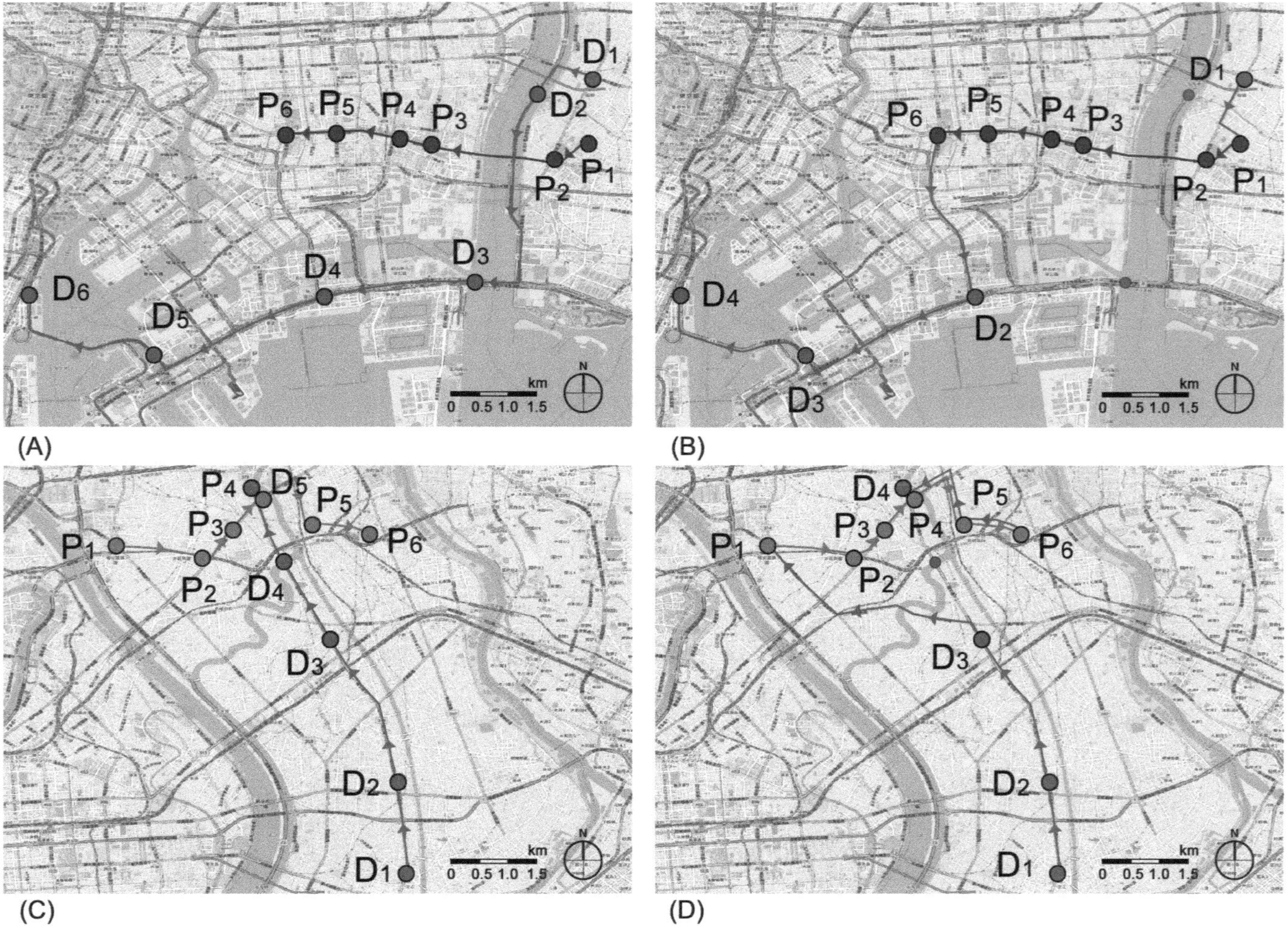

FIG. 16 Sample cases from matching result: (A) Initial trajectories of passenger and driver in case 1 (B) Matched trajectories in case 1 (C) Initial trajectories of passenger and driver in Case 2 (D) Matched trajectories in case 2.

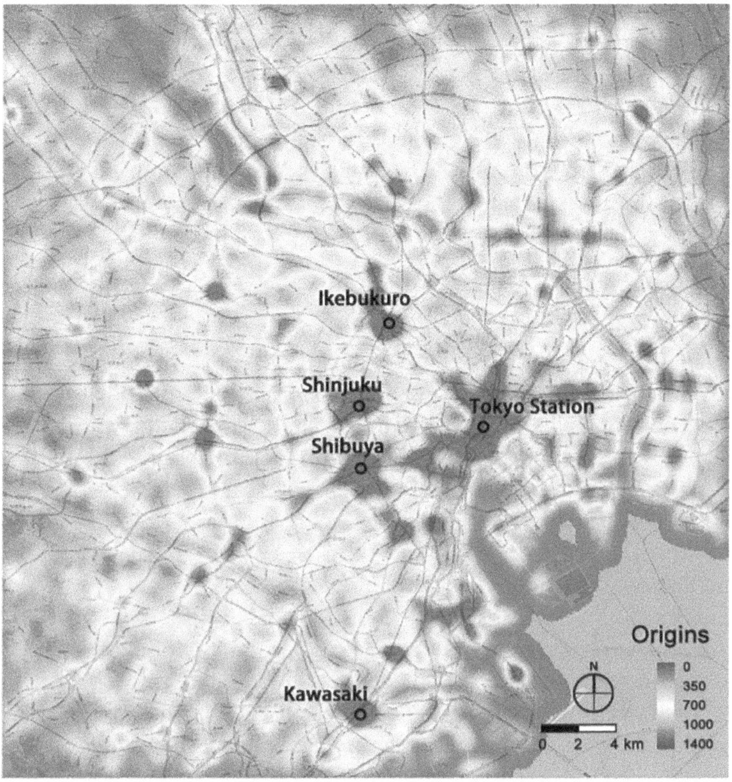

FIG. 17 Heat map of origins of trajectories.

that there are more ride-sharing potentials around these areas and ride-sharing service providers like Uber, can consider paying more attention to these places if they want to develop ride-sharing services in Tokyo area.

Emission analysis: We give the discussion on the emission results in two scenarios.

Scenario 1:

In this scenario, we assume that half of the original public transit riders will use the ride-sharing system, which refers to about 11.05% of trips. Originally, if all the trips are completed by their original travel mode, the total emission would be 41,862.36 tons of CO_2. The public transit contributed 58.97 tons, which accounts for 0.14% of total emissions. Averagely speaking, each trip emitted 24.61 kg of CO_2. Each public transit trip emitted 0.31 g of CO_2 and each vehicle trip emitted 27.63 kg of CO_2. If that 11.05% of public transit travels are shared with their potential drivers and the trips which are originally traveled with private vehicles turn to adopt ride-sharing, the total emission would be averagely reduced to 6478.45 tons of CO_2 with a reduction percentage of

84.52%, which would be 6.86 kg of CO_2 per trip. The global effect on emission reduction is great.

Here we will discuss the effect of ride-sharing on public transit travel and vehicle travel. For those 11.05% of public transit travels who turn to adopt ride-sharing along with their potential trip sharers, originally, their total emission would be 4027.69 tons of CO2. Among them, public transit travels contributed 29.13 tons and vehicle travel contributed 3998.56 tons, which also refers to 0.31 kg per public transit rider and 42.85 kg per vehicle driver. Assume they adopt the ride-sharing, the summed-up CO_2 emission would be 509.76 tons. Averaged to per trip, the quantity would be 5.42 kg; averaged to per person, the quantity would be 2.71 kg. Thus, we can see that although ridesharing brings additional emissions to original public transit riders, the beneficial effect is that emission from vehicle drivers is reduced. Overall, the reduction effect is very obvious. The total CO2 emission quantity can be reduced by 87.34%.

For the other 77.90% vehicle travels, without ride-sharing, these trips emitted about 35,798.52 tons of CO_2. Each individual trip released 27.02 kg CO_2. Suppose that these trips are shared, the total emission would be reduced to 3935.55 tons. At average level, each trip is 5.94 kg and each individual is 2.97 kg. The emission level can be reduced by 89.01%.

Scenario 2:

In this scenario, we assume that all of the original public transit riders will use the ride-sharing system, which accounts for about 22.10% of trips. If that 22.10% of trajectories that originally used public transit turn to adopt ride-sharing and all the trajectories that originally traveled with private vehicles turn to adopt ride-sharing, the total emission would be reduced from 41,862.36 to 4932.27 tons of CO_2, which would be 2.08 kg of CO_2 per trip.

In the case of original public transit riders, the total emission is reduced from 6063.83 tons to 996.72 tons, which refers to a reduction of 83.56%. The averagely distributed emission over public transit riders increases from 0.31 to 2.65 kg; one of their corresponding potential trip sharers decreases from 31.94 to 2.65 kg. Compared with the result in Scenario 1, although the emission reduction to the vehicle trip is gradually significant, it can be concluded that with the spread of ride-sharing to public transit, the transformation from public transit travel to ride-sharing doesn't necessarily mean it can provide more overall benefit. It shows a rebound effect of ride-sharing of transform from public transit travel. Our dataset reflects the benefit brought by participation from public transit riders to on-road vehicle emissions. Although in this assumed scenario, these public transit riders turn to adopt ride-sharing, the emission of public transit still takes place. In the future, data from public transit operators can be integrated from a more comprehensive analysis.

Thus, we can see that ride-sharing can truly bring an emission reduction of CO_2 to city air. In our case, when half of the public transit riders turn to adopt ride-sharing and all private car drivers adopt ride-sharing, around 84.52% of CO_2 emission can be reduced; when all of the travelers adopt ride-sharing,

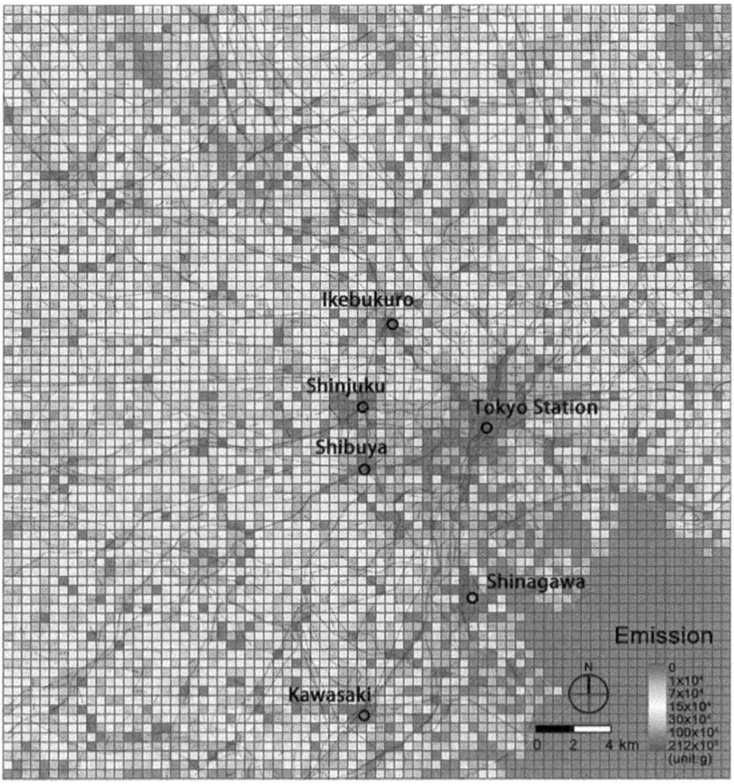

FIG. 18 Heat map of the distribution of the quantity of emission that can be reduced (Unit: g).

around 83.56% of total CO_2 emission can be reduced. Combining the result of matching that 81.29% of trajectories can be feasibly matched with another trajectory, there is a lot of potential for ride-sharing in the Tokyo area.

Here, we plot the spatial distribution of CO_2 that can be reduced in Scenarios 2 by ride-sharing in Fig. 18.

The figure mainly shows the location of the origins of exhaust emission and the total quantity of CO_2 that can be reduced by ride-sharing in the study area. According to the map, the main areas like Ikebukuro, Shinjuku, Shibuya, and Tokyo station can be improved much more in exhaust emission. This is because of the high concentration of travel origins in these areas that is explained in the previous subsection. A relatively high quantity of travel in these areas can be matched with another trip and a high potential of ride-sharing exists. Thus, ride-sharing can bring a lot of emission reduction that is caused by vehicle traffic in these areas. This also indicates that the population of ride-sharing can largely alleviate the heavy traffic congestion and exhaust emission in populated areas.

4.2.4 Conclusion

We proposed an analysis method to mine the potential of ride-sharing in an area based on quantitative historical GPS trajectories. The deep learning model in the method aims to find out how well the trajectories can be matched, thus sharing. Then we chose the case study of Tokyo area because Tokyo is a city faced with heavy traffic congestion and exhaust emission under the background of rising usage of fossil fuel in Japan. By the computation of deep learning model and exhaust emission, we found that travels in Tokyo area mainly concentrated in the center and subcenters of Tokyo, like Shibuya, Shinjuku, and Ikebukuro, which makes these places the sources of exhaust emission. If ride-sharing is adopted, the exhaust emission reduction effect by ride-sharing can perform well in these areas.

Travel segmentation and mode detection play an essential role. It helps us understand the real vehicle traveling pattern in the study area so that we can match the travel demands. What's more, it can help us recognize which trajectory is vehicle travel and if it can be the picker in the ride-sharing. The usage of the real trajectory can give a direct sense.

4.3 Map matching: Estimation of urban scale PM emission

4.3.1 Introduction

In recent years, global air pollution has been a serious problem and has become a threat to public health [57]. Braking is a common driving behavior in daily life transportation. According to some previous studies, this behavior can cause nonexhaust emission and air pollution through the wearing of braking discs and releasing material of braking discs into the air [58] as well as health problems to people [59]. Study on particulate matter emission from braking is urgent for the improvement of cleaner and fresher air. Among these studies, most of them focused on the study on individual or blurry crowd estimation from statistics and empirical evidence, which are not clear enough to observe the emission performance of braking behavior. There are few which put focus on the analysis based on massive specific records of individual driving trajectories to provide a clearer sight at an urban scale.

Based on massive GPS trajectory records with small time intervals in Tokyo area, we proposed an analytical framework on the emission behavior of braking, which will give a more comprehensive view on emission performance of on-road braking on a large urban scale. The map matching method as well as the corresponding smoothing method is utilized. We utilize millions of GPS trajectory records of vehicle travels collected by NAVITIME app in users' phones in Tokyo area with small time intervals. The work is published in JCLP [59].

4.3.2 Problem and methodology

We employed the raw GPS trajectories with the time interval of 1 s from vehicle driving and road grid data as the base of this work. To compute the quantity of PM emission and give an analysis on the emission behavior of on-road braking events, we need to overcome three challenges from our dataset:

Extraction and location of on-road driving event from raw trajectory data: The premise of detecting on-road braking event is to extract and locate the on-road driving event including accelerating behavior and deceleration behavior in the trajectory. On-road driving events always happen in a very short time interval. Thus, the trajectory data should be recorded in short time interval. In this study, we used the driving trajectory data with a time interval of 1 s. However, in the raw trajectory data, there are always biases and errors. This brings a lot of disruption to the driving event information. The first challenge is to discover this information.

Detection of on-road braking event from driving event as well as driving distance during the braking event: Once we get the on-road driving event information, we can try to detect the braking event. However, purely treating the deceleration process in the driving as the braking is not accurate. Moreover, the duration time of the deceleration event and the driving distance during the deceleration also need to be considered. A solid metric for detection and computation is necessary.

Computation of particle matter emission from the driving distance during the braking event: Based on the detected driving distance during the braking event. The transformation from driving distance to particle matter emission from braking event is also a critical problem.

To solve the challenges and problems described above, we chose the framework shown in Fig. 19.

Map Matching: Due to the error of sensors or other factors, there are always vacancies and biases in the GPS records. When there was a vacancy, the continuity and completeness of trajectory would be lost. To deal with the problem, we chose to cut the trajectories into subsets if there existed long time intervals. On the other hand, biases in the GPS sensors could cause the error in the true location of vehicles and the record of GPS location could usually be beyond the road grid. Thus, we need to match the GPS points on the road grid. Firstly, we downloaded shape file of road grid of Tokyo area from OSM (Open Street Map). The original data from OSM isn't complete. We need to use PostgreSQL to do topology on these roads and each piece of road segmentation will be given a unique ID and logical connection. The logical connection is used to indicate the geographical connection among each road segmentation. The meaning of logical connection is that after the GPS trajectory is matched to the road network model, the driving distance between the neighboring points can be computed. After topology, we have 97,235 road segmentations with unique IDs ranging from 6 to 97,240.

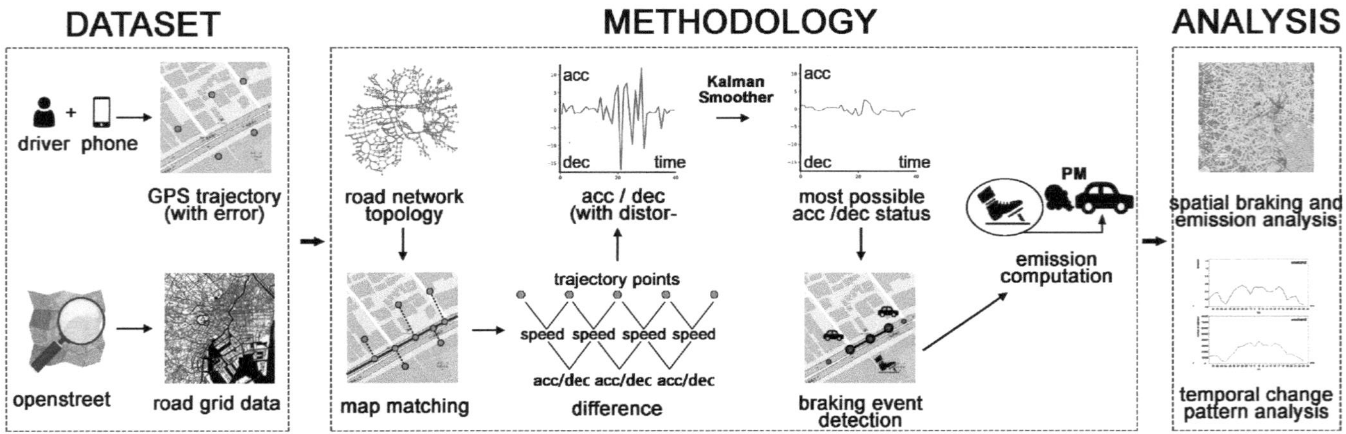

FIG. 19 Framework.

Then, we adopt map matching algorithm to find the match point of GPS trajectory point on the road grid. The principle of algorithm is to draw perpendiculars to road lines and match the point to foot of perpendicular with the least length for it is the most possible position on the road grid.

$$p_i' = \arg\min\left(verticaldis(p_i, R_i)\right), R_i \in (6, 97240) \quad (8)$$

where, p_i is the raw GPS point; p_i' is the matched point on-road grid; R_i is the road segmentation identity; *verticaldis* is the function that computes the vertical distance between the raw GPS point and road segmentation.

However, sometimes error can be enormous enough to bring about the misjudge of true position. In these cases, we would detect and get rid of these outliers in the trajectory. At these vacancies, we would do interpolation in the trajectory.

After map matching, the distance on the road grid between these continuously matched points, on-road acceleration, or deceleration can be computed.

Acceleration and Deceleration Smoothing: As discussed before, the traditional map matching method has one main flaw. As driving is a continuous process, the curve of temporal change of velocity should show a sort of continuity and smoothness. But the traditional map matching method doesn't consider the connection among the trajectory points and can probably break the continuity. The true velocity information can be undermined by the map matching method. Here is an illustration. The principal of map matching method is to match the GPS point beyond the road grid to the one on the road grid which is nearest to the original point.

As shown in Fig. 20, there are a series of six neighboring GPS points P1, P2, P3, P4, P5, P6, which are marked blue to refer to the ground truth. Because of the error of GPS record, the recorded point may be shown as the red ones in the figure. These recorded points are probably beyond the road grid line. What map matching do is to draw the perpendicular to the road grid line and consider the drop foot as the matched point on the road. Thus, it can be concluded that the map matching ignored the connection among these six points. The distance among these six points is not correctly displayed. So, there would be a lot of distortion in the computed deceleration change pattern.

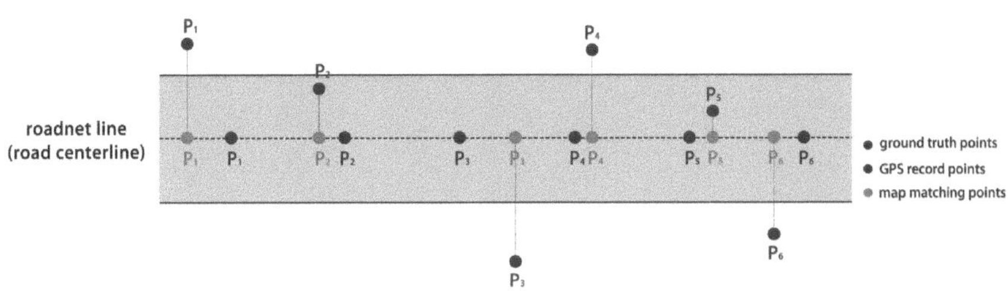

FIG. 20 Illustration of map matching method.

Thus, after applying the map matching, we also need to adopt other methods to correct the deviant states of acceleration and deceleration. In this study, we adopt the Kalman Smoother method.

Kalman Smoother uses the same computation principle as Kalman Filter but with a priori condition because all the trajectories of time series have already been observed. It is a classical algorithm to estimate the joint probability distribution of state of a system, typically physical models in real-world, over a time series with statistical noise and uncertainties in the system. The basic concept is to give the best prediction of current state of system by combining the past state information and observation of current state. Applying Kalman Smoother into the system requires a certain understanding of the system, like the covariance of observation of state and the covariance of state transition. Thus, to implement Kalman Filter to velocity correction of trajectory, we need to identify some basic information of each trajectory, mathematically estimates the state-space model parameters. In this study, we choose velocity and acceleration as the state parameter because these two are the symbols of braking event, so the state variable matrix can be written as:

$$X = \begin{bmatrix} v \\ a \end{bmatrix} \tag{9}$$

where X is the state variable matrix; v is the velocity; a is the acceleration.

Suppose we have a series of observed trajectory set after map matching $P_t{'} \in T, t = 0, 1, 2, \ldots, n$.

So, the velocity at each moment can be calculated by:

$$V_t{'} = \frac{dis(P_t{'}, P_{t+1}{'})}{\Delta t}, t = 0, 1, 2, \ldots\ldots, n-1 \tag{10}$$

where $dis(P_t{'}, P_{t+1}{'})$ is used to calculate the distance from $P_t{'}$ to $P_{t+1}{'}$ on the road grid.

As the same principle, the acceleration can be calculated by:

$$a_t{'} = \frac{V_{t+1}{'} - V_t{'}}{\Delta t}, t = 0, 1, 2, \ldots\ldots, n-2 \tag{11}$$

To make the length of v and a equal, we assume:

$$a_{n-1}{'} = 0 \tag{12}$$

Now, we can get the observed state variable series and suppose:

$$X' \sim \left(\vec{\mu'}, \vec{\sigma'} \right) \tag{13}$$

According to the relationship between velocity and acceleration, the change of state variable from time $t-1$ to t can be written as:

$$\begin{cases} V_t = V_{t-1} + a_{t-1} * \Delta t + \Delta a_{t-1} * \Delta t \\ a_t = a_{t-1} + \Delta a_{t-1} * \Delta t \end{cases} \tag{14}$$

where, V_t refers to the velocity at time t; a_t refers to the acceleration at time t; V_{t-1} and a_{t-1} refer to the same things at time $t-1$ respectively; $\triangle t$ refers to the time step.

In the form of matrix:

$$\begin{bmatrix} V_t \\ a_t \end{bmatrix} = \begin{bmatrix} 1 & \Delta t \\ 0 & 1 \end{bmatrix} \begin{bmatrix} V_{t-1} \\ a_{t-1} \end{bmatrix} + \begin{bmatrix} \Delta t \\ \Delta t \end{bmatrix} \Delta a_{t-1} = AX_{t-1} + FU \tag{15}$$

This is the equation for posteriori prediction of state variables. In this equation, term FU refers to the input of system from outside.

Simultaneously, the covariance of prediction needs to be updated by:

$$E_t = ACov(X_{t-1})A^T + Q_t \tag{16}$$

where, E_t refers to the posteriori error covariance matrix used to estimate the error of state variable prediction; Q_t refers to the external uncertainty; $Cov(X_{t-1})$ is the covariance matrix of state variable at time $t-1$.

Since we don't know the state of variables before the observation data. In this study, we set the initial state variable as the first observed state and set the initial covariance as zero:

$$\begin{bmatrix} v_0 \\ a_0 \end{bmatrix} = \begin{bmatrix} v_0^r \\ a_0^r \end{bmatrix} \tag{17}$$

$$Cov(X_0) = \begin{bmatrix} 0 & 0 \\ 0 & 0 \end{bmatrix} \tag{18}$$

v^r and a^r refer to the velocity and acceleration computed from raw trajectory data.

The above equations are prediction part of Kalman Filter. The principle of Kalman Smoother is combining prediction and observation. The observation data we use is the velocity computed from raw trajectory. Thus, the observation matrix is:

$$H = \begin{bmatrix} 1 & 0 \\ 0 & 0 \end{bmatrix} \tag{19}$$

Suppose the combined final result satisfies:

$$\hat{X}_t \sim N\left(\vec{\hat{\mu}}, \vec{\hat{\sigma}}\right) \tag{20}$$

The best prediction of state variable from Kalman Filter can be written as:

$$\vec{\hat{\mu}} = HX_t + K\left(\vec{\mu'} - HX_t\right) \tag{21}$$

$$\vec{\hat{\sigma}} = (E-K)ACov(X_{t-1})A^T \tag{22}$$

$$\hat{X}_t = X_t + K\left(\vec{\mu'} - HX_t\right) \tag{23}$$

$$\hat{E}_t = E_t(I - KH) \tag{24}$$

$$K = ACov(X_{t-1})A^T H^T \left(HE_t H^T + \vec{\sigma'}\right)^{-1} \tag{25}$$

Here, we still have some parameters like FU, Q_t have not been determined. Generally, offline Kalman Smoother aims to find the maximal possibility of state variables with priori estimation:

$$\max P\left(X_t | X'_{0,1,2,\ldots,n-1}\right) \tag{26}$$

Thus, the rest of the parameters can be identified by applying EM algorithm. It is used to compute and estimate the parameter of model that are unobservable latent variables.

In this study, the EM algorithm can be employed to find:

$$\left(FU, \vec{\mu'}, \vec{\sigma'}\right) = argmaxP\left(X'_{0,1,2,\ldots,n-1}\right) \tag{27}$$

The algorithm involves two steps in each iteration: E(Expectation) step and M(Maximization) step. At the beginning of iteration, the algorithm employee expectation step to create a function that uses the current estimated system parameters to expect the evaluation of log-likelihood. Then, the maximization step is utilized to compute the system parameters that can maximize the expected likelihood estimation function created in the expectation step. The iteration process will come to a convergence until the differences among the results from each iteration process are small enough. According to Eq. (15), we need U to estimate the state variable. However, we cannot directly obtain the input from raw data. So, in this study, we apply the EM algorithm combined with observed $a^{r}_{0,1,2,\ldots,n-2}$ to estimate the U.

By applying the EM algorithm before filtering the raw velocity curve, the optimal parameter of Kalman Filter can be determined. But before implementing Kalman Filter, we need to remove the outliers in the raw velocity data because Kalman Filter assumes the noise and presence of observed data can be described with a Gaussian distribution. However, error of observation brought by map matching and bad internet connection during the record can be so enormous that there could be outliers that don't satisfy this condition. This can corrupt the performance of Kalman Filter [60]. Thus, we need to compute the acceleration of each second and remove the unreasonable ones first and then apply EM algorithm and Kalman Filter.

Braking Event Detection: Braking event in the GPS trajectory cannot be purely defined by a decrease in the velocity curve of traveling because turning can also cause deceleration.

According to some previous researches, despite special events, such as traffic accidents, the braking event in the GPS trajectory can be detected by the deceleration in the driving process over a certain time series. According to previous research [61], if there is an integral of deceleration over at least 4 s more than −2.5 m/s, the driving event can be thought of as braking. According to this metric, we can find the GPS trajectory point referring to the braking event. Thus, the driving distance during braking can be calculated.

Particulate Matter Emission Computation: In this research, we mainly consider the emission quantity of PM10. We employed the brake wear PM10 emission factors proposed by MOVES2014 (Motor Vehicle Emission Simulator 2014). MOVES is an emission modeling system released by the US Environmental Protection Agency and it is considered to perform well in emission estimation [62, 63]. Based on some generalized assumptions [64], it conducted various tests under the near driving condition. The emission factor concluded in the study is suitable for big data computation. Under this circumstance, we assume that the emission factor of each vehicle in the same vehicle type is the same. The average PM10 emission rate of braking events of LDV is 17.6 mg/mi/vehicle and the one for HDV is 24.13 mg/mi/vehicle, which can be converted to 11 mg/km/vehicle for LDV and 15 mg/km/vehicle for HDV. The distance is the distance of on-road braking event.

4.3.3 Result illustration and analysis

Study area: used the dataset of GPS trajectories in the Tokyo area with the time interval of 1 s with the mark of vehicle type distinguished by LDV and HDV.

Spatial Braking and Emission Analysis: After applying the smoothing method, we adopted the metric introduced in the methodology part to detect the braking event in the trajectory. We found 729,743 braking events in these trajectories according to this braking event detection metric from all matched vehicle travel trajectories of 24,846 independent navigation identities. The spatial distribution of all braking events is plotted in Fig. 21. The braking events also usually happened in the city center and popular crowded areas due to the large traffic volume. Also, these areas are the origins and destinations of vehicle traveling. On the expressway or highway, the braking events mainly happened on the main trunk roads, like Shibuya Line, Shinjuku Line, and Ikebukuro Line. These roads link the popular and crowded commercial centers with the outside and play an important role in transportation paths with heavy traffic. Thus, these links are also frequent areas of braking and high traffic emission.

The distribution of PM 10 braking emission can be shown in Fig. 22. Similarly, the emission of braking events is mainly located in the popular areas and primary trunk roads in the city. These areas also usually suffer more pollution from the particulate matter of exhaust. So, the effort from the government

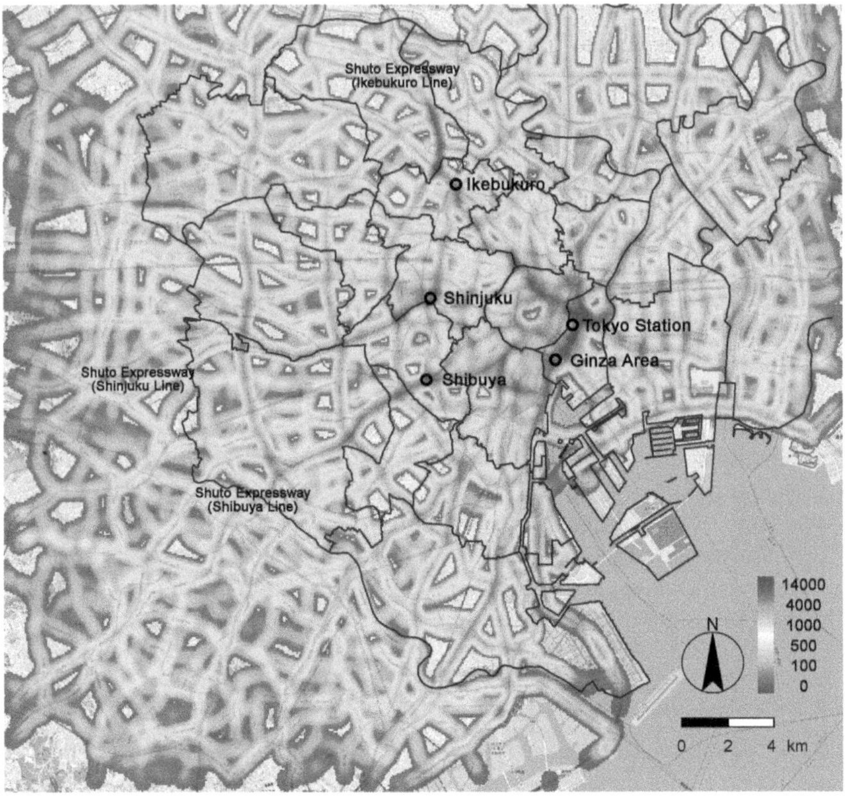

FIG. 21 The spatial distribution of braking events.

should focus on the volume limitation of vehicle traffic in these areas as well as the traffic dredge to reduce the braking events.

Temporal Change Pattern Analysis: The whole study case across 1 month, so we chose a weekday and weekend for the temporal change pattern analysis. The temporal pattern in 2 days is shown in Fig. 23.

We could see from the figure that the temporal change pattern between weekdays and weekends is quite different. On weekdays, the braking emission and traffic volume are continuously high during the day time but low in the evening and deep night. on weekends, the temporal change pattern is much steadier compared with weekdays. The total quantity of braking emissions on weekends is lesser than weekdays and the change pattern remains at a stable level during the urban active time period.

We also plotted the spatial distribution of particle matter emission in the determined time periods. They are the two peak hours on weekdays and daytime on weekends. They are shown in Fig. 24.

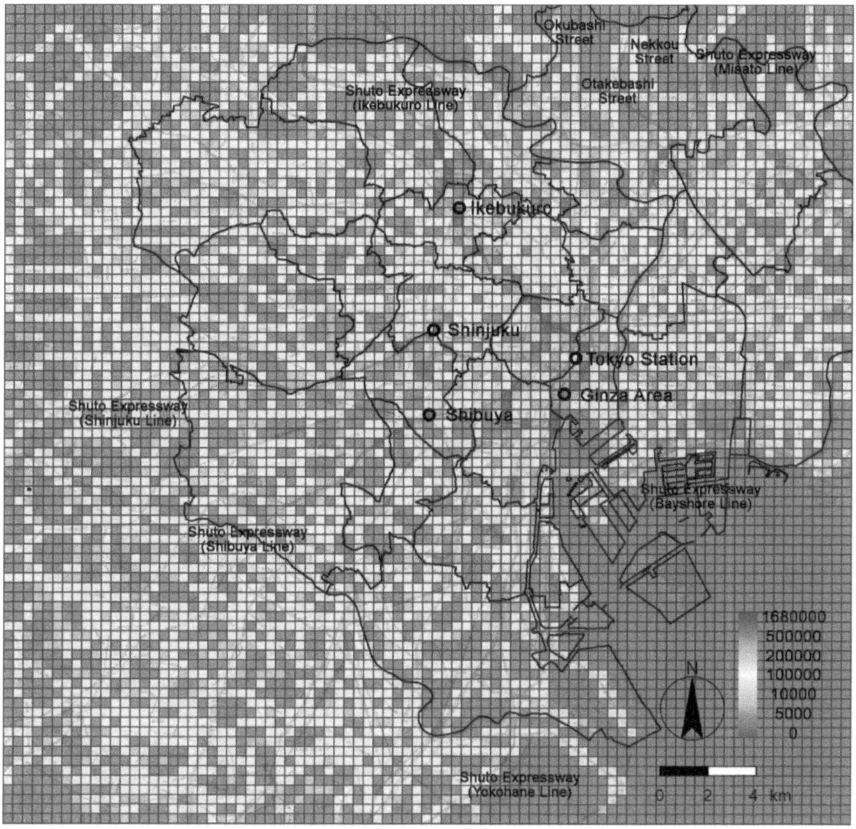

FIG. 22 The distribution of PM10 emission of braking events.

From the figure, we can see that from 6 to 8 a.m. on weekdays, the particle matter emission of braking events is mainly distributed in the city center and subcenters. The quantity of particle matter emission is descending from the main areas of Tokyo to the surrounding areas. This spatial pattern indicates that most of mobility is crowding to the city center. From 15 to 17 p.m. on weekdays, the emission turns to focus in the surrounding areas. The traffic congestion and mobility transfer to these places. This is also similar to the mobility pattern of commuting vehicles. Thus, in the evening, the main focus should be supporting the large traffic volume in the surrounding areas. During the daytime on weekends, the high concentration areas of emission simultaneously appear in the city center and surrounding areas of the city due to the various mobility activities on weekends like shopping or picnic.

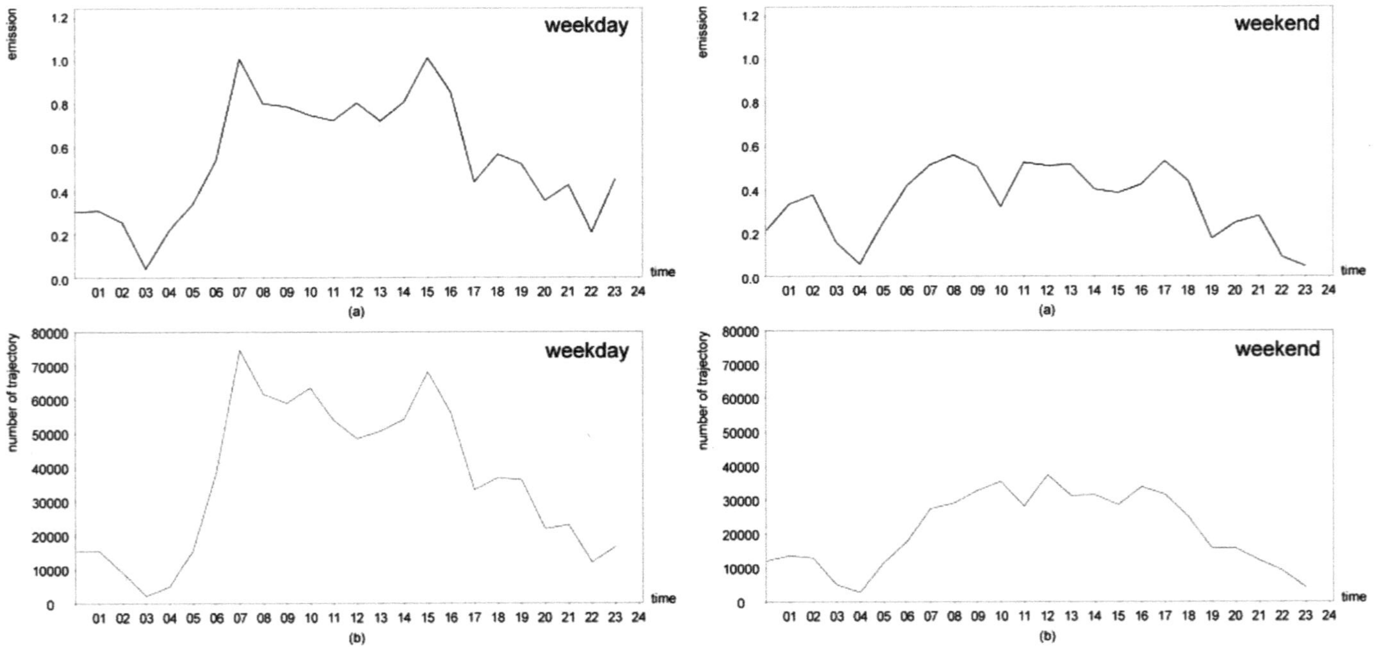

FIG. 23 The temporal change of emission and quantity of trajectory record in the study area (Unit: kg).

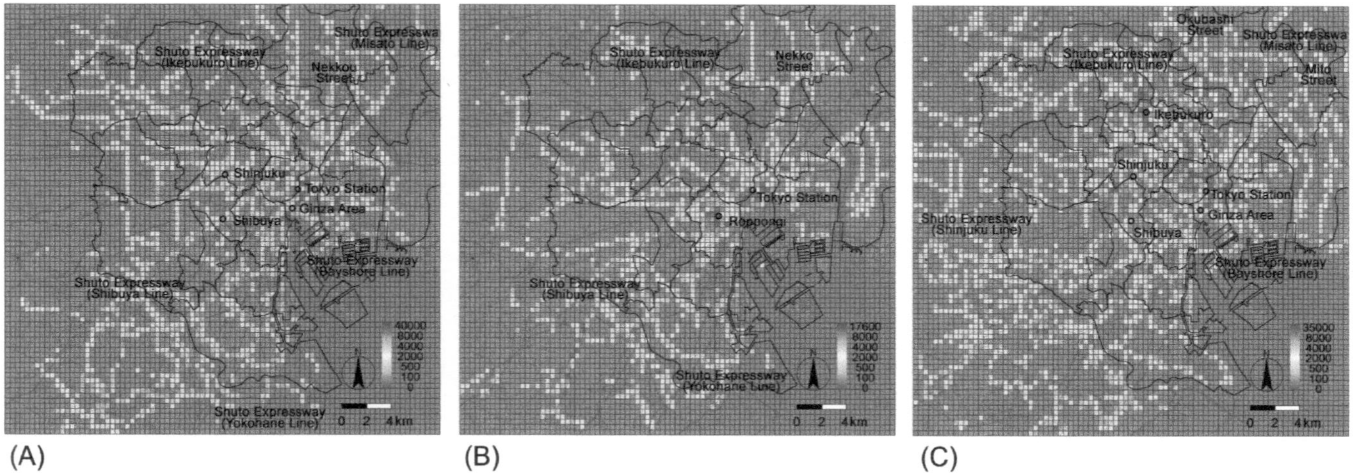

FIG. 24 The spatial distribution of emission in the study area. (A) The spatial distribution of particle matter emission from 6 to 8 a.m. on weekdays (Unit: mg). (B) The spatial distribution of particle matter emission from 15 to 17 p.m. on weekdays (Unit: mg). (C) The spatial distribution of particle matter emission during day time on weekends (Unit: mg).

4.3.4 Conclusion

This case tried to recover the particulate matter emission behavior from on-road braking events by data mining on large-scale vehicle trajectory data. From the perspective of spatial distribution of braking events and emissions, we found that the braking event usually happened in the areas of city center, crowded and popular areas, and main trunk roads of expressways. So as the braking emission. The result shows that over the Tokyo area, the particulate matter emission can account for at least 33.03% of on-road exhaust emission and the combined emission of braking and on-road exhaust of HDV is 2.33 times of LDV.

Map matching method helps us recover the probable trajectory of vehicle. Based on it, we can use generalized emission model to compute the on-road PM emission. In addition, the Kalman Filter method is also used to help correct and improve the result of map matching.

5 Conclusion

Data preprocessing is essential in the analysis of spatio-temporal mobility data. Raw GPS trajectory data is not suitable as the input of MaaS algorithm. In this chapter, we introduce several key technologies in data preprocessing as well as their workflow and sequence. Generally, the workflow consists of outlier removement, stay location detection, travel segmentation, travel mode detection as map matching. The workflow is not just an entity. Each step also has its meaning in the unique utility in different studies. To illustrate their utility, in the case study, we separately introduced three works that independently adopted stay location detection, travel segmentation, and mode detection as well as map matching and showed their essential functions in the study. The whole chapter gives a comprehensive introduction to the spatio-temporal data preprocessing.

References

[1] Hong Z, Chen Y, Mahmassani HS, Xu S. Commuter ride-sharing using topology-based vehicle trajectory clustering: methodology, application and impact evaluation. Transp Res Part C Emerg Technol 2017;85:573–90.

[2] Christian K, Cho S-H, Kho S-Y, Kim D-K. Bayesian models with spatial autocorrelation for bike sharing ridership variability based on revealed preference GPS trajectory data. IET Intell Transp Syst 2019;13:1658–67.

[3] Cai H, Wang X, Adriaens P, Xu M. Environmental benefits of taxi ride sharing in Beijing. Energy 2019;174:503–8.

[4] Yu Q, Zhang H, Li W, Sui Y, Song X, Yang D, et al. Mobile phone data in urban bicycle-sharing: market-oriented sub-area division and spatial analysis on emission reduction potentials. J Clean Prod 2020;254:119974.

[5] Yu Q, Zhang H, Li W, Song X, Yang D, Shibasaki R. Mobile phone GPS data in urban customized bus: dynamic line design and emission reduction potentials analysis. J Clean Prod 2020;272:122471.

[6] Cottrill CD. MaaS surveillance: privacy considerations in mobility as a service. Transp Res Part A Policy Pract 2020;131:50–7.
[7] Huo X, Wu X, Li M, Zheng N, Yu G. The allocation problem of electric car-sharing system: a data-driven approach. Transp Res Part D: Transp Environ 2020;78:102192.
[8] Zhang H, Song X, Xia T, Zheng J, Haung D, Shibasaki R, et al. MaaS in bike-sharing: smart phone GPS data based layout optimization and emission reduction potential analysis. Energy Procedia 2018;152:649–54.
[9] Hwang S, Evans C, Hanke T. Detecting stop episodes from GPS trajectories with gaps. In: Seeing cities through big data. Springer; 2017. p. 427–39.
[10] Birant D, Kut A. ST-DBSCAN: an algorithm for clustering spatial–temporal data. Data Knowl Eng 2007;60:208–21.
[11] Rocha JAM, Times VC, Oliveira G, Alvares LO, Bogorny V. DB-SMoT: A direction-based spatio-temporal clustering method. 2010 5th IEEE international conference intelligent systems. IEEE; 2010114–9.
[12] Ankerst M, Breunig M, Kriegel H, Ng R, Sander J. Ordering points to identify the clustering structure, In: Proc ACM SIGMOD; 2008.
[13] Gong H, Chen C, Bialostozky E, Lawson CT. A GPS/GIS method for travel mode detection in new York City. Comput Environ Urban Syst 2012;36:131–9.
[14] Wu L, Yang B, Jing P. Travel mode detection based on GPS raw data collected by smartphones: a systematic review of the existing methodologies. Information 2016;7:67.
[15] Xiao G, Juan Z, Zhang C. Travel mode detection based on GPS track data and Bayesian networks. Comput Environ Urban Syst 2015;54:14–22.
[16] Nguyen MH, Armoogum J. Hierarchical process of travel mode imputation from GPS data in a motorcycle-dependent area. Travel Behav Soc 2020;21:109–20.
[17] Zhou C, Jia H, Gao J, Yang L, Feng Y, Tian G. Travel mode detection method based on big smartphone global positioning system tracking data. Adv Mech Eng 2017;9: 1687814017708134.
[18] Li L, Zhu J, Zhang H, Tan H, Du B, Ran B. Coupled application of generative adversarial networks and conventional neural networks for travel mode detection using GPS data. Transp Res A Policy Pract 2020;136:282–92.
[19] Jahangiri A, Rakha H. Developing a support vector machine (SVM) classifier for transportation mode identification by using mobile phone sensor data, In: Transportation research board 93rd annual meeting; 2014.
[20] Xiao G, Juan Z, Gao J. Travel mode detection based on neural networks and particle swarm optimization. Information 2015;6:522–35.
[21] Shafique MA, Hato E. Travel mode detection with varying smartphone data collection frequencies. Sensors 2016;16:716.
[22] Byon Y-J, Abdulhai B, Shalaby A. Real-time transportation mode detection via tracking global positioning system mobile devices. J Intell Transp Syst 2009;13:161–70.
[23] Stenneth L, Wolfson O, Yu PS, Xu B. Transportation mode detection using mobile phones and GIS information, In: Proceedings of the 19th ACM SIGSPATIAL international conference on advances in geographic information systems; 2011. p. 54–63.
[24] Widhalm P, Nitsche P, Brändie N. Transport mode detection with realistic smartphone sensor data, In: Proceedings of the 21st International Conference on Pattern Recognition (ICPR2012): IEEE; 2012. p. 573–6.
[25] Xiao G, Cheng Q, Zhang C. Detecting travel modes from smartphone-based travel surveys with continuous hidden Markov models. Int J Distrib Sens Netw 2019;15: 1550147719844156.

[26] Zong F, Yuan Y, Liu J, Bai Y, He Y. Identifying travel mode with GPS data. Transp Plan Technol 2017;40:242–55.

[27] Rezaie M, Patterson Z, Yu JY, Yazdizadeh A. Semi-supervised travel mode detection from smartphone data, In: 2017 International Smart Cities Conference (ISC2): IEEE; 2017. p. 1–8.

[28] Shaheen S, Cohen A. Mobility on demand (MOD) and mobility as a service (MaaS): early understanding of shared mobility impacts and public transit partnerships. In: Demand for emerging transportation systems. Elsevier; 2020. p. 37–59.

[29] Jang S, Caiati V, Rasouli S, Timmermans H, Choi K. Does MaaS contribute to sustainable transportation? A mode choice perspective. Int J Sustain Transp 2021;15:351–63.

[30] Storme T, De Vos J, De Paepe L, Witlox F. Limitations to the car-substitution effect of MaaS. Findings from a Belgian pilot study. Transp Res A Policy Pract 2020;131:196–205.

[31] Singh S, Singh J, Sehra SS. Genetic-inspired map matching algorithm for real-time GPS trajectories. Arab J Sci Eng 2020;45:2587–603.

[32] Jagadeesh GR, Srikanthan T. Online map-matching of noisy and sparse location data with hidden Markov and route choice models. IEEE Trans Intell Transp Syst 2017;18:2423–34.

[33] Hsueh Y-L, Chen H-C. Map matching for low-sampling-rate GPS trajectories by exploring real-time moving directions. Inform Sci 2018;433:55–69.

[34] Tu W, Wei C, Zhao T, Li Q, Zhong C, Li Q. Uncovering online sharing vehicle mobility patterns from massive GPS trajectories. In: Spatial synthesis. Springer; 2020. p. 413–29.

[35] Li W, Wang S, Zhang X, Jia Q, Tian Y. Understanding intra-urban human mobility through an exploratory spatiotemporal analysis of bike-sharing trajectories. Int J Geogr Inf Sci 2020;34:2451–74.

[36] Kou Z, Cai H. Understanding bike sharing travel patterns: an analysis of trip data from eight cities. Phys A: Stat Mech Appl 2019;515:785–97.

[37] Zheng Y, Capra L, Wolfson O, Yang H. Urban computing: concepts, methodologies, and applications. ACM Trans Intell Syst Technol 2014;5:1–55.

[38] Song X, Guo R, Xia T, Guo Z, Long Y, Zhang H, et al. Mining urban sustainable performance: millions of GPS data reveal high-emission travel attraction in Tokyo. J Clean Prod 2020;242:118396.

[39] Fu C, McKenzie G, Frias-Martinez V, Stewart K. Identifying spatiotemporal urban activities through linguistic signatures. Comput Environ Urban Syst 2018;72:25–37.

[40] Li W, Zhang H, Chen J, Li P, Yao Y, Shibasaki M, et al. Effective metagraph-based life pattern clustering with big human mobility data. arXiv preprint arXiv:210411968.

[41] Tomasev N, Radovanovic M, Mladenic D, Ivanovic M. The role of hubness in clustering high-dimensional data. IEEE Trans Knowl Data Eng 2013;26:739–51.

[42] Khan K, Rehman SU, Aziz K, Fong S, Sarasvady S. DBSCAN: Past, present and future, In: The fifth international conference on the applications of digital information and web technologies (ICADIWT) 2014IEEE; 2014. p. 232–8.

[43] Lee DD, Seung HS. Learning the parts of objects by nonnegative matrix factorization. Nature 1999;401:788–91.

[44] Ye Y, Zheng Y, Chen Y, Feng J, Xie X. Mining individual life pattern based on location history. 2009 tenth international conference on mobile data management. In: Systems, services and middleware. IEEE; 2009. p. 1–10.

[45] Zhang H, Song X, Long Y, Xia T, Fang K, Zheng J, et al. Mobile phone GPS data in urban bicycle-sharing: layout optimization and emissions reduction analysis. Appl Energy 2019;242:138–47.

[46] Sun L, Wang S, Liu S, Yao L, Luo W, Shukla A. A completive research on the feasibility and adaptation of shared transportation in mega-cities–a case study in Beijing. Appl Energy 2018;230:1014–33.
[47] Hu Y, Zhang Y, Lamb D, Zhang M, Jia P. Examining and optimizing the BCycle bike-sharing system–a pilot study in Colorado, US. Appl Energy 2019;247:1–12.
[48] Hietanen S. Mobility as a Service. the new transport model. 2–4.
[49] Sun H, Wang H, Wan Z. Model and analysis of labor supply for ride-sharing platforms in the presence of sample self-selection and endogeneity. Transp Res B Methodol 2019;125:76–93.
[50] Zhang H, Chen J, Li W, Song X, Shibasaki R. Mobile phone GPS data in urban ride-sharing: an assessment method for emission reduction potential. Appl Energy 2020;269:115038.
[51] Zhang H, Song X, Xia T, Yuan M, Fan Z, Shibasaki R, et al. Battery electric vehicles in Japan: human mobile behavior based adoption potential analysis and policy target response. Appl Energy 2018;220:527–35.
[52] Van Dyk DA, Meng X-L. The art of data augmentation. J Comput Graph Stat 2001;10:1–50.
[53] LeCun Y, Ranzato M. Deep learning tutorial, In: Tutorials in international conference on machine learning (ICML'13)Citeseer; 2013. p. 1–29.
[54] Foster I, Kesselman C. Computational grids. In: The future of high performance distributed computing. Morgan Kaufmann Publishers; 1998.
[55] Kingma DP, Adam BJ. A method for stochastic optimization. arXiv preprint arXiv:14126980.
[56] Ministry of Land I, Transport and Tourism,. 大都市交通センサス首都圏報告書. 2012.
[57] McNeill VF. Addressing the global air pollution crisis: Chemistry's role. Trends Chem 2019;1:5–8.
[58] Chiappino G, Pellissetti D, Moretto O, Picchi O. Asbestos risk in the textile industry: braking systems on machinery used until the 1990's. Med Lav 2005;96:250–7.
[59] Chen J, Li W, Zhang H, Jiang W, Li W, Sui Y, et al. Mining urban sustainable performance: GPS data-based spatio-temporal analysis on on-road braking emission. J Clean Prod 2020;270:122489.
[60] Kautz T, Eskofier BM. A robust Kalman framework with resampling and optimal smoothing. Sensors 2015;15:4975–95.
[61] Mohan P, Padmanabhan VN, Ramjee R. Nericell: rich monitoring of road and traffic conditions using mobile smartphones, In: Proceedings of the 6th ACM conference on Embedded network sensor systems: ACM; 2008. p. 323–36.
[62] Bai S, Eisinger D, Niemeier D. MOVES vs. EMFAC: A comparative assessment based on a Los Angeles County case study. Task Orders No 9 and53.
[63] Guensler R, Liu H, Xu Y, Akanser A, Kim D, Hunter MP, et al. Energy consumption and emissions modeling of individual vehicles. Transp Res Rec 2017;2627:93–102.
[64] Simulator MVE. User Guide for MOVES2014. US Environmental Protection Agency; 2014.

Chapter 3

Travel similarity estimation and clustering

Yuhao Yao, Ryosuke Shibasaki, and Haoran Zhang
Center for Spatial Information Science, The University of Tokyo, Kashiwa, Chiba, Japan

1 Introduction

Traveling is the most significant activity of human beings. It not only refers to making a long journey such as migration and going vacationing but also includes all the daily movements like commuting, going shopping, or even taking a walk. In this sense, understanding human mobility of travel provides sufficient information for urban planning [1], transportation management [2–4], ride-sharing system [5], emergency management [6], and prevention of infectious diseases [7], which exerts an enormous impact on serving for Mobility-as-a-Service (MaaS) and improving urban mobility, accessibility, and reliability.

The mobility expression of traveling is various. As Fig. 1 shows, at the individual-level, full-time trajectory is the most fundamental way to perfectly record an individual's travel mobility. The trajectory as spatial-temporal data records the position of the individual at different times and could be processed into nearly all other mobility metrics. However, such a high-precision expression usually requires large storage space and lots of time to process. Therefore, individual trajectory data is usually processed into travel patterns such as most frequented locations and motifs, which is more proper in some circumstances. On the other hand, high precision is not necessary for many applications that just need aggregated statistics. Therefore, at the population level, some metrics are more widely used such as Origin-Destination (OD) matrices.

No matter what kind of mobility expression is utilized, the mobility similarity comparison plays a critical role in modeling dynamic human mobility. Basically, mobility similarity is the key indicator to evaluate the quality of mobility observing datasets. A variety of mobility observing datasets has been widely utilized to extract large-scale population flow information (including GPS [8], Bluetooth [9], GSM [10], automatic number plate recognition (ANPR) sensors [11, 12], and smartphone call detailed record (CDR) [13, 14]). Due to the different observing mechanisms, gaps exist in the qualities of these data

Big Data and Mobility as a Service. https://doi.org/10.1016/B978-0-323-90169-7.00004-X

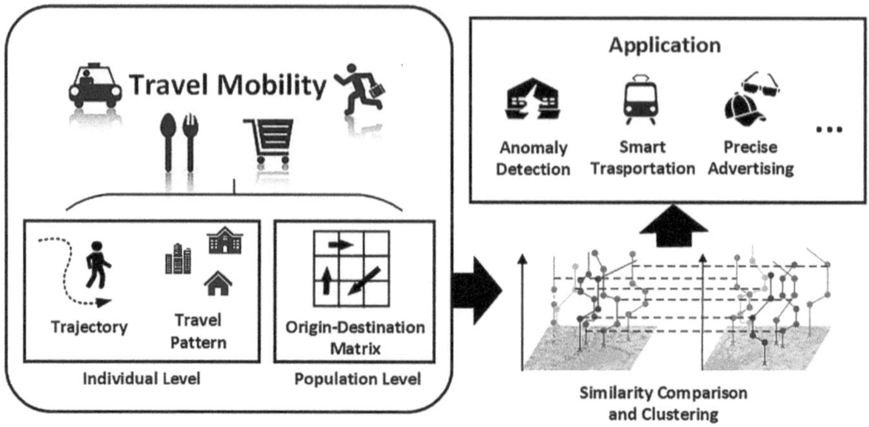

FIG. 1 Framework of travel similarity.

sources. An example of an estimated OD trajectory from CDR data is demonstrated in Fig. 2 CDR data is generated when the mobile device has any data interaction with a neighboring cell tower; it records the position of the cell tower and the timestamp. Since the location of CDR data is the position of the cell tower, there is usually a distance between CDR data and the real location. When we utilize a grid-based origin-destination pair to describe the traveling, we will find it incorrect. And when individuals' trajectories aggregate into population density map, the error will also accumulate. Mobility similarity comparison is a high-dimensional and homogeneous method to evaluate these gaps.

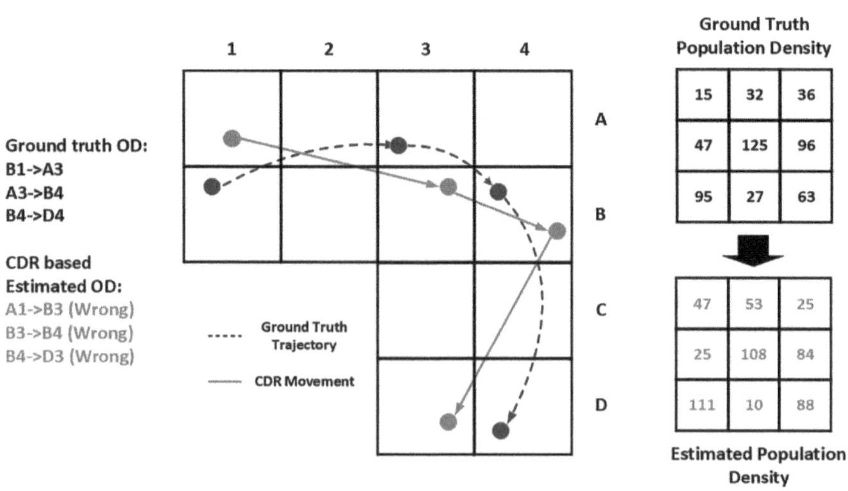

FIG. 2 Comparison between estimated mobility and ground truth.

Additionally, mobility similarity comparison helps to match the mobility between different regions from a policymaker's perspective. For first-tier cities, there are usually abundant data of human mobility and full-fledged data-driven application such as public transportation management [15], traffic congestion avoiding plan [16], car-sharing system [17], etc. However, such human mobility data collection and data-driven services are much less in small cities. A well-designed mobility similarity measurement can provide abundant useful information for transferring mature applications from one city into another by mobility features.

Moreover, mobility similarity comparison can be naturally used for mobility clustering, which distills general rules from messy mobility information [18] and mines the coupling relationship of urban subspace and dynamic causes [19].

For mobility clustering, to be more specific, it could be divided into three types: clustering for different periods, clustering for different regions, and clustering for different populations.

Clustering for different periods is usually utilized to extract temporal mobility rules in one region. For example, a daily mobility clustering in a city could basically reflect weekdays and weekends, and further indicates special days like a vacation or huge event days.

Clustering for different regions could extract spatial mobility rules in different regions and help for cross-city mobility-related applications transferring like what is introduced above.

Clustering for different populations provides sufficient information to identify population groups, which is essential in abundant of commercial applications such as advertising. For example, by analyzing the human mobility of different populations in a metro station, the properties such as age, occupation, gender of the major metro users are clear, the station advertising can therefore have a definite object in view.

In this chapter, we will introduce methods of similarity estimation and clustering for different typical kinds of mobility expressions. Trajectory similarity as individual level will be firstly introduced in Section 2. Then Origin-Destination (OD) matrices similarity will be introduced in Section 3. In Section 4, we will introduce mobility patterns similarity. Two case studies will be demonstrated in Section 5 while one is a similarity comparison between CDR-based estimated mobility and the ground truth, the other is a metro usage pattern clustering.

2 Trajectory similarity

Trajectory refers to a set of points that describe the path one individual passes through in the target period, which includes both positions and corresponding timestamps. For example, a standard two-dimensional trajectory T, contains a series of n timestamped points l_i, $i \in [1, n]$, will be $T = \{(t_1, l_1), (t_2, l_2), \ldots, (t_n, l_n)\}$ where t_i is the corresponding timestamp of l_i and $t_i < t_i + 1$.

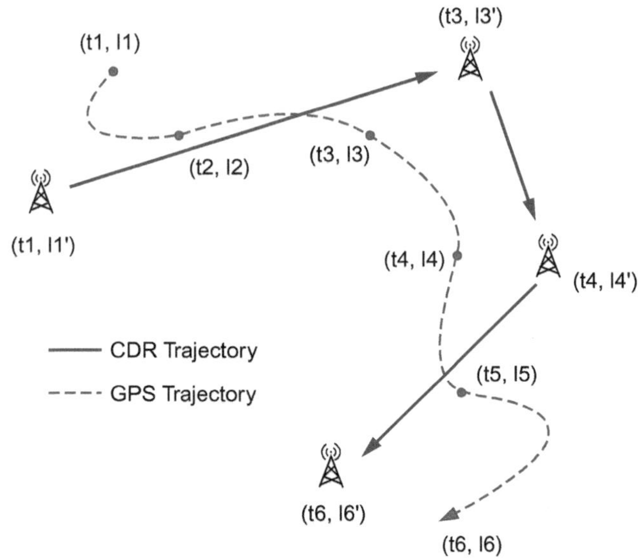

FIG. 3 Comparison between CDR trajectory and GPS trajectory.

A dense trajectory can highly describe the traveling of an individual and can be used to compute other metrics such as velocity and traveling direction. However, since the data collecting mechanisms are manifold, the time interval and positioning accuracy are not fixed. Fig. 3 demonstrates an example of the comparison between GPS trajectory and CDR trajectory of the same individual, while GPS records have shorter time intervals and higher accuracy.

The similarity between two trajectories is usually measured by a set of distance between trajectories' points. Therefore, the point-to-point distance metric is also one of the key points of trajectory similarity measurement. It is worth noting that some of the similarity functions were originally proposed for time series data. Since trajectory is a kind of particular time series in multidimensional space, these similarity functions are also suitable for trajectory data.

In this section, we will first introduce several point-to-point distance metrics in Section 2.1, then introduce the typical similarity functions for trajectory in Section 2.2, and finally introduce the standard process of trajectory clustering and a specific algorithm for example in Section 2.3.

2.1 Point-to-point distance metric

- Euclidean distance

Euclidean distance (also known as Euclidean Metric) is the most common definition of distance, which refers to the real distance between two points in

n-dimensional space, or the natural length of a vector, which is the distance between the point and the origin.

For two given points $A_1 = (x_1, y_1)$ and $A_2 = (x_2, y_2)$ in a two-dimensional space, we have:

$$D_{Euclidean}(A_1, A_2) = \sqrt[2]{(x_1 - x_2)^2 + (y_1 - y_2)^2} \tag{1}$$

- Manhattan distance

Manhattan distance (also known as Taxi Geometry) is proposed by Herman Minkowski in the 19th century. It is used in geometric space to indicate the sum of the absolute wheelbase of two points in the coordinate.

For two given points $A_1 = (x_1, y_1)$ and $A_2 = (x_2, y_2)$ in a two-dimensional space, we have:

$$D_{Manhattan}(A_1, A_2) = |x_1 - x_2| + |y_1 - y_2| \tag{2}$$

Manhattan distance comes from the shortest driving path in a city that buildings are planned as square blocks such as Manhattan, where most of the roadways there are with north-south or east-west directions and there are just several inclined lanes. In this kind of city, the driving distance from one point to another equals the sum of the distance from the north-south direction and the distance from the east-west direction. Therefore, the Manhattan distance is also called Taxi Geometry. Obviously, Manhattan distance is not a distance invariant, when the coordinate axis changes, the distance between two points will be different. In the early computer graphics, coordinates of points were all integers. It was too hard for computing the Euclidean distance between two points to avoid a floating-point operation, which was expensive, slow, and inaccurate at that time. By using Manhattan distance, the computation just needs addition and subtraction, which avoids floating-point operation and can greatly improve the computation speed and accuracy.

- Chebyshev distance

Chebyshev distance represents the maximum absolute wheelbase of two points in the coordinate system. For two given points $A_1 = (x_1, y_1)$ and $A_2 = (x_2, y_2)$ in a two-dimensional space, we have:

$$D_{\text{Chebyshev}}(A_1, A_2) = \max\left(|x_1 - x_2|, |y_1 - y_2|\right) \tag{3}$$

If we place the chessboard in a two-dimensional coordinate system, in which the side length of the grid is defined as 1, the X and Y axes of the coordinate are parallel to the chessboard grid, then the number of steps required for the King to move from one position to another just equals to the Chebyshev distance between these two positions, so the Chebyshev distance is also called chessboard distance. In practice, Chebyshev distance is often used in warehouse logistics, because the time for a crane to move objects just fits the rule of it.

- Minkowski distance

Minkowski distance is a group of distance definitions rather than a single kind of distance. For two given points in a two-dimensional space $A_1=(x_1,y_1)$ and $A_2=(x_2,y_2)$, we have:

$$D_{\text{Minkowski}}(A_1,A_2)=\left(|x_1-x_2|^p+|y_1-y_2|^p\right)^{\frac{1}{p}} \tag{4}$$

where p is a constant.

It is not hard to see that Minkowski distance includes all the above three distances. When $p=1$, Minkowski distance represents the Manhattan distance. When $p=2$, Minkowski distance represents the Euclidean distance. When $p=\infty$, Minkowski distance represents the Chebyshev distance. It should be noticed that Minkowski distance is related to the dimension of character parameters, so the Minkowski distance of character parameters with different dimensions is usually meaningless.

2.2 Similarity function of trajectory

Compared with the distance measurement between points or between points and trajectories, the similarity measurement between trajectories is more complex and needs to consider more factors, such as the sampling rate of trajectories, the timestamp of different points, and the noise. Basically, similarity functions of the trajectory can be divided into several types by what they lay particular emphasis on:

1. Point-based function measures the similarity by the distance between point pairs of two trajectories, such as Euclidean, DTW, LCSS, EDR, and ERP.
2. Shape-based function focuses on the shape similarity of two trajectories, such as Hausdorff distance, Frechet distance.
3. Segment-based function divides trajectories into different pieces and measures the similarity by distances between pieces, such as One Way Distance, LIP distance.

We will serially introduce these functions.

- Euclidean distance

As a naïve similarity measure, Euclidean distance for trajectories is simple and clear, which is the mean value of the sum of distances between corresponding points of two trajectories. For two given trajectories $T_A=\{A_1,A_2,\ldots,A_n\}$ and $T_B=\{B_1,B_2,\ldots,B_n\}$ where A_i and B_i are the ith points of T_A and T_B. We have:

$$D_{\text{Euclidean}}(T_A,T_B)=\frac{1}{n}\sum_{i=1}^{n}D_{\text{Euclidean}}(A_i,B_i) \tag{5}$$

Similarly, there is another measure named Sum-of-Pairs distance, which is the sum of distances between corresponding points of two trajectories. Although these two measures are simple and effective, which just have $O(n)$ computational complexity, they have obvious disadvantages. They struggle with different sampling rates, outliers, and require trajectories of different lengths to be cut to equal size [20]. They are also sensitive to noises.

- Dynamic time warping

As we see, the major limitation of Euclidean distance is that it requires two trajectories to have the same length, which is rare in practice. An ideal similarity measure should be flexible in the length of two trajectories. Dynamic time warping (DTW) distance is the first method based on this motivation [21]. The basic idea of DTW is to allow some points to be repeated for the best alignment. By using DTW, the trajectories are "warped" in a nonlinear way to measure similarity while allowing for varying sampling rates [22]. For two given trajectories T_A and T_B with length m and n relatively, we have:

$$D_{DTW}(T_A, T_B) = \begin{cases} 0, & \text{if } m=0 \text{ and } n=0 \\ \infty, & \text{if } m=0 \text{ or } n=0 \\ D(Head(T_A), Head(T_B)) + \min \begin{cases} D_{DTW}(T_A, Rest(T_B)) \\ D_{DTW}(Rest(T_A), T_B) \\ D_{DTW}(Rest(T_A), Rest(T_B)) \end{cases}, & \text{otherwise} \end{cases} \quad (6)$$

where function D can be any point-to-point distance metric, *Head* refers to the first point of the trajectory and *Rest* refers to the rest of the trajectory except the first point.

- Longest common subsequence

Longest Common Subsequence (LCSS) comes from a classical algorithmic problem that is to find the LCSS of two sequences. Obviously, the similarity between two trajectories can just be represented by a kind of LCSS. LCSS distance [20] represents the number of points that can be regarded as the same one in two trajectories, that is, the number of point pairs that satisfy the minimum distance threshold in two trajectories. For two given trajectories T_A and T_B with length m and n relatively, we have:

$$D_{LCSS}(T_A, T_B) = \begin{cases} 0, & \text{if } m=0 \text{ or } n=0 \\ 1 + D_{LCSS}(Rest(T_A), Rest(T_B)), & \text{if } D(Head(T_A), Head(T_B)) \le \varepsilon \\ \max\left(D_{LCSS}(Rest(T_A), T_B), D_{LCSS}(T_A, Rest(T_B))\right), & \text{otherwise} \end{cases} \quad (7)$$

Same with DTW distance, function D can be any point-to-point distance metric, *Head* and *Rest* functions refer to the first point of the trajectory and the rest of the trajectory except the first point relatively. ε is the set distance threshold.

Compared with Euclidean and DTW, LCSS is much more robust towards noise because the basic idea of LCSS is to allow points to be skipped instead of rearranged, so noise point that is far away from the real position will be ignored. However, it faces the problem that threshold ε is hard to determined, sometimes for an inappropriate ε there may be no similarity between two trajectories with high LCSS distance at all.

- Edit distance on real sequence

Although LCSS can deal with trajectories with noise, it cannot distinguish trajectories with similar common subsequences. In order to cover it better, a new similarity function called edit distance on real sequence (EDR) is proposed [23]. The fundamental idea of EDR is to count the minimum number of edits required to transfer one trajectory into the other. For two given trajectories T_A and T_B with length m and n relatively, we have:

$$D_{\mathrm{EDR}}(T_A, T_B) = \begin{cases} n, & \text{if } m = 0 \\ m, & \text{if } n = 0 \\ \min\begin{cases} D_{\mathrm{EDR}}(Rest(T_A), Rest(T_B)) + subcost(T_A, T_B) \\ D_{\mathrm{EDR}}(Rest(T_A), T_B) + 1 \\ D_{\mathrm{EDR}}(T_A, Rest(T_B)) + 1 \end{cases}, & \text{otherwise} \end{cases} \tag{8}$$

where

$$subcost(T_1, T_2) = \begin{cases} 0, & \text{if } D(Head(T_1), Head(T_2)) \le \varepsilon \\ 1, & \text{otherwise} \end{cases} \tag{9}$$

Same with LCSS distance, function D can be any point-to-point distance metric, *Head* and *Rest* functions refer to the first point of the trajectory and the rest of the trajectory except the first point relatively. ε is the set distance threshold.

Compared with LCSS, EDR punishes the result based on the gap length between two matching subtrajectories, which makes it more accurate. There is a basic transformation between LCSS and EDR as:

$$D_{\mathrm{EDR}}(T_A, T_B) = num(T_A) + num(T_B) - 2 * D_{\mathrm{LCSS}}(T_A, T_B) \tag{10}$$

where function *num* refers to the number of points in the trajectory.

- Edit distance with real penalty

Edit distance with real penalty (ERP) is another edit distance-based measure. All the trajectory similarity functions discussed above cannot solve the problem of supporting time shift and being metric together. To address this challenge, ERP is proposed [24].

By analyzing DTW distance, we may find that the reason why DTW is not a metric is that when the point in one trajectory is missed to match the other trajectory, it will repeat the previous point, so the distance between this point pair depends on the previous point. On the contrary, ERP uses a real penalty value for the distance between two matched points and a constant value for the distance between two mismatched points. Therefore, ERP can support time shift and is a metric. For two given trajectories $T_A = \{A_1, A_2, \ldots, A_m\}$ and $T_B = \{B_1, B_2, \ldots, B_n\}$ where A_i and B_i are the ith points of T_A and T_B. We have:

$$D_{\mathrm{ERP}}(T_A, T_B) = \begin{cases} \sum_1^n |B_i - R|, & \text{if } m = 0 \\ \sum_1^m |A_i - R|, & \text{if } n = 0 \\ \min \begin{cases} D_{\mathrm{ERP}}(Rest(T_A), Rest(T_B)) + D(Head(T_A), Head(T_B)) \\ D_{\mathrm{ERP}}(Rest(T_A), T_B) + D(Head(T_A), R) \\ D_{\mathrm{ERP}}(T_A, Rest(T_B)) + D(Head(T_B), R) \end{cases}, & \text{otherwise} \end{cases} \tag{11}$$

where function D can be any point-to-point distance metric, *Head* refers to the first point of the trajectory and *Rest* refers to the rest of the trajectory except the first point. R is a random point.

- Hausdorff distance

Hausdorff distance is a shape-based similarity measure, which refers to the maximum distance between the closest points of two trajectories. For two given trajectories T_A and T_B with length m and n relatively, we have:

$$D_{\mathrm{Hausdorff}}(T_A, T_B) = \max\left(h(T_A, T_B), h(T_B, T_A)\right) \tag{12}$$

where

$$h(T_1, T_2) = \max_{p \in T_1}\left(\min_{q \in T_2}(D(p, q))\right) \tag{13}$$

Function h is an unidirectional Hausdorff distance and function D can be any point-to-point distance metric.

- Frechet distance

The Frechet distance is amongst the most popular trajectory similarity measures [25]. It was first defined by Frechet [26] and can be applied to both continuous directed curves as well as the discretized trajectories.

An intuitive interpretation for Frechet distance is that a person is walking a dog with a dog leash. The dog will not always move on the person's trajectory. Both the dog and the person are able to change the moving speed and even stop, but obviously, at any time the dog will not be more distant from the person than the length of the leash. The Frechet distance between the trajectories of the person and the dog is just the minimum length of the leash required to complete the traversal of both trajectories. For two given trajectories $T_A = \{A_1, A_2, \ldots, A_m\}$

and $T_B = \{B_1, B_2, \ldots, B_n\}$ where A_i and B_i are the ith points of T_A and T_B. We have:

$$D_{\text{Frechet}}(T_A, T_B) = \begin{cases} \max\limits_{1 \le i \le n} D(A_i, B_1), & \text{if } m = 1 \\ \max\limits_{1 \le i \le m} D(A_1, B_i), & \text{if } n = 1 \\ \max \begin{cases} D(A_m, B_n) \\ \min \begin{cases} D_{\text{Frechet}}(Sub(T_A), T_B) \\ D_{\text{Frechet}}(T_A, Sub(T_B)) \\ D_{\text{Frechet}}(Sub(T_A), Sub(T_B)) \end{cases} \end{cases}, & \text{otherwise} \end{cases} \tag{14}$$

where function D can be any point-to-point distance metric and function *sub* represents the subtrajectory which removes the last point.

Frechet distance provides us with a simple and intuitive way to measure trajectory similarity, which can also achieve as good results as other methods. However, it cannot deal with noise points at all because every point of two trajectories is used in the calculation. If a point in one trajectory deviates a lot due to the noise, the Frechet distance will be much larger than it should be, which is far away from the ground truth.

- One way distance

One way distance (OWD) [27] is a segment-based function. The basic idea of OWD distance is to measure the similarity between two trajectories based on the area enclosed by them. Obviously, when the area is large, the distance between trajectories is far, which means the similarity is low. On the contrary, if the enclosed area equals to zero, two trajectories are exactly the same one and the similarity is the highest. For two given trajectories T_A and T_B with length m and n relatively, we have:

$$D_{\text{OWD}}(T_A, T_B) = \frac{1}{2}(o(T_A, T_B), o(T_B, T_A)) \tag{15}$$

where

$$o(T_1, T_2) = \frac{1}{|len(T_1)|} \int_{p \in T_1} d(p, T_2) dp \tag{16}$$

Here function o is a unidirectional OWD distance. Function *len* refers to the length of the trajectory and function d of a point and a trajectory refers to the shortest distance between the point and the trajectory, which can be any distance metric.

- Locality in-between polylines

Locality in-between polylines (LIP) [28] method is also a segment-based function. It has a very similar basic idea with OWD, which is to measure the similarity between two trajectories based on the area enclosed by them. This time, it assigns different weights to enclosed regions by the proportion of their perimeters of the total length of two trajectories, which counteracts the interference of noise points to a certain extent. When the perimeter of a region accounts for a large proportion of the total length of two trajectories, the weight will be large. When the weight of the region is large, the gap between the two trajectories is large, so the LIP distance is large. When the area of the region equals zero, obviously there is no gap between the two trajectories, so the LIP distance is 0. For two given trajectories T_A and T_B with length m and n relatively, R_i represents each region enclosed by the two trajectories, we have:

$$D_{LIP}(T_A, T_B) = \sum S(R_i)^* w_i \tag{17}$$

where

$$w_i = \frac{C(R_i)}{len(T_A) + len(T_B)} \tag{18}$$

Here w_i is the weight function of R_i. S function is the area of the region and C function is the perimeter of the region.

2.3 Trajectory clustering

By measuring similarity between trajectories as a kind of distance, it is possible to cluster trajectories by some traditional data clustering methods such as K-means [29], BIRCH [30], DBSCAN [30], OPTICS [31], STING [32], etc. In this way, each trajectory just belongs to one cluster.

However, in some scenarios, this kind of global-based trajectory clustering will lose some details. Fig. 4 demonstrates an example of three trajectories.

Obviously, the parts of the three trajectories in the red circle are very similar and should be in one cluster. However, the left parts of the three trajectories are totally different, which move into different directions. If we apply global-based clustering algorithm, three trajectories may belong to different clusters, which is improper.

In the real case, trajectories are usually longer and more complex than what the example shows. If we can divide trajectory into multiple segments and cluster segments instead of the original trajectory, each trajectory can then belong to multiple clusters and the result will be much better. To achieve it, trajectory segmentation is necessary.

Trajectory segmentation is to divide the trajectory into multiple segments by selecting proper feature points from the original trajectory. In this process, accuracy and simplicity should be guaranteed. Accuracy means that feature points should be able to describe the original trajectory; Simplicity means that feature points should use as few points as possible to summarize features of the

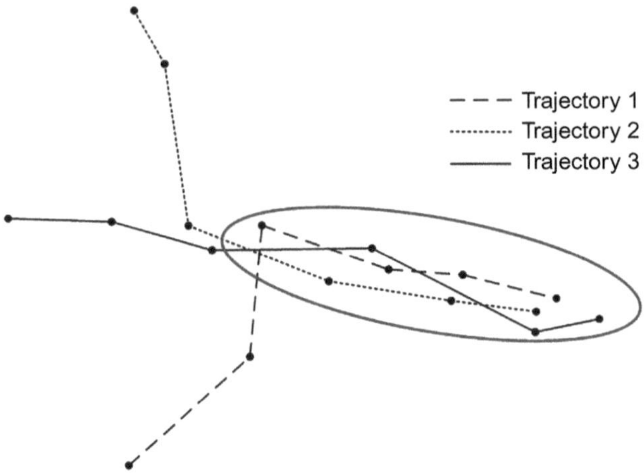

FIG. 4 An example of a common subtrajectory.

trajectory. A good segmentation algorithm should be able to balance these two contradictory characteristics.

The trajectory segmentation method can be basically divided into three types:

The first type segment is the trajectory based on time interval. For example, a simple way is to segment the trajectory once if the interval between timestamps of two points is too long. It can also segment the trajectory according to the time period.

The second type is to segment the trajectory based on the shape. For example, it is easy to compute the moving direction of the trajectory, thus the moving direction changing angle is an ideal indicator. If the changing angle exceeds a certain threshold, it means the moving state of this individual is changed dramatically, so the segmentation could be performed.

The last type is to segment trajectory based on semantic information, such as stay points and travel modes. For example, when the trajectory shows that the individual stays at a position for a long time, it has a high probability for the individual to change the moving state here, so we can segment the trajectory. The moving velocity can also indicate the moving mode of the individual like low velocity represents walking and high velocity represents moving by car. Then a dramatic change of moving velocity can also be a symbol to segment the trajectory.

Here we take a classical trajectory clustering algorithm named TRACLUS [33] as an example to show the whole process of trajectory clustering.

The first step is to determine the similarity function.

For two given lines that have the same length and are parallel to each other, the lines of two start points and lines of two end points are both perpendicular to these two lines, the vertical distance $d_{\perp}$ can be used to represent their distance or similarity. The closer the distance is, the more similar they are.

When two lines have different lengths or one line is dislocated along its direction, it is obvious that only vertical distance is not enough to represent their similarity, which should be measured by the difference in the horizontal direction, that is, the horizontal distance $d_{\|}$.

Further, when two lines have different directions, the greater the angle between their directions is, the smaller the similarity is. Therefore, an angle distance d_θ should also be considered.

Based on this idea, TRACLUS introduced a similarity function of two trajectory segment L_i and L_j as:

$$D(L_i, L_j) = w_{\perp} * d_{\perp}(L_i, L_j) + w_{\|} * d_{\|}(L_i, L_j) + w_\theta * d_\theta(L_i, L_j) \tag{19}$$

$$d_{\perp}(L_i, L_j) = \frac{l_{\perp 1}^2 + l_{\perp 2}^2}{l_{\perp 1} + l_{\perp 2}} \tag{20}$$

$$d_{\|}(L_i, L_j) = \min(l_{\|1}, l_{\|2}) \tag{21}$$

$$d_\theta(L_i, L_j) = \begin{cases} \|L_j\| * sin\theta, & \text{if } 0 \le \theta < 90^\circ \\ \|L_j\|, & \text{if } 90^\circ \le \theta \le 180^\circ \end{cases} \tag{22}$$

The geometric relations between $l_{\perp 1}$, $l_{\perp 2}$, $l_{\|1}$, $l_{\|2}$, θ and L_i, L_j are shown in Fig. 5. $w_{\perp}$, $w_{\|}$, w_θ are weights of three parts, which are all set as 1.

To segment the trajectory, MDL (minimum description length) principle is adopted, which is widely used for information compression.

The principle of minimum description length (MDL) was proposed by Rissane for general coding. The basic idea is that for a given data D, in order to save the storage space, it is usually encoded and compressed by a model M into D', where D' is much smaller than D. Since the compressed data needs to be recovered correctly, the model M also needs to be saved. Therefore, the necessary data length to save D equals to the data length of compressed data D' and model M, which is called description length. The principle of MDL is to select the model with the minimum total description length. Similarly, selecting feature points for segmentation in a trajectory can also obey this principle.

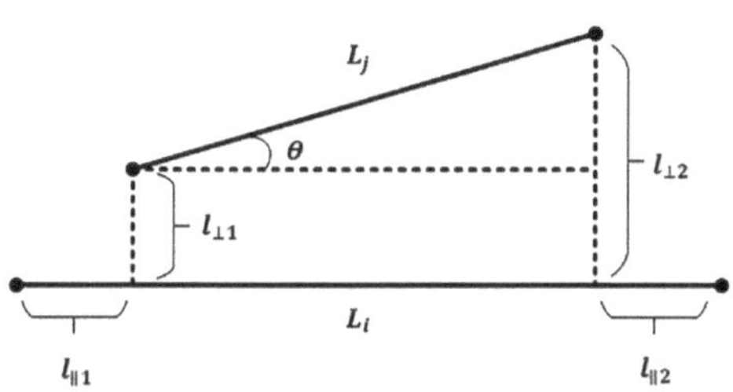

FIG. 5 Three components of the distance function for line segments.

The MDL principle consists of two parts [34], one is $L(H)$, which represents the length that required to describe the compression model, the other is $L(D|H)$, which describes the length of the data encoded by the compression model.

For a given trajectory $T=\{p_1,p_2,\ldots,p_n\}$, assume that m feature points are selected as $F=\{p_{c_1},p_{c_2},\ldots,p_{c_m}\}$.In TRACLUS, the definition of two parts is as follows:

$$L(H)=\sum_{i=1}^{m-1}\log_2(len(p_{c_i}p_{c_{1+1}})) \tag{23}$$

$$L(D|H)=\sum_{i=1}^{m-1}\sum_{j=c_i}^{c_{i+1}-1}\left\{\log_2\left(d_\perp\left(p_{c_i}p_{c_{1+1}},p_jp_{j+1}\right)\right)+\log_2\left(d_\theta\left(p_{c_i}p_{c_{1+1}},p_jp_{j+1}\right)\right)\right\} \tag{24}$$

For a given trajectory, to select feature points equals to find the proper set F that has the minimum $L(H)+L(D|H)$, so we just need to select feature points by checking whether the point makes the MDL value smaller one by one.

Finally, segments are clustered by an algorithm that is similar to DBSCAN. Different from DBSCAN, the algorithm will finally check the trajectory cardinality of each cluster, which refers to the number of different trajectories that segments come from in the clusters. If its trajectory cardinality is below the threshold, the algorithm filters out the corresponding cluster because they do not explain the behavior of a sufficient number of trajectories.

3 Travel pattern similarity

Travel pattern describes the individual's mobility regularity and general travel style in a target time period. Since it is a recapitulative mobility expression, different from trajectory or OD matrix, it does not have a fixed data structure. What it focuses on mainly depends on the research purpose. Some studies may also refer it in terms of "mobility pattern" [35–37], "activity pattern" [38] or "commute pattern" [39, 40]. But the common thing is that all the research about it discusses how humans travel and conduct daily activities. The routine mobility between home and workplaces are commonly discussed, other places include tourist attraction [41], shopping [42] and catering [43] also be subdivided into some studies.

To extract travel pattern from an individual's mobility data is quite similar to semantic information based trajectory segmentation, which shares some common steps such as stay point detection and moving mode detection, but there are still several differences: Travel patterns usually pay more attention to the moving mode, origin-destination pair, travel purpose, instead of the trajectory itself. In addition, travel pattern usually describes several routes and assign a frequency to each route based on the individuals' mobility regularity, while semantic information based trajectory segment is just a route.

In this section, we will first introduce some basic methods to help with mobility pattern extraction in Section 3.1, then introduce a kind of travel pattern structure in Section 3.2 and finally introduce a kind of travel pattern clustering method in Section 3.3.

3.1 Travel pattern extraction

Although travel pattern usually includes numerous mobility features, there are two basic features that are usually utilized: stay point and travel mode.

Stay point provides OD pair information for OD matrix and is regarded as a significant separator for human activity. For example, in one day, an individual may stay at home and sleep until 8:00, then he goes to the workplace to work until 12:00. He leaves the workplace to have lunch and then goes back to continue working until 17:00. Finally, he goes back home and stays there. On this day, his activity is perfectly divided by stay points in his trajectory: home, workplace, and restaurant.

Since stay point means that the individual is staying here and just moves in a small range, a sliding window can simply detect the stay state.

Assume that the trajectory of the individual is $T=\{r_1, r_2, \ldots, r_n\}$, where r is one position records that contain the timestamp t and the position p.We apply a sliding window $W=\{r_k, r_{k+1}, \ldots, r_{k+l}\}$ where $k \geq 1$, $k+l \leq n$, $t_{r_{k+l}} - t_{r_k} < \Delta t$, Δt is the set minimum stay time threshold.

We start W with $k=1$ and define a maximum cluster distance threshold Δd and the formula (26).

$$\max distance\left(p_{r_i}, p_{r_j}\right) < \Delta d \forall k < i, \;\; j < k+l \tag{25}$$

If the records in the sliding window fit the formula, it means records in the window form a stay point. Then we start to temporally expand the sliding window until the formula is no longer established. All the points in the sliding window form a stay point.

Travel mode detection is used to segment the trajectory into segments based on the travel mode change and detect the corresponding travel mode. For example, we can divide a moving mode into four types: *Stay*, *Nonmotorized*, *Motor*, and *Metro*. Here *Nonmotorized* type means the individual moves without a motor vehicle, usually by foot or bicycle. *Motor* type means the individual moves by motor vehicle, usually by own car or bus. *Metro* means the individual moves by metro, including underground and ground.

If we use different numbers to code travel modes, the travel mode of a trajectory can be represented by a sequence with time interval Δt, in which the ith number represents the moving mode of the ith Δt time period. If we describe moving mode changing as a Markov process, the moving mode sequence could be regarded as a hidden states sequence and we need to estimate it based on the observable information. This could be achieved by Hidden Markov Model (HMM).

HMM is a statistical model that describes a Markov process with unknown parameters. Its states cannot be observed directly but could be indirectly observed through the observation vector sequence. Each observation vector is expressed as various states through some probability density distribution, and each observation vector is generated by a state sequence with corresponding probability density distribution. Therefore, HMM is a dual stochastic process, i.e., a set of hidden Markov chains with a certain number of states and a set of dominance random functions.

For mode detection, observable states can be any features of the trajectory that has a relationship with moving mode, such as moving velocity, moving direction, time, stay point, etc.

3.2 Travel pattern expression

We introduce a method [44] to use a support graph to express the travel pattern of different individuals and the way to convert support graphs of different individuals into a topology-attribute matrix (T-A matrix).

Assume that we want to cluster the daily travel pattern of a group of people by their one-month mobility data, in which we just focus on where they are. For each user, we first build an individual graph that represents the location of each hour within one day. The nodes of the individual graph assume to begin from the top and one layer of the individual graph represents 1 h, each node of the graph represents one significant place for a user within 1 h. For users that stay at multiple significant places within one hour, we select the one with the longest duration. There is only one edge between the same source and target nodes. The direction of the edge is always from the node in the higher layer to the node in the lower layer.

Secondly, a support graph is constructed in the way that any of the individual graphs can always be found in the support graph to have the same topology. In the support graph, each edge is assigned a unique index in an ascending manner from top to bottom and from left to right. Then, the edges of the individual graph will be respectively assigned one index which corresponds to the index of the edges with the same topology in the support graph by definition.

Then, we use T-A matrix to represent the graphs. Let n denotes the number of edges in the support graph, a T-A matrix of one user one day will be generated with its row number being equal to the total index number of the support graph edges n. The first row of the T-A matrix corresponds to the first edge of the support graph, the second row to the second edge, and so forth. For individual each day, the element in the ith row is assigned the value 1 if the ith edge of the support graph is also contained in the individual graph; otherwise, 0 is assigned to the element in the ith row. Let $T_u = \{T_{(u,1)}, T_{(u,2)}, \ldots, T_{(u,m)}\}$ corresponding to T-A matrix of user u in day m, by computing the average value of T_u, we can obtain the average probability matrix $T_{\text{average } u}$ of user u with m days.

3.3 Travel pattern clustering

Since we can use an average probability T-A matrix to represent the mobility pattern of a population, we can conduct graph clustering in the matrix space M. Once we identify the bases of M, each $T_{\text{average } u}$ can be decomposed into a linear combination of these bases. Then, the similarity between a pair of graphs can be defined based on the coefficients of the linear combination.

Considering that the value of individual graph edges and the value of probability T-A matrix are all nonnegative, the nonnegative matrix factorization (NMF) is applied [45]. For any given nonnegative matrix V, it can find a nonnegative matrix W and a nonnegative matrix H, satisfying the condition $V=W \times H$.Then a nonnegative matrix is decomposed into two nonnegative matrices. W matrix is called base matrix and H matrix is called coefficient matrix or weight matrix. By replacing the original matrix with the coefficient matrix H, the dimension reduction of the original matrix can be achieved, and the dimension reduction matrix of the data features can be obtained, so as to reduce the storage space.

NMF does not make the orthogonality assumption, thus it adapts to a wide range of matrix manifolds [46]. The bases obtained from NMF have physical meanings [47], called metagraph in the context of this study. NMF decomposes the total T-A matrix as $T_{n\times u} \approx W_{n\times k} \cdot H_{k\times u}$, where n represents edges and u represents individuals. Note that the columns of W are actually the metagraphs and the axis of the metagraph space M. The columns of H are the corresponding coefficients of the linear combination of these metagraphs for the corresponding users, also can be seen as the corresponding coordinate point of the corresponding individual graph in the metagraph space M. The origin of the metagraph is an empty graph. Then travel patterns could be clustered by any Euclidean space clustering method.

4 Origin-destination matrix similarity

Origin-destination (OD) matrix is an effective and the most widely used form to describe the high-dimensional information of human mobility, which describes the travel demand distributed between different origin and destination locations of a study site. OD matrix emphasizes the volume of population flow between different region pairs and exerts an enormous function on numerous fields.

Since OD matrix is a kind of matrix, simply focusing on the value difference can somehow measure the similarity between different OD matrices. However, people flow usually consists of numerous basic patterns, including commuting, working, and commercial. Such complex spatial-temporal data is extremely difficult to extract useful information directly by the volume similarity provided by OD matrix, thus requires high-dimensional abstract features to describe them [48]. To further introduce the mobility similarity measurement problem, we introduce a concept named mobility tableau [49].

Mobility tableau is an aggregated tableau representation to the population flow distributed between different location pairs of a study site. A mobility tableau can be described by a vector graph, in which the grid represents the map of the study site while vectors represent the volume of population flow and the corresponding origin/destination regions. Compared with traditional OD matrix comparison, mobility tableau is geographic correlation related, so the comparison between two mobility tableaus can not only reflect the volume similarity of each population flow as OD matrix but also demonstrate several significant physical properties, including:

- Spatial similarity and dissimilarity
- Mass inclusiveness and exclusiveness
- Structural similarity and dissimilarity

Three cases are shown in Fig. 6 to further illustrate the mobility tableau and these properties. Assume that all study sites are 3×3 grid, and each cell represents a region. There is always an omitted regular population flow between any two regions. By using an arrow as a vector to represent huge population flow, a vector graph can be constructed to describe the mobility tableau.

Spatial similarity and dissimilarity represent whether two mobility tableaus are similar by taking geographic correlation into account. In Fig. 6A, though

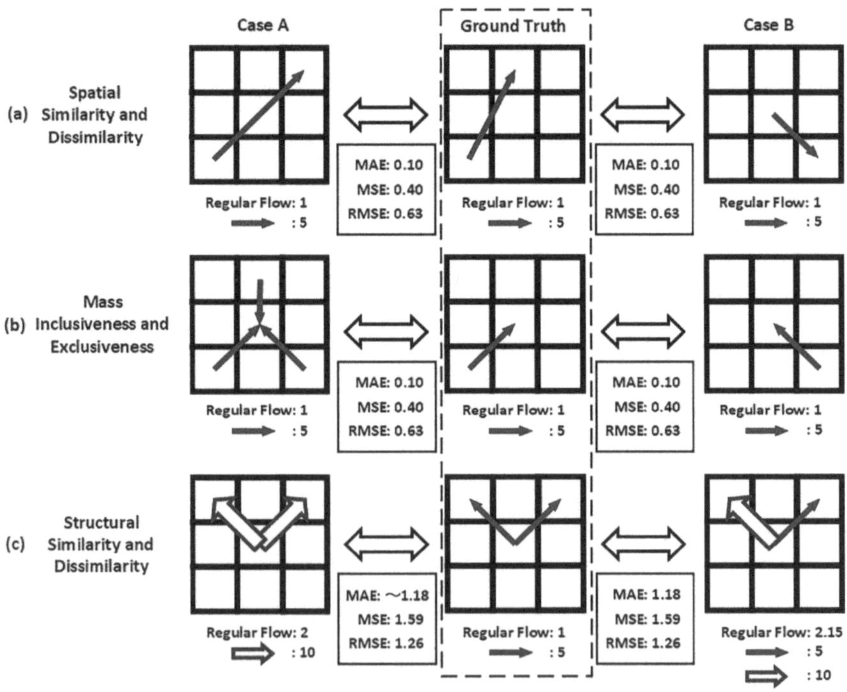

FIG. 6 Demonstration of three physical properties of mobility similarity.

compared with ground truth all traditional metrics are the same for case A and case B, without no doubt case A is much closer to ground truth because the only huge population flow is closer in space. Therefore, case A has much more spatial similarity to ground truth than case B.

Mass inclusiveness and exclusiveness represent the extent to one mobility tableau to cover the other. In Fig. 6B, though compared with ground truth all traditional metrics are the same for case A and case B, three huge population flows in case A totally cover the only huge population flow in ground truth, while in case B the huge population flow is exclusive from the ground truth. Therefore, case A has much more mass inclusiveness to ground truth than case B.

Structural similarity and dissimilarity represent the (dis)similarity of the proportion of population flows from the same origin region to different regions between two mobility tableaus. In Fig. 6C, though compared with ground truth all traditional metrics are similar for case A and case B, case A is just double the population flow of ground truth, which indicates they have the same structure, while the structure of case B is totally different. Therefore, case A has much more structural similarity to ground truth than case B.

In this chapter, we divide mobility tableau similarity measurement methods into three types based on their key ideas and introduce them serially: volume difference focused measure, image-based measure, and transforming distance-based measure. Finally, we mainly introduce a distance-based method in detail for example.

4.1 Volume difference focused OD similarity measure

The key idea among volume difference focused measures is to regard population flow of different origin-destination pairs as an independent statistic, then traditional statistical metrics can be applied to measure the volume (dis)similarity of two mobility tableau.

Here several notable measures are listed: root mean square error (RMSE) [50–52]; normalized root mean square error (RMSN) [53]; mean square error (MSE) [54]; mean absolute error ratio (MAER) [55]; mean absolute percent error (MAPE) [56]; goodness of Theil's fit (GU) [57]; maximum possible absolute error (MPAE) [58]; relative error (RE) [59]; total demand deviation (TDD) [60]; R-squared (R^2) [61], and entropy measure (E) [62].

Assume that X and Y are two mobility tableaus, $X_{i,j}$ and $Y_{i,j}$ represents population flows from region i to region j in X and Y, where $i, j \in K$, K is the total number of regions. Eqs. (26)–(28) show formulations of three classical measures:

$$\mathrm{RMSE}(X,Y) = \sqrt{\frac{1}{K^2}\sum_{i,j\in K}\left(X_{i,j}-Y_{i,j}\right)^2} \tag{26}$$

$$\mathrm{GU}(X,Y)=\frac{\sqrt{\frac{\sum_{i,j\in K}\left(X_{i,j}-Y_{i,j}\right)^2}{K^2}}}{\sqrt{\frac{\sum_{i,j\in K}X_{i,j}{}^2}{K^2}}+\sqrt{\frac{\sum_{i,j\in K}Y_{i,j}{}^2}{K^2}}} \tag{27}$$

$$\mathrm{E}(X,Y)=\sum_{i,j\in K}\left(X_{i,j}\log\left(\frac{X_{i,j}}{Y_{i,j}}\right)-X_{i,j}+Y_{i,j}\right) \tag{28}$$

Although the aforementioned measures have simple mathematical formulations and compute deviations between individual population flows, they can just simply reflect the volume similarity of mobility tableaus and ignore all other properties.

4.2 Image-based OD similarity measure

Mean structural SIMilarity index (MSSIM) [63] is the prototype to compare the structural degradation between two images by comparing pixels. Since cells of matrix are similar to pixels in an image, Djukic et al. [64] applied MSSIM on OD matrices.

Three different components—luminance l, contrast c, and structure s are introduced in *SSIM*, which can be calculated by mean (μ_x and μ_y), standard deviation (σ_x and σ_y) and covariance (σ_{xy}). Here, x and y represent the group of OD pairs in the same $n \times n$ scale window in OD matrices X and Y.

$$SSIM(x,y)=\frac{\left(2\mu_x\mu_y+C_1\right)\left(2\sigma_{xy}+C_2\right)}{\left(\mu_x^2+\mu_y^2+C_1\right)\left(\mu_x^2+\mu_y^2+C_2\right)} \tag{29}$$

$$MSSIM(X,Y)=\frac{1}{\overline{W}}\sum_{w\in W}SSIM(x,y) \tag{30}$$

Eq. (29) demonstrates the simplified form of *SSIM*. The constants C_1, C_2 are meant to stabilize the result when either mean or standard deviation is close to zero, which are respectively suggested to be 10^{-10} and 10^{-2} by Pollard et al. [65]. The average of all SSIM values (over $\overline{W}$ local windows) yields MSSIM as Eq. (30).

Based on MSSIM, several improvements have been made:

- In 4D-MSSIM proposed by Van Vuren and Day-Pollard [66], neighborhood OD pairs are identified using spatial proximity between OD pairs by using Euclidian distance.
- In GSSI proposed by Behara et al. [67], geographical boundaries of higher-level zones are used to define the local windows.

- In SpSSIM proposed by Jin et al. [68], a series of spatial weight matrices are used to take place in the original moving window.

Although the problem about the sensitivity of MSSIM towards local window size and geographical adjacency of zones have been addressed, some problems still remain, especially issues related to stability constants and network-specific parameters selection.

4.3 Transforming distance-based OD similarity measure

The main idea of distance-based measure is to measure mobility similarity by computing the least cost of transforming one mobility tableau into the other.

- Wasserstein metric

One of the transforming distance-based measures utilizes Wasserstein distance [69], which used to be widely applied in mass transportation problems such as optimal cost required to transfer iron-ore from many mining locations to several factories etc. In the context of mobility tableaus, the Wasserstein distance between two mobility tableaus are defined as the minimum total travel time required to assign the trips between origin-destination pairs of mobility tableau X and Y.

For example, for an OD pair $\tau = (\tau_o, \tau_d)$, if we want to transform mobility tableau X to Y, the difference of τ's population flow needs to be assigned into another OD pair $\gamma = (\gamma_o, \gamma_d)$, the cost of which can be computed by:

$$d(\tau, \gamma) = d(\tau_o, \gamma_o) + d(\tau_d, \gamma_d) \tag{31}$$

Here $d(a, b)$ represents the average travel time cost of moving to a from b. By creating virtual population flow to boundary, X and Y can have the same total population flows. Assume that v_γ^τ represents the population flow that needs to be assigned from τ to γ, of course, we have $v_\gamma^\tau > 0$. Then Wasserstein distance between X and Y can be computed by:

$$\text{Wasserstein}(X, Y) = \min\left(\sum_{\tau, \gamma} v_\gamma^\tau d(\tau, \gamma)\right) \tag{32}$$

Although this metric can measure the structural similarity by taking geographic correlation into account, it has two fatal limitations: First, as an optimization problem, when the number of regions is too large, it is computationally very expensive; second, when the total population flow number of two mobility tableau is different, there is no definite way to balance the difference, so the result is not unique.

- Levenshtein metric

Another measure selects Levenstein distance [70], which used to be a measure of proximity between two strings. To be more specific, Levenstein distance calculates the least expensive set of insertions, deletions or substitutions that are required to transform one string into another.

For a mobility tableau X, using $D^n_{X_i}$ and $A^n_{X_i}$ to respectively represent the i-th preferred destination and its corresponding demand value from nth origin of X, the sorted set of destination IDs and corresponding demand from nth origin is expressed as $R^n_X = (D^n_X, A^n_X) = [(D^n_{x_1}, A^n_{x_1}), \ldots (D^n_{x_i}, A^n_{x_i}), \ldots (D^n_{x_m}, A^n_{x_m})]$.

Let $S = (S_0, S_1, \ldots S_k, \ldots S_s)$ be the sequence of edit operations to transform R_Y into R_X, and the cost associated with each edit operation are $\beta = (\beta_0, \beta_1, \ldots \beta_k, \ldots \beta_s)$, respectively. While the LODn is an absolute comparison of nth rows, we can have a relative comparison by normalizing LODn with the trip productions (sum of origin flows) for nth row from both matrices. This normalized version of LODn is referred to as NLODn as Eq. (34):

$$\mathrm{LOD}n\left(R^n_Y, R^n_X\right) = \min_S \left(\sum_{k=0}^{k=s} \beta_k \right) \tag{33}$$

$$\mathrm{NLOD}n\left(R^n_Y, R^n_X\right) = \frac{LODn\left(R^n_Y, R^n_X\right)}{\left(\sum_{j=1}^{j=M} A^n_{x_j} + \sum_{j=1}^{j=M} A^n_{y_j} \right)} \tag{34}$$

$$\mathrm{LOD}\left(R^n_Y, R^n_X\right) = \frac{\sum_{n=1}^{n=N} LODn\left(R^n_Y, R^n_X\right)}{N} \tag{35}$$

$$\mathrm{NLOD}\left(R^n_Y, R^n_X\right) = \frac{\sum_{n=1}^{n=N} \mathrm{NLOD}n\left(R^n_Y, R^n_X\right)}{N} \tag{36}$$

The overall comparison between the mobility tableaus is obtained through mean Levenshtein distance named LOD, which is the average of all LODn values, and the mean normalized Levenshtein distance named NLOD, which is the average of all NLODn as expressed in Eqs. (35), (36), respectively.

This metric clearly distinguishes the structural similarity from the volume similarity of mobility tableaus. However, the geographic correlation of different OD pairs is ignored, which will cause the result to be worse when mobility tableaus have high spatial similarity.

4.4 OD tableau similarity measure: Mobsimilarity

Since the OD tableau of an OD matrix can be represented by a vector map, measuring similarity between two OD matrices equals to measure the similarity between two vector sets, which can be transformed into computing the cost of the transformation between two vector sets. An operator "→" is defined for vector or vector set named *Transformation*. $S_X \rightarrow S_Y$ means the transformation from vector graph S_X to vector graph S_Y.

To conduct the transformation, three basic operations are defined as Fig. 7 shows: $0 \rightarrow \vec{\delta}$ named *Add*, $\vec{\delta} \rightarrow \vec{\delta}\prime$ named *Shift* and $\vec{\delta} \rightarrow 0$ named *Delete*. Manhattan distance is selected as the cost function and an operator "$\|\|$" is defined to represent the cost of each operation:

- **Add:** Add represents directly adding a new vector into the graph. The cost equals the Manhattan distance between added vector's start point and end point, i.e.,

$$\left\|0 \rightarrow \vec{\delta}_{(ox,\, oy),(dx,\, dy)}\right\| = |dx - ox| + |dy - oy| \tag{37}$$

For example, as Fig. 7A shows, when $S_X = \left\{\vec{\delta}_{(1,2),(2,1)}\right\}$, $S_Y = \left\{\vec{\delta}_{(1,2),(2,1)}, \vec{\delta}_{(3,1),(3,3)}\right\}$, to conduct $S_X \rightarrow S_Y$, the red vector $\vec{\delta}_{(3,1),(3,3)}$ is added. The cost $\left\|0 \rightarrow \vec{\delta}_{(3,1),(3,3)}\right\| = |3-3| + |3-1| = 2$.

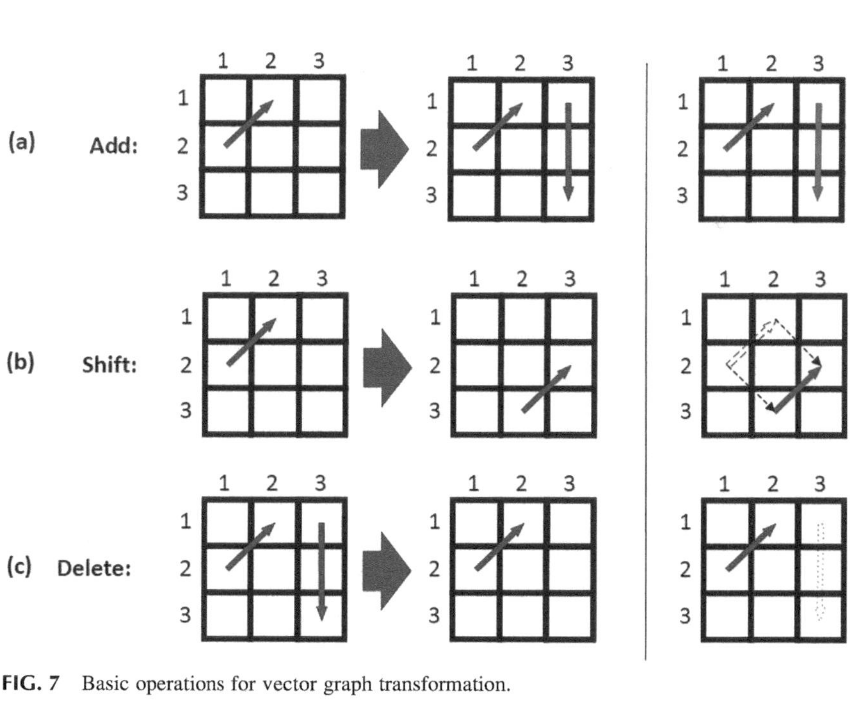

FIG. 7 Basic operations for vector graph transformation.

- **Shift:** Shift represents shifting a vector from its original position to another, which changes the vector from one into another. In Shift operation, two-shift vectors are generated between the previous start point and new start point, and the previous end point and new end point. The cost of shift operation equals the sum of these two vectors' Manhattan distance between the start point and end point, i.e.,

$$\left\| \vec{\delta}_{(ox,oy),(dx,dy)} \rightarrow \vec{\delta}_{(ox',oy'),(dx',dy')} \right\| = |ox' - ox| + |oy' - oy| + |dx' - dx| + |dy' - dy| \tag{38}$$

For example, as Fig. 7B shows, when $S_X = \left\{ \vec{\delta}_{(1,2),(2,1)} \right\}$, $S_Y = \left\{ \vec{\delta}_{(2,3),(3,2)} \right\}$, to conduct $S_X \rightarrow S_Y$, the vector $\vec{\delta}_{(1,2),(2,1)}$ is shifted from $(1,2) \rightarrow (2,1)$ to $(2,3) \rightarrow (3,2)$. The cost $\left\| \vec{\delta}_{(1,2),(2,1)} \rightarrow \vec{\delta}_{(2,3),(3,2)} \right\| = |2-1| + |3-2| + |3-2| + |2-1| = 4$.

- **Delete:** Delete represents directly deleting a vector from the set.

$$\left\| \vec{\delta}_{(\mathrm{ox},\mathrm{oy}),(\mathrm{dx},\mathrm{dy})} \rightarrow 0 \right\| = |\mathrm{dx} - \mathrm{ox}| + |\mathrm{dy} - \mathrm{oy}| \tag{39}$$

For example, as Fig. 7C shows, when $S_X = \left\{ \vec{\delta}_{(1,2),(2,1)}, \vec{\delta}_{(3,1),(3,3)} \right\}, S_Y = \left\{ \vec{\delta}_{(1,2),(2,1)} \right\}$, to conduct $S_X \rightarrow S_Y$, the vector $\vec{\delta}_{(3,1),(3,3)}$ is deleted. The cost $\left\| \vec{\delta}_{(3,1),(3,3)} \rightarrow 0 \right\| = |3-3| + |3-1| = 2$.

Function $V(S_X \rightarrow S_Y)$ is defined as the least cost of a transformation $S_X \rightarrow S_Y$. The least cost $V(S_X \rightarrow S_Y)$ can be computed iteratively:

$$V(S_X \rightarrow S_Y) = \min_{S_X' \subseteq S_X} \min_{S_Y' \subseteq S_Y} \min \begin{cases} V(S_X' \rightarrow S_Y') + V((S_X - S_X') \rightarrow (S_Y - S_Y')) \\ V(S_X' \rightarrow \emptyset) + V((S_X - S_X') \rightarrow S_Y) \\ V(\emptyset \rightarrow S_Y') + V(S_X \rightarrow (S_Y - S_Y')) \end{cases} \tag{40}$$

where

$$V(S \rightarrow \emptyset) = \sum_{\vec{\delta}_i \in S} \left\| \vec{\delta}_i \rightarrow 0 \right\| \tag{41}$$

$$V(\emptyset \rightarrow S) = \sum_{\vec{\delta}_i \in S} \left\| 0 \rightarrow \vec{\delta}_i \right\| \tag{42}$$

$$V\left(\left\{\vec{\delta}\right\}\rightarrow\left\{\vec{\delta'}\right\}\right)=\min\left(\left\|\vec{\delta}\rightarrow\vec{\delta'}\right\|,\left\|\vec{\delta}\rightarrow 0\right\|+\left\|0\rightarrow\vec{\delta'}\right\|\right) \tag{43}$$

Computing least cost can be converted into a bipartite graph minimum weight matching problem and solved by Kuhn-Munkres algorithm [71].

By computing least cost $V(S_X \rightarrow S_Y)$, all the similarity metrics can be achieved as follow:

- Volume similarity

Obviously, the least cost $V(S_X \rightarrow S_Y)$ reflects the volume difference between two vector graphs S_X and S_Y of mobility tableaus by taking geographical correlation into account. Considering about $V(S_X \rightarrow S_Y)$ is affected by cell width w, Manhattan distance for mobility tableaus (MD) is computed by:

$$MD(X,Y)=V(S_X \rightarrow S_Y)^*w \tag{44}$$

To measure the mass similarity, mass difference is normalized by computing the cost from empty set to S_X and S_Y to get normalized Manhattan distance for mobility tableaus (NMD) by:

$$\text{NMD}(X,Y)=1-\frac{V(S_X \rightarrow S_Y)}{V(\emptyset \rightarrow S_X)+V(\emptyset \rightarrow S_Y)} \tag{45}$$

- Spatial similarity

For a transformation $\left\{\vec{\delta}\right\}\rightarrow\left\{\vec{\delta}\,'\right\}$, there will be two possible implementation ways: $\vec{\delta}\rightarrow\vec{\delta}\,'$ and $\left(\vec{\delta}\rightarrow 0\right)+\left(0\rightarrow\vec{\delta}\,'\right)$. Assume that $\vec{\delta}$ is the vector from point A to point B while $\vec{\delta}\,'$ is the vector from point A' to point B'. Measuring the cost of two ways is equivalent to comparing the Manhattan length between two pairs of opposite sides of the quadrilateral formed by $ABA'B'$, i.e. $AB+A'B'$ and $AA'+BB'$. When $AB+A'B'$ is longer than $AA'+BB'$, the transformation prefers *Shift* operation because δ and δ' are close in space. On the other hand, when $AA'+BB'$ is longer, the transformation prefers *Add* and *Delete* operation because $\vec{\delta}$ and $\vec{\delta}\,'$ are quite far away. The total cost of *Shift* operation and *Add/Delete* operation just correspondingly indicates that whether the error comes from Spatial Similarity and Spatial Dissimilarity.

Therefore, decompose the function V into $V_{add/delete}$ and V_{shift} where $V_{add/delete}$ is the total cost of *Add/Delete* operation and V_{shift} is the total cost of *Shift* operation, a shift proportion (SP) can be computed as:

$$\text{SP}(X,Y)=\frac{V_{shift}(S_X \rightarrow S_Y)}{V_{shift}(S_X \rightarrow S_Y)+V_{add/delete}(S_X \rightarrow S_Y)} \tag{46}$$

SP illustrates the proportion of mobility tableaus difference that can be corrected by spatial similarity.

- Mass inclusiveness

For vector sets S_X and S_Y, if we put them into a Euclidean space, in which the distance between two vector sets equals to the least cost of Manhattan distance between them and use point A, B to represent them correspondingly, a triangle can be formed by introducing an empty set as point O as Fig. 8 shows. Obviously, the angle of $\angle AOB$ can reflect the direction difference between S_X and S_Y. Here direction refers to the process to build the mobility tableau. The same direction means building one mobility tableau needs the same process of building the other one while the contrary direction means there is no common process between building two mobility tableaus.

A normalized mass angle (NMA) that represents the mass inclusiveness is defined as:

$$\mathrm{NMA} = 1 - \frac{\arccos \dfrac{V(\emptyset \rightarrow S_X)^2 + V(\emptyset \rightarrow S_Y)^2 - V(S_X \rightarrow S_Y)^2}{2V(\emptyset \rightarrow S_X)^* V(\emptyset \rightarrow S_Y)}}{\pi} \tag{47}$$

- Structural similarity

The structural (dis)similarity metric can be computed in the same way with NMA by normalizing two mobility tableaus.

Assume that X' and Y' are normalized mobility tableaus for X and Y. Then normalized structural angle (NSA) can be defined similarly with Eq. (47) by:

$$\mathrm{NMA} = 1 - \frac{\arccos \dfrac{V(\emptyset \rightarrow S'_X)^2 + V(\emptyset \rightarrow S'_Y)^2 - V(S'_X \rightarrow S'_Y)^2}{2V(\emptyset \rightarrow S'_X)^* V(\emptyset \rightarrow S'_Y)}}{\pi} \tag{48}$$

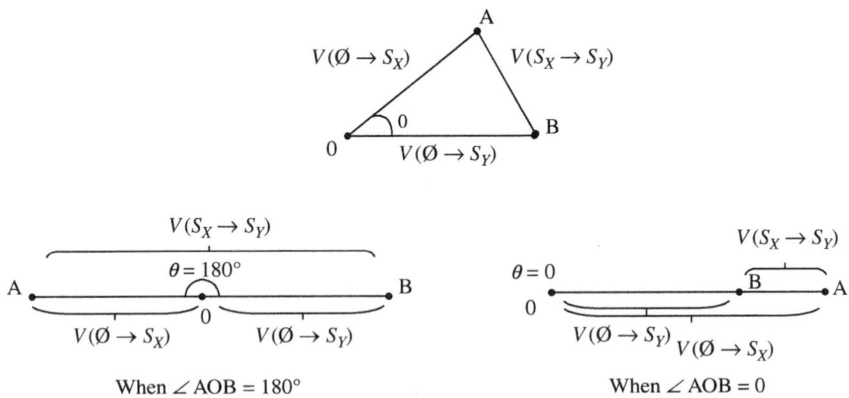

FIG. 8 Manhattan distance triangle for two mobility tableaus.

However, in the real case, NSA = 0 will almost never happen because when the total population flow of mobility tableau is large, two mobility tableaus seldom have no spatial similarity. Therefore, NSA will be usually very high. Considering about two random mobility tableaus will also have a relatively stable NSA μ, randomness removed normalized structural angle (RRNSA) is defined as:

$$\mathrm{RRNSA} = \begin{cases} \frac{\mathrm{NSA}-\mu}{1-\mu}, & \frac{\mathrm{NSA}-\mu}{1-\mu} > 0 \\ 0, & \frac{\mathrm{NSA}-\mu}{1-\mu} \leq 0 \end{cases} \tag{49}$$

5 Case study

In this section, we demonstrate two case studies, while one is a similarity comparison between CDR-based estimated mobility and the ground truth, the other is a metro usage pattern clustering.

In these two case studies, a GPS dataset in Tokyo, Japan is utilized, which is a part of Konzatu-Tokei (R) provided by Zenrin DataCom INC. Konzatsu-Tokei (R) Data refers to people flow data collected by individual's location data sent from a mobile phone under utilizers' consent, through applications provided by NTT DOCOMO, INC. These data are processed collectively and statistically in order to conceal private information. Original location data is GPS data (latitude, longitude) sent in about every 5 min and does not include the information to specify an individual. We randomly select 10,000 users in January 2011.

For the first case study, a simulated CDR dataset is generated based on the GPS dataset as ground truth. The cell tower distribution map is constructed from *Opencellid*, which is the largest Open Database of Cell Towers and their locations.

5.1 CDR-based travel estimation accuracy analysis

As we have introduced in Section 1, CDR is a kind of mobile phone data, which could be utilized as a great data source for many human mobility-related studies and applications [72–74].

However, the position of the cell tower only indicates a course range of the mobile terminator device's position, which results in the massive difference between CDR trajectory and the real movement. For different operators, their CDR sampling conditions are different. These different sampling conditions, such as cell tower density and sampling interval, may lead to different quality of restoration results. Therefore, evaluating the accuracy of the restoration result in different conditions is a significant issue.

In this case study, we test the ability to estimate human mobility based on CDR data in different sampling conditions by measuring the similarity between the estimated result and the ground truth. We mainly change two CDR sampling conditions: the cell tower density and the sampling time interval. The original GPS trajectory is regarded as the ground truth.

We first test the accuracy of CDR-based estimated trajectory, which is processed by a classical HMM-based map-matching method [75]. For cell tower density, we fix the sampling time interval as 5 min and change the cell tower density in [120/km^2, 40/km^2, 10/km^2, 2.5/km^2, 0.5/km^2], which means the number of cell towers per kilometers. For the sampling time interval, we fix the cell tower density as 40/km^2 and change the sampling time interval in [5 min, 10 min, 15 min, 30 min, 1 h]. We select two trajectory similarity metrics: DTW distance and Fréchet Distance, which have been introduced in Section 2. The result is shown in Table 1.

From the result, we may find that the sparser the cell tower distribution is and the longer the time interval is, the worse the estimated result will be, which is very reasonable. Although two sampling conditions both affect the accuracy of estimation result, from DTW distance we could see that the estimated trajectory is still close to the ground truth even in bad sampling conditions, which means the method is robust to conditions changing if we ignore the time shift.

Then we test the estimated OD matrices. *R*-squared, RMSE, and the method introduced in Section 4.4 are utilized to measure the similarity. Since OD matrix is very sensitive, when cell tower distribution is sparse, the result will be too bad

TABLE 1 Trajectory Similarity in Different Conditions.

	DTW distance	Fréchet distance
5 min, 120/km^2	45.58 m	296.8 m
5 min, 40/km^2	52.89 m	425.2 m
5 min, 10/km^2	101.64 m	547.72 m
5 min, 2.5/km^2	173.73 m	785.82 m
5 min, 0.5/km^2	217.88 m	1071.89 m
5 min, 40/km^2	52.89 m	425.2 m
10 min, 40/km^2	83.44 m	675.04 m
15 min, 40/km^2	122.08 m	814.34 m
30 min, 40/km^2	189.82 m	1039.67 m
1 h, 40/km^2	250.65 m	1285.36 m

to analyze similarity, so this time cell tower density is fixed as 25/km^2. The sampling time interval varies in [10 min, 15 min, 30 min, 1 h, and 3 h]. The result is shown in Table 2.

Compared with 10-min interval, when the time interval gets longer, the loss of trips information will cause volume similarity reduces. When the time interval gets longer than 1 h, the changing of stay point detection method causes the number of trips grow up to form a rebound. From the table we can conclude findings as follows:

- The changing tendency with an interval for NMD is similar with R^2 and RMSE because they all represent the volume similarity of mobility tableaus. However, RMSE is disturbed by long-distance OD pairs that are all zeros, so it is too small to help us judge whether the result is good or not. R^2 cannot reflect the geographical correlation, so the value shows the correlation is not high, but NMD indicates that the volume similarity is considerable indeed.
- When NMD is high, SP is also high, which means the major differences are spatial error. When NMD is low at 1-h time interval, SP is close to zero because at this time differences are caused by information loss instead of spatial error.
- NMA is always close to 100% for 1 h, 15 min, 10 min time intervals, because the information loss will not cause exclusiveness. For 3-h and 2-h time intervals, since the stay point detection is not accurate, NMA slightly gets lower.
- RRNSA is quite different from volume similarity. Although volume similarity is lowest at 1-h time interval, the structural similarity is the highest because less records bring less spatial error from cell towers, but the general structure is the same. For 3- and 2-h time intervals, though volume similarity is not low, RRNSA is much lower because inaccurate stay points break the original structure.

TABLE 2 OD matrix similarity in different conditions.

	3 h[a]	2 h[a]	1 h	15 min	10 min
R-squared	0.57	0.59	0.31	0.68	0.82
RMSE	0.0981	0.0951	0.1001	0.0692	0.0557
NMD (volume similarity)	92.11%	87.67%	70.40%	93.96%	97.99%
SP (spatial similarity)	75.19%	43.77%	2.32%	16.74%	71.43%
NMA (mass inclusiveness)	95.46%	96.97%	98.21%	98.66%	98.81%
RRNSA (structural similarity)	84.61%	82.29%	96.47%	94.48%	91.54%

[a]*Stay point detection method is different.*

5.2 Metro usage pattern clustering

With the development of the transportation system, metro becomes more and more important in city dweller's daily life, especially for commuting. Under this circumstance, metro station advertising has become a great strategy. However, in order to get a better advertising effect, understanding different station's metro usage pattern is significant, in which the first step is metro usage pattern clustering.

In this case study, we focus on a metro line named Yamanote Line, which is one of the most important commuter routes in Tokyo, Japan, to cluster the metro usage pattern.

The data has been preprocessed before to remove users who have not taken Yamanote Line before. The travel pattern extraction, expression, and clustering methods introduced in Section 4 are adopted for this case study. Different from the normal travel pattern, since we just focus on metro usage, nodes in the support graph are all stations instead. To achieve a better result, the time interval is set as 3 h, so the support graph just has eight layers.

After NMF decomposition, each user's metro usage pattern can be represented by a 3-dimensional vector in a Euclidean space. We pick *K*-means as the clustering method to incorporate with this metagraph space. Fig. 9 shows the result of the clustering that classifies all the users into 14 groups.

From the result we can see that some groups are not so ideal like group 9, the population of this group do not have a clear metro usage pattern and get on and off the metro in multiple stations. Since the metro usage will be affected by too many factors, the data duration is not long enough to extract a stable pattern. However, it still successfully catches some important features and groups. For example, group 12 describes the population who prefer to get on or off the metro in Shinjuku station in the afternoon or morning, who have a high probability work there. Group 4 describes the population who prefer to get on or off the metro in Tokyo station in the afternoon or morning, who also have a high probability work there.

6 Conclusion and future directions

In this chapter, we mainly introduce methods of similarity estimation and clustering for three typical kinds of mobility expressions: individual-level trajectory, population-level origin-destination matrices, and individual-level travel pattern. We also demonstrate two case studies, while one is a similarity comparison between CDR-based estimated mobility and the ground truth, the other is a metro usage pattern clustering.

Although the similarity measurement and clustering algorithm for some simple mobility structures like trajectory have already been mature enough to be utilized for various applications, there are still some problems remained to be solved for complex mobility structures.

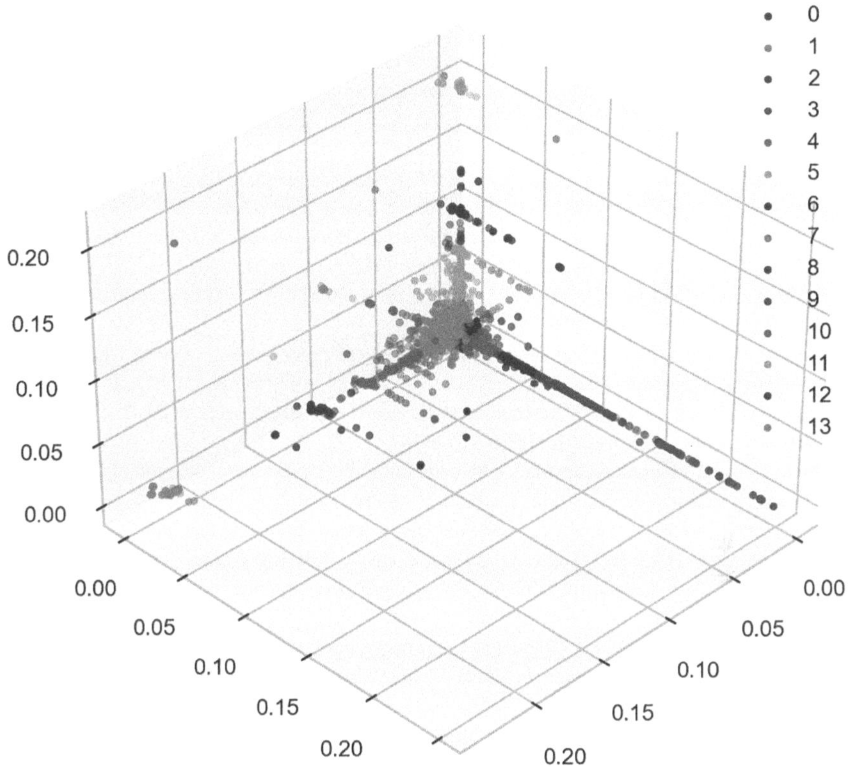

FIG. 9 Clusters by *K*-means in the metagraph space.

For OD matrix, it is hard to measure all the similarity properties in a short time. Algorithm with rapid computation speed cannot comprehensively measure the different similarity introduced before, while the algorithm that introduced in Section 4.4 face with high computational complexity in some circumstances. On the other hand, all the similarity measurements pay more attention to a static similarity comparison, while a temporal measure is also important.

For travel patterns, it is hard to construct a unified model for different purposes. The similarity measurement is also seriously affected by some parameters setting, such as time interval. Travel pattern clustering still has great development space.

No matter in what kind of mobility expression, despite the dataset contains records of massive users, it usually just accounts for a very small part of the real population. This could bring a huge degree of bias and cause a significant impact on numerous applications. In the future, how to combine travel mobility with demographic information to get better estimation and analysis is a major challenge.

References

[1] He T, et al. What is the human mobility in a new city: transfer mobility knowledge across cities. In: Proceedings of the web conference 2020; 2020.

[2] Chen J, et al. Mining urban sustainable performance: GPS data-based spatio-temporal analysis on on-road braking emission. J Clean Prod 2020;270:122489.

[3] Wang J-P, Liu T-L, Huang H-J. Tradable OD-based travel permits for bi-modal traffic management with heterogeneous users. Trans Res E: Logist Transp Rev 2018;118:589–605.

[4] Zhang Q, et al. Origin-destination-based travel time reliability under different rainfall intensities: an investigation using open-source data. J Adv Transp 2020;2020.

[5] Ma J, et al. Ridesharing user equilibrium problem under OD-based surge pricing strategy. Transp Res B Methodol 2020;134:1–24.

[6] Wang Y, et al. Using mobile phone data for emergency management: a systematic literature review. Inf Syst Front 2020;1–21.

[7] Kraemer MU, et al. The effect of human mobility and control measures on the COVID-19 epidemic in China. Science 2020;368(6490):493–7.

[8] Huo E, et al. Mining massive truck GPS data for freight OD estimation: case study of liaoning province in China. In: CICTP 2020; 2020. p. 86–98.

[9] Behara K, Bhaskar A, Chung E. A methodological framework to explore latent travel patterns and estimate typical OD matrices: a case study using Brisbane Bluetooth multi-density OD database; 2020. Quteprints: 199363. [Working Paper] (Submitted (not yet accepted for publication)).

[10] Forghani M, Karimipour F, Claramunt C. From cellular positioning data to trajectories: steps towards a more accurate mobility exploration. Transp Res C: EmergTechnol 2020;117:102666.

[11] Katranji M, et al. Deep multi-task learning for individuals origin-destination matrices estimation from census data. Data Min Knowl Disc 2020;34(1):201–30.

[12] Dey S, Winter S, Tomko M. Origin-destination flow estimation from link count data only. Sensors 2020;20(18):5226.

[13] Iqbal MS, et al. Development of origin-destination matrices using mobile phone call data. Transp Res C: EmergTechnol 2014;40:63–74.

[14] Mamei M, et al. Evaluating origin-destination matrices obtained from CDR data. Sensors 2019;19(20):4470.

[15] Barua L, Zou B, Zhou Y. Machine learning for international freight transportation management: a comprehensive review. Res Transp Bus Manag 2020;34:100453.

[16] Liu C, et al. Time-dependent vehicle routing problem with time windows of city logistics with a congestion avoidance approach. Knowl-Based Syst 2020;188:104813.

[17] Shaheen S, Cohen A. Innovative mobility: carsharing outlook carsharing market overview, analysis, and trends. UC Berkeley: Transportation Sustainability Research Center; 2020. Retrieved from: https://escholarship.org/uc/item/9jh432pm.

[18] Xiang Q, Wu Q. Tree-based and optimum cut-based origin-destination flow clustering. ISPRS Int J Geo Inf 2019;8(11):477.

[19] Guo X, et al. An OD flow clustering method based on vector constraints: a case study for Beijing taxi origin-destination data. ISPRS Int J Geo Inf 2020;9(2):128.

[20] Vlachos M, Kollios G, Gunopulos D. Discovering similar multidimensional trajectories. In: Proceedings 18th international conference on data engineering, IEEE; 2002.

[21] Berndt DJ, Clifford J. Using dynamic time warping to find patterns in time series. In: KDD workshop, Seattle, WA, USA; 1994.

[22] Yuan Y. Image-based gesture recognition with support vector machines. University of Delaware; 2008.
[23] Chen L, Özsu MT, Oria V. Robust and fast similarity search for moving object trajectories. In: Proceedings of the 2005 ACM SIGMOD international conference on management of data; 2005.
[24] Chen L, Ng R. On the marriage of lp-norms and edit distance. In: Proceedings of the thirtieth international conference on very large data bases-Volume 30; 2004.
[25] Gudmundsson J, Laube P, Wolle T. Computational movement analysis. In: Springer handbook of geographic information. Springer; 2011. p. 423–38.
[26] Fréchet MM. Sur quelques points du calcul fonctionnel. Rendiconti Circolo Matematico Palermo (1884–1940) 1906;22(1):1–72.
[27] Lin B, Su J. Shapes based trajectory queries for moving objects. In: Proceedings of the 13th annual ACM international workshop on Geographic information systems; 2005.
[28] Pelekis N, et al. Similarity search in trajectory databases. In: 14th international symposium on temporal representation and reasoning (TIME'07), IEEE; 2007.
[29] Lloyd S. Least squares quantization in PCM. IEEE Trans Inf Theory 1982;28(2):129–37.
[30] Tian Z, Ramakrishnan R, Birch LM. An efficient data clustering method for very large databases. ACM SIGMOD Rec 1996;25:103–14.
[31] Ankerst M, et al. Ordering points to identify the clustering structure. In: Proc ACM SIGMOD; 2008.
[32] Wang W, Yang J, Muntz R. STING: a statistical information grid approach to spatial data mining. In: VLDB; 1997.
[33] Lee J-G, Han J, Whang K-Y. Trajectory clustering: a partition-and-group framework. In: Proceedings of the 2007 ACM SIGMOD international conference on management of data; 2007.
[34] Grünwald PD, Myung IJ, Pitt MA. Advances in minimum description length: theory and applications. MIT Press; 2005.
[35] Jiang S, Ferreira J, González MC. Activity-based human mobility patterns inferred from mobile phone data: a case study of Singapore. IEEE Trans Big Data 2017;3(2):208–19.
[36] Liu K, Murayama Y, Ichinose T. A multi-view of the daily urban rhythms of human mobility in the Tokyo metropolitan area. J Transp Geogr 2021;91:102985.
[37] Tu W, et al. Uncovering online sharing vehicle mobility patterns from massive GPS trajectories. In: Spatial synthesis. Springer; 2020. p. 413–29.
[38] Hafezi MH, Liu L, Millward H. A time-use activity-pattern recognition model for activity-based travel demand modeling. Transportation 2019;46(4):1369–94.
[39] Acheampong RA, et al. Mobility-on-demand: An empirical study of internet-based ride-hailing adoption factors, travel characteristics and mode substitution effects. Transp Res C: EmergTechnol 2020;115:102638.
[40] Bai H, et al. Measurement of the differential cross sections and angle-integrated cross sections of the 6Li (n, t) 4He reaction from 1.0 eV to 3.0 MeV at the CSNS Back-n white neutron source. Chinese Phys C 2020;44(1):014003.
[41] Phithakkitnukoon S, et al. Understanding tourist behavior using large-scale mobile sensing approach: a case study of mobile phone users in Japan. Pervasive Mobile Comput 2015;18:18–39.
[42] Hu T, et al. Mining shopping patterns for divergent urban regions by incorporating mobility data. In: Proceedings of the 25th ACM international on conference on information and knowledge management; 2016.
[43] Zhang F, et al. Exploiting dining preference for restaurant recommendation. In: Proceedings of the 25th international conference on World Wide Web; 2016.

[44] Li W, et al. Effective metagraph-based life pattern clustering with big human mobility data. arXiv preprint arXiv:2104.11968; 2021.

[45] Lee DD, Seung HS. Learning the parts of objects by non-negative matrix factorization. Nature 1999;401(6755):788–91.

[46] Lu N, Miao H. Clustering tree-structured data on manifold. IEEE Trans Pattern Anal Mach Intell 2015;38(10):1956–68.

[47] Donoho D, Stodden V. When does non-negative matrix factorization give a correct decomposition into parts? In: Advances in neural information processing, vol. 16. MIT Press; 2003.

[48] Fan Z, Song X, Shibasaki R. CitySpectrum: a non-negative tensor factorization approach. In: Proceedings of the 2014 ACM international joint conference on pervasive and ubiquitous computing; 2014.

[49] Yao Y, et al. Mobsimilarity: vector graph optimization for mobility tableau comparison. arXiv preprint arXiv:2104.13139; 2021.

[50] Ashok K, Ben-Akiva ME. Estimation and prediction of time-dependent origin-destination flows with a stochastic mapping to path flows and link flows. Transp Sci 2002;36(2):184–98.

[51] Barceló Bugeda J, et al. A Kalman-filter approach for dynamic OD estimation in corridors based on bluetooth and Wi-Fi data collection. In: 12th world conference on transportation research WCTR, 2010; 2010.

[52] Tamin O, Willumsen L. Transport demand model estimation from traffic counts. Transportation 1989;16(1):3–26.

[53] Antoniou C, Ben-Akiva M, Koutsopoulos HN. Incorporating automated vehicle identification data into origin-destination estimation. Transp Res Rec 2004;1882(1):37–44.

[54] Cascetta E. Estimation of trip matrices from traffic counts and survey data: a generalized least squares estimator. Transp Res B Methodol 1984;18(4-5):289–99.

[55] Kim S-J, Kim W, Rilett LR. Calibration of microsimulation models using nonparametric statistical techniques. Transp Res Rec 2005;1935(1):111–9.

[56] Cools M, Moons E, Wets G. Assessing the quality of origin-destination matrices derived from activity travel surveys: results from a Monte Carlo experiment. Transp Res Rec 2010;2183 (1):49–59.

[57] Barceló J, et al. Robustness and computational efficiency of Kalman filter estimator of time-dependent origin-destination matrices: exploiting traffic measurements from information and communications technologies. Transp Res Rec 2013;2344(1):31–9.

[58] Yang H, Iida Y, Sasaki T. An analysis of the reliability of an origin-destination trip matrix estimated from traffic counts. Transp Res B Methodol 1991;25(5):351–63.

[59] Gan L, Yang H, Wong SC. Traffic counting location and error bound in origin-destination matrix estimation problems. J Transp Eng 2005;131(7):524–34.

[60] Bera S, Rao K. Estimation of origin-destination matrix from traffic counts: the state of the art. Eur Transp/Trasp Eur 2011;49:3–23.

[61] Tavassoli A, et al. How close the models are to the reality? Comparison of transit origin-destination estimates with automatic fare collection data. In: Proc 38th Australas Transp Res Forum (ATRF); 2016.

[62] Ros-Roca X, et al. Investigating the performance of SPSA in simulation-optimization approaches to transportation problems. Transp Res Procedia 2018;34:83–90.

[63] Wang Z, et al. Image quality assessment: from error visibility to structural similarity. IEEE Trans Image Process 2004;13(4):600–12.

[64] Djukic T, Hoogendoorn S, Van Lint H. Reliability assessment of dynamic OD estimation methods based on structural similarity index. Transportation Research Board 92nd Annual Meeting; 2013.

[65] Pollard T, et al. Comparing the quality of OD matrices in time and between data sources. In: Proceedings of the European transport conference, AET Frankfurt, Germany; 2013.
[66] Day-Pollard T, van Vuren T. When are origin-destination matrices similar enough?; 2015.
[67] Behara KN, Bhaskar A, Chung E. Geographical window based structural similarity index for OD matrices comparison; 2020.
[68] Jin C, et al. Similarity measurement on human mobility data with spatially weighted structural similarity index (SpSSIM). Trans GIS 2020;24(1):104–22.
[69] Ruiz De Villa A, Casas J, Breen M. OD matrix structural similarity: Wasserstein metric. Transportation Research Board 93rd Annual Meeting; 2014.
[70] Behara KN, Bhaskar A, Chung E. A novel approach for the structural comparison of origin-destination matrices: Levenshtein distance. Transp Res C: EmergTechnol 2020;111:513–30.
[71] Kuhn HW. The Hungarian method for the assignment problem. Berlin, Heidelberg: Springer Berlin Heidelberg; 2009. p. 29–47.
[72] Pei T, et al. A new insight into land use classification based on aggregated mobile phone data. Int J Geogr Inf Sci 2014;28(9):1988–2007.
[73] Shin D, et al. Urban sensing: using smartphones for transportation mode classification. Comput Environ Urban Syst 2015;53:76–86.
[74] Xu Y, et al. Human mobility and socioeconomic status: analysis of Singapore and Boston. Comput Environ Urban Syst 2018;72:51–67.
[75] Newson P, Krumm J. Hidden Markov map matching through noise and sparseness. In: Proceedings of the 17th ACM SIGSPATIAL international conference on advances in geographic information systems; 2009.

Chapter 4

Data fusion technologies for MaaS

Yi Sui[a,b,c], Haoran Zhang[b], Wenxiao Jiang[b], Rencheng Sun[a,b,c], and Fengjing Shao[c]
[a]College of Computer Science and Technology, Qingdao University, Qingdao, China, [b]Center for Spatial Information Science, The University of Tokyo, Kashiwa, Chiba, Japan, [c]Institute of Smart City and Big Data Technology, Qingdao, China

1 Introduction

With the emergence of the concept and practice of Mobility as a Service (MaaS), urban mobility has undergone profound changes [1, 2]. As a new transport solution, MaaS integrates different available transport modes into one platform and provides an ease-of-use transport option with a higher level of availability and reliability. The rapid growth of Transportation Network Companies (TNCs), such as Uber, Didi Chuxing, Lyft, Grab, Careem, and Ola, has allowed MaaS to expand rapidly. Various online ride services, such as ride-sourcing/ride-hailing, ride-sharing, ride-splitting, bike-sharing, and minibus shared-ride service, have grown into immense popularity around the world, which poses a great challenge to traditional transportation modes, and reshapes the urban transportation system [3]. For promoting sustainable urban transport, optimizing MaaS is vital, encompassing numerous research problems from diverse perspectives [4].

In general, MaaS system consists of three parts: service providers (such as taxis, private cars, or bikes), service users (such as passengers), and service platforms (such as TNCs) [5]. The service platform needs to employ the right operational strategies to balance users' demand and providers' supply over time and space that is to minimize the gap between users' demand and providers' supply. For example, based on demand prediction, service platforms can manage and distribute their fleet (ride-hailing) to reduce empty mileages and improve their service quality. With rapid developments of information and pervasive sensing technologies, a wide array of data sourced from service providers and users are collected. As shown in Fig. 1, on the providers' side, data include vehicle

Big Data and Mobility as a Service. https://doi.org/10.1016/B978-0-323-90169-7.00005-1

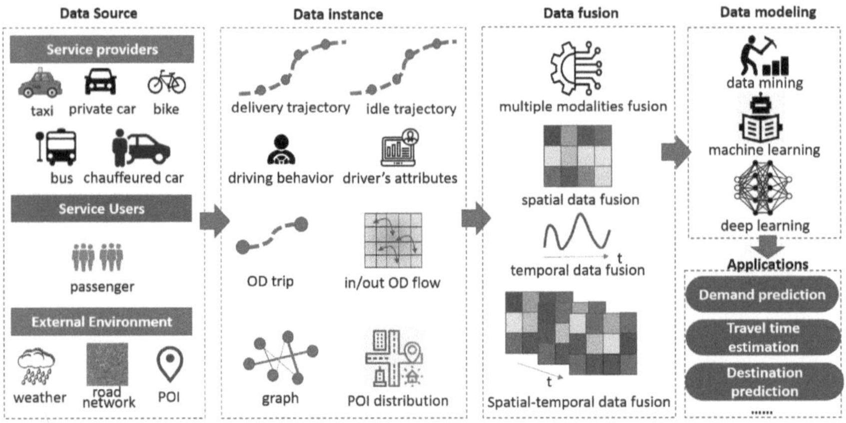

FIG. 1 Different data sources in MaaS system and their roles in decision-making.

trajectory, drivers' behaviors (traffic accident), drivers' attributes (age, gender, service rate); on the users' side, data include passengers' OD trip, travel mode choice, and demographical attributes. The service platform also collects external environmental data, such as road network, traffic volume, weather, and POI (points of interest), which have a significant influence on drivers' and users' travel behaviors. All of these data (from drivers, passengers, and environment) are very useful and valuable for the decision-making of the service platform. They are first integrated and then further processed through data mining and machine learning technology for different applications, such as spatio-temporal demand prediction of passengers, travel time estimation, ride-splitting strategies optimization, and layout optimization of bike-sharing stations.

However, how to integrate these data from different sources for supporting decision-making in the MaaS system is an uneasy task. First, data from different sources have multi-modal characteristics, each of which has a different formula. For example, global position system (GPS) records travel positions of vehicles, which is usually expressed in time sequences; a road network is a graph structure composed of nodes and edges; Origin-Destination (OD) trip data is usually regarded as raster data in form of a matrix. Fusing data with different modalities poses a significant challenge, and this requires advanced data fusion methods that can deal with many different types of input [6]. Second, solutions to many research problems in the MaaS system depend on accurate analysis of spatio-temporal correlation. For example, when forecasting the riding demand of an area, it is important to extract the historical demand pattern of the area (time dependence) and study the influence of neighboring areas and distant areas (spatial dependence) on the demand of the area. Potential time dependence and spatial dependence interact with each other, and together determine the future trend of ride demand. The spatio-temporal correlation is difficult to obtain through human's prior knowledge. Therefore, this calls data fusion methods to learn from a large amount of historical data and have the ability to extract spatial and temporal information and merge them effectively.

The primary goal of this chapter is to review the advanced data fusion methodologies and their application in the MaaS system and discuss the directions for future research in this area. In Section 2, we summarize the data types in the MaaS system and their input formula of data fusion methods. In Section 3, the categories of data fusion methods in the MaaS system and the reasons for such classification are introduced. In Section 4, we first discuss the deep learning-based feature extraction and then summarize the feature fusion strategies in recent works. In Section 5, we introduce decomposition-based methods and their applications. Challenging problems and future study trends are discussed in Section 6.

2 Data formula

When fusing data from different sources for further mining, the first task is to figure out the formula of the input data. Different data fusion methods usually have different requirements for the format of the input. Tuple, sequence, vector, matrix/tensor, and graph are widely used data formulas. Fig. 2 presents common data instances in the MaaS system and their possible formulas. In the following, we discuss each data instance and its corresponding formulas and mining applications.

2.1 Attribute and event data

Drivers' and users' attributes (i.e., age, gender, income, education, car-ownership) and their behaviors' events (driver's traffic accident, user's travel mode choice) can be recorded in the formula of a tuple. For example, event data can be defined as a tuple (id, t, l, e), indicating the occurring time t and location l of $id's$ event e. Correlation analysis of attributes and behaviors can help us understand the influence mechanism of target events.

It was found that the road crash rate among online ride-hailing drivers was higher than taxi drivers and private car drivers, which had aroused research interests in the aberrant driving behaviors of ride-hailing drivers [7]. Based on attributes data of drivers, influencing factors such as gender, income, driving

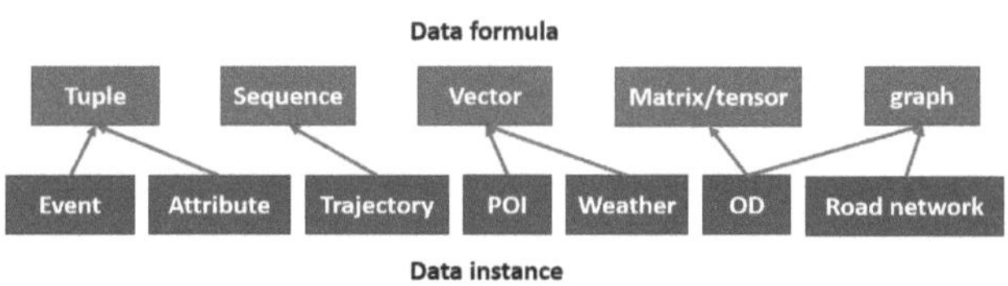

FIG. 2 Correlation between data instance and data representation.

style, and work-condition factors were found to be the significant predictors of abnormal driving behaviors of ride-hailing drivers [8].

Comparative analysis of the characteristics of MaaS users and other groups can help us find the socioeconomic factors that affect the usage of MaaS. Previous works concluded that ride-hailing/ride-sharing users tended to be younger, well-educated, and higher incomes [9–11].

2.2 Trajectory data

A trajectory is a sequence of points recording the trace of a moving object over time. Trajectory data comprise discrete locations and times. Each trajectory $Traj = [(t_1, l_1), \ldots, (t_n, l_n), \ldots]$ can be represented as a sequence of tuples, where $l_n = (x_n, y_n)$ is the two-dimensional location of the moving object at a time t_n.x_n, y_n often refer to longitude and latitude, respectively, which can be extended to more than two dimensions. GPS data often includes instantaneous travel speed v, moving direction θ. These data can be included in *Traj* and then obtain a time sequence composed of four-tuple $l_n = (x_n, y_n, v_n, \theta_n)$.

Trajectory data often come from Global Position System (GPS) which are installed on the vehicle (taxis, buses, private cars) or smartphone. The length of trajectory *Traj* depends on the sample intervals of the sensor. For example, trajectory data of ride-hailing vehicles from Didi Chuxing is sampled at a time interval of 3s, which means the passing time between two adjacency points in a trajectory sequence is 3s. Data with shorter time intervals can reflect a more realistic trajectory shape on a two-dimensional plane.

Trajectory data is mostly used for destination prediction, giving a trajectory prefix and predicting the destination. The key to destination prediction is a good recognition of spatio-temporal correlation between adjacent locations recorded in prefix trajectory. As shown in Fig. 3, trajectory data can be expressed as a sequence, which can be used as the input of a model that can extract sequence correlation [12, 13], or is transformed into a grid image in which each pixel of the grid is represented whether the corresponding location is visited in the trajectory. The grid image can be used for the input of models that can extract spatial correlation features [14].

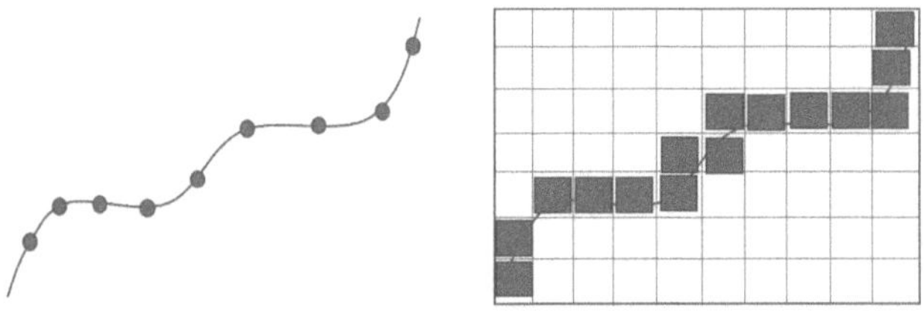

FIG. 3 Sequence and grid image formula of trajectory data.

2.3 Origin-destination (OD) trip data

A trip data can be defined as a tuple Trip $= [id, (t_{up}, l_{up}), (t_{off}, l_{off})]$, where $(t_{up}, l_{up})(t_{off}, l_{off})$ represent pick-up and pick-off information (time and location) of passenger *id*, respectively. For online ride service, trip data can be divided into two categories: demand trip and answer trip. When passengers send a ride request to the online ride system, demand trip data is generated. When a driver answers this demand, the finished trip is regarded as an answer trip. OD flow at a citywide scale can be calculated by aggregating long time (one month or one year) trip data. Based on the OD flow, the in–out-flow of each area at a given time span can be summarized.

As shown in Fig. 4, in-flow of each area can be defined as a 3D matrix F_{in} with m rows and n columns, where the value $F_{in}(i,j)$ is a vector length of $m \times n$. Each element in $F_{in}(i,j)$ indicates the number of trips from other locations to location (i,j) at time span t. The value of an element can be obtained by aggregating trajectory data according to origin and destination. Out-flow of each area can also be defined as F_{out} that has a similar structure with F_{in}.

OD trip data can be used for demand prediction that learns where and when a trip will start and end. Through demand prediction, service platform can manage and distribute their fleet (ride-hailing, bike-sharing) to match the demand and the supply. Previous works suggested that demand prediction depended not only on historical OD trips, but also the environmental data such as traffic, weather, time, and date. Therefore, OD trips data and environmental data are usually input to the model separately for feature extraction and then features are combined for the further process [15–20].

2.4 Correlation network

Given entities and their relation, a graph can be constructed for describing correlation among entities. In the MaaS system, many types of data can be abstracted as graphs. For example, according to latitude and longitude, an urban area can be divided into a grid, in which a cell can be regarded as a node and the geographical connection between cells is an edge. A grid network with four-connected neighbors (such as an image) is a special case in graphs.

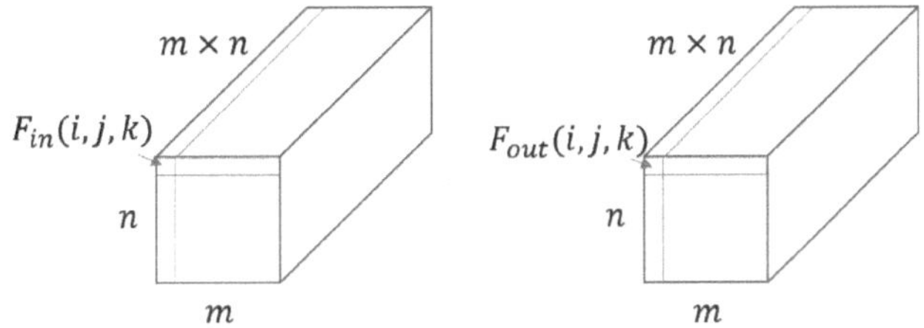

FIG. 4 Formula of OD trip data.

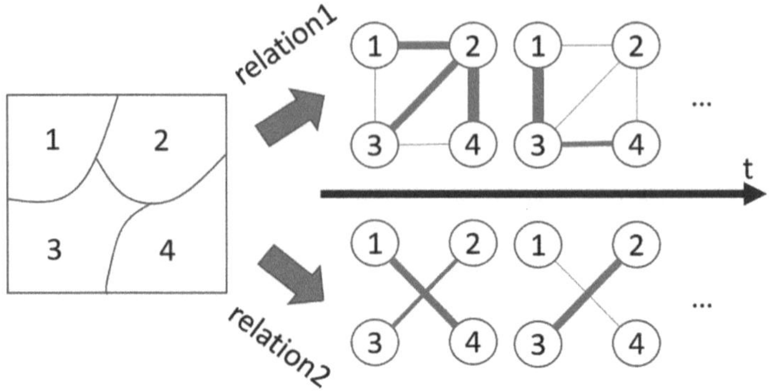

FIG. 5 Graph formula of OD trip data.

In practice, an edge can be generated according to the given relation. For example, as shown in Fig. 5, an edge e_{ij} between node v_i, v_j can be generated when $dis(v_i, v_j) < \tau$, where $dis(v_i, v_j)$ is a function measuring the distance between node v_i and v_j. The distance can be Euclidean distance, indicating the neighbor relationship between block areas; or transport connectivity, such as travel time by public transport or private car; functional similarities, such as residential district or office district. OD trip between v_i and v_j can be regarded as weight w_{ij} of edge e_{ij}. When considering time dimension, we can obtain a sequence of OD trip graphs over different timestamps. This formula integrates geographical information and OD trip to demonstrate the spatio-temporal travel structure of passengers.

Dividing the area into equal subareas based on latitude and longitude results in the loss of functional information, which has a negative influence on problem-solving, such as demand prediction. For example, area with the same function (business area, education area) is divided into different subareas, which leads to redundant data. For solving this problem, subarea with the same function can be abstract as a node through clustering methods [21–23]. Then multiple relations between subareas can be constructed for better mining spatio-temporal correlation between subarea for supporting demand prediction of each subarea [17, 24].

2.5 Environmental data

We refer to environmental data as POI, weather, date, and time data. Function of location can be recorded as POI, such as hospital, school, metro, park, factory, etc. POI of location i is defined as a vector $P_i = (p_i^1, \ldots, p_i^n, \ldots, p_i^k)$, where p_i^n indicates the amounts of type j facility in area i and k is the total number of facilities ($n \in \{1, 2, \ldots k\}$). Weather of location i at time t is defined as a vector $w_i = (t, w_i^1, \ldots, w_i^n, \ldots, w_i^m)$, where $w_i^n (n \in \{1, 2, \ldots m\})$ is one kind of weather data, such as temperature, humidity, air pressure, air quality, etc. Date and time data are categorical data, which can be expressed as one-hot code as model input.

When using environmental data, they are often first transformed into a low-dimensional real vector (embedding code in latent space) by the embedding method and then are fused with other feature vectors for further processing [25–28]. In general, let $v \in R^{1 \times I}$ denotes environmental data, where I is the vocabulary size of input categorical value. Embedding representation $u \in R^{1 \times O}$ of v can be calculated by $u = f(Wv + b)$, where $W \in R^{I \times O}$ is a parameter matrix and $b \in R^{1 \times O}$ is a bias term. Function f is a nonlinear activation function. Usually $O \ll I$. One important property of the embedding method is similar vectors are usually very similar in embedding space. In MaaS application, environmental data is used together with other data. For example, as mentioned in Section 2.3, environmental data is used with OD trip data for demand prediction. In addition, environmental data and GPS trajectory data were combined as the input of learning model for destination prediction [12, 14], origin-destination prediction [26], and traffic flow prediction [28, 29].

3 Categories of data fusion methods in MaaS

There are many definitions of data fusion in the literature. One of the most popular definitions appeared in the 1990s by the Joint Directors of Laboratory and the U.S. Department of Defense [30]: "*A process dealing with the association, correlation, and combination of data and information from single and multiple sources to achieve refined position and identity estimates, and complete and timely assessments of situations and threats as well as their significance.*" A more general definition of data fusion is reported by [31]: "*Data fusion is a formal framework in which are expressed means and tools for the alliance of data originating from different sources. It aims at obtaining information of greater quality; the exact definition of 'greater quality' will depend upon the application.*" In general, data fusion technique aims to integrate multiple data sources represented different parts of the scene that allows achieving more complete information to specific research targets.

Data fusion technique has attracted research interest from many applications, such as intelligent transportation system [32], urban computing [33], remote sensing [34], and Internet of Thing [35]. Each area focuses on a different motivation of data fusion. In general, the available data fusion techniques can be classified into three categories [36]: (i) low-level fusion, where raw data are integrated, such as data imperfection and data correlation [37]; (ii) mid-level fusion, where features extraction from raw data are fused for further processing. In this category, machine learning techniques are usually used for feature extraction [38]; and (iii) high-level fusion, where preliminary decisions are provided as input of data fusion methods to obtain joint inference decisions. Probabilistic approaches, such as Bayesian network [39] and state-space models (Markov Random Field) [40] are widely used for this category of data fusion.

In the MaaS system, we are faced with a large amount of data with different formulas (sequence, matrix, graph, tuple). The solution of many problems depends on complex spatio-temporal correlation mining from these data. This requires data fusion methods to deal with a large amount of historical data with multiple modalities and has the ability to extract complex, nonlinear spatio-temporal correlation from the data. One widely used idea of data fusion in MaaS is to handle data separately and obtain their latent representation (feature extraction) for further fusion processing. Learning feature representation can not only unify the data formula in latent space but also remove data redundancy and retain useful data information (spatial and temporal characteristics). Therefore, in this chapter, we focus on mid-level fusion and divided it into two subcategories: deep learning methods and decomposition methods.

Deep learning is a data-driven paradigm, usually used in situations where complex, nonlinear models are required to solve a problem. Different from traditional machine learning methods that require manual features, deep learning model can automatically learn feature representations from data. The multilayer structure of deep network enables it to extract low, medium, and high-level features from data. High-level features (more abstract features) of the data are the result of the nonlinear function of low-level and mid-level features. Moreover, various architectures give deep learning models the ability to process different formulas of data, such as vector, matrix, tensor, and graph. Therefore, deep learning methods have become widely used in data fusion for MaaS applications.

In general, there are two kinds of deep learning-based data fusion workflow. As shown in Fig. 6, one is there are multiple data sources and each dataset is used as the input of the deep learning model. After extracting the features of each dataset, they are fused together. The other one is to have a single dataset and use it as the input of multiple deep learning models. Different features can be extracted and then be fused.

In Section 4, several widely used deep learning models for feature extraction are introduced. After obtaining data features, different strategies and models are then used to fuse them. Choice of data fuse strategies and models depends on specific research problems. We will briefly introduce preliminary strategies that are widely used in MaaS data fusion.

Although deep learning is associated with powerful learning ability, it is still a black-box method with weak interpretability and is difficult to intuitively understand by humans [41]. Compared with deep learning methods, the latent features obtained by the decomposition method are more intuitive and have better interpretability. The main idea of decomposition-based methods is to extract latent features by leveraging matrix factorization. Matrix is one most common data formulas in MaaS. Therefore, we review decomposition methods used in MaaS application in Section 5.

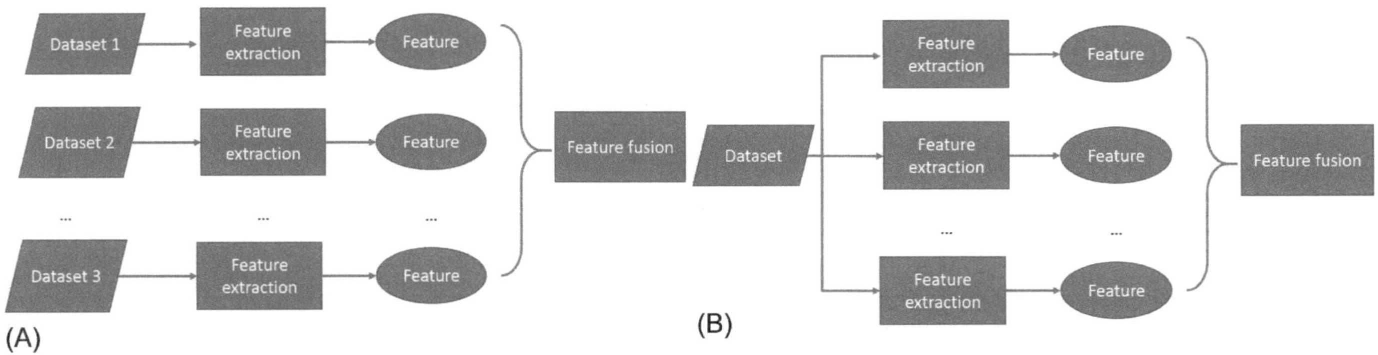

FIG. 6 Data fusion workflow based on deep learning. (A) Feature fusion from multiple datasets and (B) feature fusion from one dataset.

4 Data fusion based on deep learning

In Section 4.1, we briefly introduce the fundamental building blocks of deep learning networks first, and then discuss their applications in feature extraction in MaaS. Feature fusion strategies are summarized in Section 4.2.

4.1 Fundamental building units of deep learning network

4.1.1 CNN

Convolutional neural network (CNN) is a class of feedforward deep neural networks. CNN is originally designed to process image data. Classical CNN models include LeNet-5 [42], AlexNet [43], VGG [44], GoogleNet [45], FCN [46], ResNet [47], etc. With its power ability on capturing spatial correlations and patterns, it is widely used in many data processing tasks, such as spatio-temporal data, which has been achieved state-of-the-art results.

As shown in Fig. 7, the basic architecture of CNN consists of three types of layers, namely convolutional layer, pooling layer, and fully connected layer. Aiming for learning the feature representation of inputs, the convolutional layer is composed of several convolution kernels for calculating different feature maps. Each neuron of the feature map is connected to the local area of neighboring neurons in the previous layer, which allows the feature map to learn local patterns. The learned feature representation has spatial invariance (i.e., invariance to translation, invariance to scale, etc.).

For each feature map, all spatial locations of the input associate with the same kernel. The output of neuro located at (i,j) in feature map k of the lth layer can be calculated by:

$$z_{i,j,k} = a\left(w_k x_{i,j} + b_k\right)$$

where w_k and b_k are the weight vector and bias term of the kernel k of the lth layer, and $x_{i,j}$ is the input at location (i,j), respectively. Function a is a nonlinear

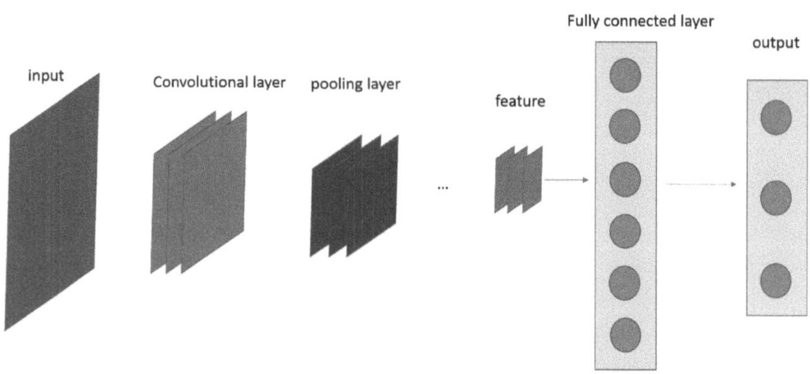

FIG. 7 The basic architecture of CNN.

activation function for detecting nonlinear features. Typical activation functions are sigmoid, tanh, and ReLU. Pooling layer aims to downsample the input feature map, that is, to reduce the size of the feature map. Max-pooling and average pooling are two kinds of widely used pooling layers.

Usually, CNN architecture stacks several convolutional layers and pooling layers to form a hierarchical structure, however, pooling layer is not necessary. As the number of layer increases, the size of the feature map becomes smaller and the number of feature maps increases. After several convolutional and pooling layers, a feedforward neural network (including several fully connected layers) is introduced for final prediction tasks. Feature maps at shallow layers indicate the low-level features of the input data, while the feature maps at the top-layer can capture the high-level or abstract feature composed by low-level features.

CNN is commonly used for extracting spatial correlation features of data. When using CNN, formula of inputs must be matrix or raster data with several channels. In general, road network and trajectory data can be easily transformed into matrix and then used as input of CNN. In work [48], for estimating travel speed of road segment a time-space matrix M was constructed by taxis GPS trajectory data, in which row i indicated road intersection, column j indicated time and entity m_{ij} was traffic speed on road intersection i at time j. A depth-4 CNN model with 256, 128, 64 convolutional layers (kernel filters of size 3×3 and max-pooling of size 2×2 and fully connected layers) was constructed for extracting features of M and then predicted the travel speed in the next time interval. Similar data transformation was performed in [49], in which taxi speed was predicted based on an improved CNN model by incorporating recurrent layers.

Some studies converted each trajectory into a two-dimensional image I by splitting the city area into grids. The difference between them was the value assignment of $I(i,j)$. In [14], if the destination of the trajectory located at (i, j), the value of $I(i,j)$ was designated as 1; and if the trajectory passed through (i,j), the value of $I(i,j)$ was designated as 0.5, and if the trajectory did not pass, its value was 0. Then the data was input into a CNN architecture with two convolution-pooling stacks for destination prediction. In [50], the value of pixel $I(i,j)$ was proportional to car's stay time and when the stay time of a pixel was larger than a given threshold, the color was set to 255. Authors adopted a simplified version of CNN model (AlexNet) and Random Forest to detect ride-sourcing cars by co-train strategies.

4.1.2 RNN

Recurrent neural network (RNN) is a neural network that can process sequence with arbitrary length. Unlike feedforward neural networks that only predict the output based on the current input, RNN has a memory that can store the past inputs and can predict output based on information stored in memory and current inputs. In RNN, information stored in memory is called a hidden state. The basic structure of RNN is shown in Fig. 8A.

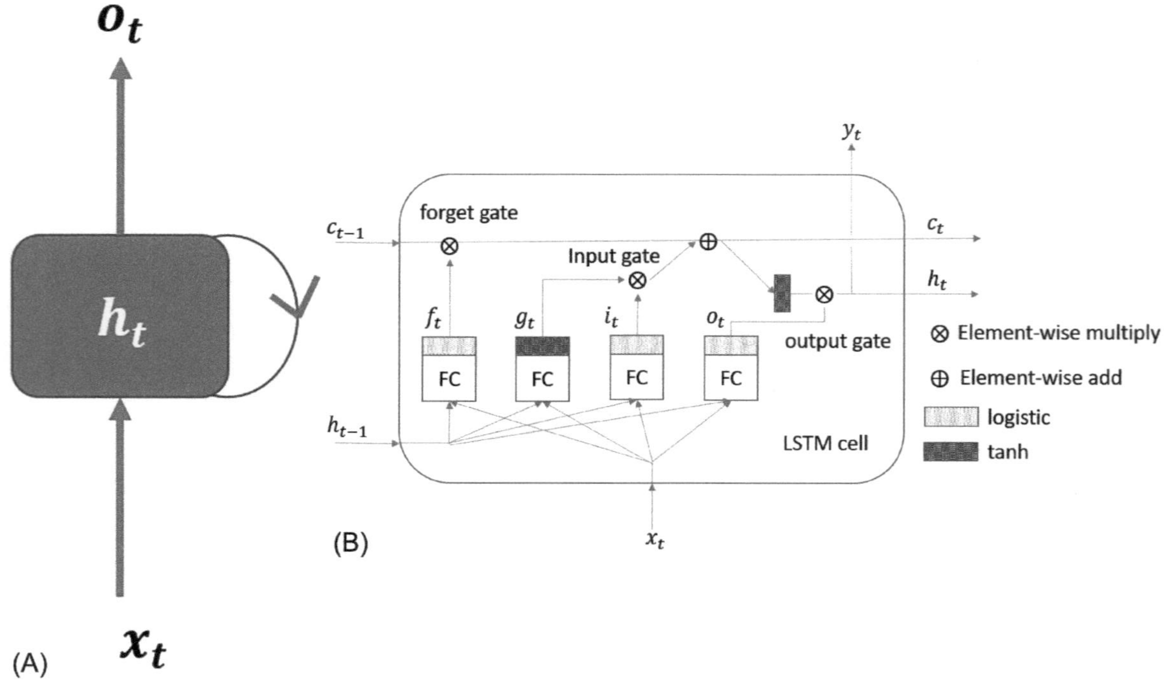

FIG. 8 The basic structure of RNN and LSTM: (A) RNN and (B) LSTM.

RNN processes the inputs x_t, produces the hidden state h_t and gives outputs o_t at each time step t. Hidden state and outputs can be calculated as $h_t = f_h(W_x x_t + W_h h_{t-1})$, and $o_t = f_o(W_o h_t)$, respectively, where W_x, W_h, W_o are learnable weights that are shared among each time stamp, and f_h, f_o are a nonlinear function, such as tanh, ReLu.

When the length of the sequence is long, vanilla RNN faces the problem of gradient vanishing and gradient exploding. Long short-term memory (LSTM) cell is put forward to solve this problem by introducing a gating mechanism [51]. As shown in Fig. 8B, LSTM cell state is divided into a short-term state h_t and a long-term state c_t. Two gates are used to control the value of long-term state c_t. One is the forget gate, which determines how much of past long-state c_{t-1} to remember in c_t; the other is input gate, which determines how much of the current input to be save in c_t. Output gate is used to control how much of the current long-term state c_t to output. Calculation equations in LSTM cells are:

$$f_t = \sigma\left(W_f x_t + U_f h_{t-1} + b_f\right)$$

$$i_t = \sigma(W_i x_t + U_i h_{t-1} + b_i)$$

$$o_t = \sigma(W_o x_t + U_o h_{t-1} + b_o)$$

$$g_t = \tanh\left(W_g x_t + U_g h_{t-1} + b_g\right)$$

$$c_t = f_t \otimes c_{t-1} + i_t \otimes g_t$$

$$y_t = h_t = o_t \otimes \tanh(c_t)$$

where W_f, W_i, W_o, W_g are weight matrix for input x_t, U_f, U_i, U_o, U_g are weight matrix for short-term state h_t, and b_f, b_i, b_o, b_g are basic terms. σ is a sigmoid function and $\otimes$ is an element-wise multiple operation.

Gated Recurrent Unit (GRU) is a simplified version of LSTM, which is another widely used RNN unit [52]. It turns the input gate, forget gate, and output gate into two gates: update gate z and reset gate r; merges the unit hidden state and the output as h. Update gate determines how much of the past state to remember and reset gate determines how much of the past state to forget. The hidden state h is then updated iteratively using the following equations:

$$z_t = \sigma(W_Z x_t + U_Z h_{t-1} + b_Z)$$

$$r_t = \sigma(W_r x_t + U_r h_{t-1} + b_r)$$

$$g_t = \tanh\left(W_g x_t + U_h(r_t \otimes h_{t-1}) + b_g\right)$$

$$h_t = (1 - z_t) \otimes g_t + z_t \otimes h_{t-1}$$

where W_Z, W_r, W_g, U_Z, U_r, U_h are learnable weight matrix, and b_f, b_i, b_o, b_g are learnable basis terms. σ is a sigmoid function and $\otimes$ is an element-wise multiple operation.

RNN models have been widely used to process time sequence data. Through RNN, time-dependencies of prediction tasks can be obtained, such as periodic characteristics for short-term travel time estimation [53], destination prediction [12, 13], traffic demand prediction [15, 16].

In [15, 16], entire city was divided into grids and converted the past taxi pick-up data into week-long data sequences. For each time step, the data sequence consisted of the number of pickups in each area and time information (day of the week, hours, minutes and other impacting factors). The authors constructed LSTM model with mixture density network on top to train the sequence data for forecasting future pickups in each subarea. In [54], a mixture deep LSTM was proposed for predicting traffic flow after traffic accidents, in which LSTM can learn the traffic patterns by "memorizing" historical traffic as well as interruption caused by accidents. The deep LSTM contains two hidden layers of size 64 and 32 cells. In [55], bidirectional LSTM was trained for travel time estimation, in which one LSTM processed the sequence from pick-up location to pick-off location and another LSTM from the opposite direction, and finally the output of the two LSTMs were concatenated. In [56], LSTM and DNN were combined to predict travel time. One best model was selected from different 16 hyperparameters including the number of LSTM cells and layers and of deep fully connected layers.

4.1.3 ConvLSTM

In LSTM, the input and hidden state are completely connected to four gates, resulting in weak spatial information extraction capacities. When the input are spatio-temporal data, i.e., 3D OD flow data shown in Fig. 4, due to this shortcoming, the model cannot learn the spatial correlation between regions. ConvLSTM [57] was proposed for handling this problem, in which all the inputs $\mathcal{X}_1,\ldots,\mathcal{X}_t$, cell outputs $C_1, \ldots, C_t$, hidden states $\mathcal{H}_1,\ldots,\mathcal{H}_t$ and gates i_t, f_t, o_t were 3D tensors whose last two dimensions are spatial dimensions (row and columns). In other words, each cell's state in the grid depends on the inputs and past states of its local neighbors. Calculation equations in ConvLSTM cell are:

$$i_t = \sigma(W_{xi} * \mathcal{X}_t + W_{hi} * \mathcal{H}_{t-1} + W_{ci} \circ C_{t-1} + b_i)$$

$$f_t = \sigma\left(W_{xf} * \mathcal{X}_t + W_{hf} * \mathcal{H}_{t-1} + W_{cf} \circ C_{t-1} + b_f\right)$$

$$C_t = f_t \circ C_{t-1} + i_t \circ \tanh\left(W_{xc} * \mathcal{X}_t + W_{hc} * \mathcal{H}_{t-1} + b_c\right)$$

$$o_t = \sigma(W_{xo} * \mathcal{X}_t + W_{ho} * \mathcal{H}_{t-1} + W_{co} \circ C_t + b_o)$$

$$\mathcal{H}_t = o_t \circ \tanh(C_t)$$

where * denotes the convolutional operator and ∘ is an element-wise multiple operation.

Forecasting of passenger demand or online ride services in each region associates with time dependence and spatial dependence. In [26], taxi demand between all region pairs in a future interval was studied, in which the convolutional result of OD matrix and DO matrix was feed with weather embedding code (obtained by Multiple-Layer Perceptron, MLP) into ConvLSTM for spatial-temporal feature extraction. In [27], for predicting the short-term demand of free-floating bike-sharing, authors divided Shanghai's old central area into 5 × 5 grids, in which the number of bike trips in each grid cell in every 15, 20, 30 min was summarized based on Mobike data. Then the image-like sequence input was processed by multiple-layer neural networks with ConvLSTM unit to extract spatio-temporal correlation of the trip map.

4.1.4 Autoencoder (AE)

An autoencoder is a type of neural network attempting to reproduce its input by learning data code in an unsupervised manner. As shown in Fig. 9, given input $x \in R^d$, AE first creates a latent code $z = \Phi(x)$ to describe the input by encoder function Φ and then creates a reconstruction $\hat{x} = \Psi(z)$ of the input x by the decoding function Ψ. AE are trained to adjust the weights by minimizing reconstruction errors $L(x, \hat{x}) = \|x - \hat{x}\|^2$. AE can be used for dimension reduction of input and the low-dimension code z can be regarded as a latent feature of the input. Stacked AE (SAE) is formed by stacking each hidden layer of an AE on the top of others, which is a powerful tool for automatic feature extraction. AE/SAE is a universal structure where encode and decode functions can be MLP, CNN, and RNN/LSTM/GRU.

In [58], an encoder–decoder framework based on convolutional and ConvLSTM units to identify complex features that capture spatio-temporal influences and pick-up–dropoff interactions on citywide passenger demand. The encoder was composed of convolutional units in the lower layers and ConvLSTM units in higher layers. Input sequence was converted into fixed dimensional representations by the encoder. The hidden representations can capture the temporal information flow in the sequence while retaining spatial correlations.

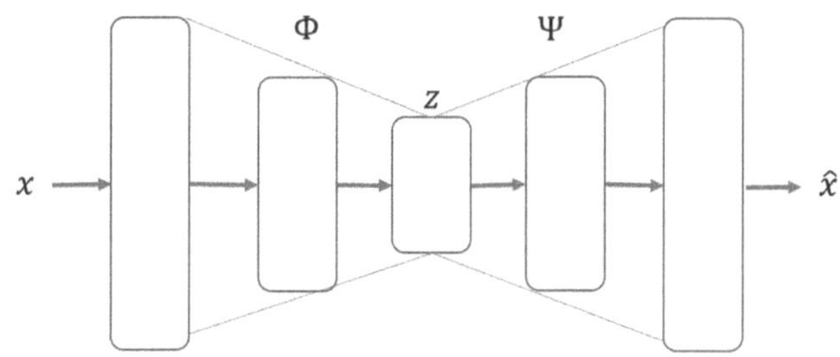

FIG. 9 Basic structure of AE.

4.1.5 Convolution graph neural network (ConvGNN)

The graph convolutional network is proposed for processing data in the form of graph, which has been widely used in node/network classification, link prediction, or community detection. There are two kinds of graph convolutional network models. One is to extend convolutions to graph in spectral-domain by finding corresponding Fourier basis, named spectral-based ConvGNN. GCN is one of the classical works in this type of model [59]. The other one is to perform convolution directly on graph based on local connectivity structure, named spatial-based ConvGNN. Typical works are GAT [60] with attention mechanism and Graph SAGE [61] that generates node embedding by sampling and aggregating features. In this chapter, we introduce the general concept of spatial-based ConvGNN. Readers who are interested in GNN can refer to a comprehensive review in [62].

Consider a graph $\mathcal{G}=(V,A)$, where V is the set of all nodes and $A\in\mathbb{R}^{|V|\times|V|}$ is the adjacency matrix of $\mathcal{G}$. Let $L=I-D^{-1/2}AD^{-1/2}$ denotes the graph Laplacian matrix, where D is the degree matrix, and feature vector $X\in\mathbb{R}^{|V|\times d}$ where d is the dimensions of a feature vector, latent feature of nodes can be calculated as

$$H_1=f(LXW_1)$$

where W_1 is trainable weight matrix of size $d\times d'$ (i.e., d' output features of nodes at layer-1), and function f can be any nonlinear activation function. The main idea of graph convolution is to take the weighted average of all neighbors' node features (including itself): low-degree nodes get larger weights. The output of layer-1 (i.e. node's feature vector) is calculated with the connectivity information from its neighbors. When stacking another layer on top of the layer-1, calculation of hidden representation of nodes at layer-2 can be repeated, but this time, the neighbors already have information about their own neighbors (i.e., 2-hop neighbor). Latent feature of nodes H_2 can update using $H_2=f(LH_1W_2)$, where W_2 is trainable weight matrix of size $d'\times d''$ (i.e. d'' output features of nodes at layer-2). Therefore, by stacking multiple graph convolutional layers, the latent feature of a node can contain global connectivity information of the graph. The basic structure of spatial-based ConvGNN is shown in Fig. 10. After obtaining the latent feature of nodes, we can feed them into feedforward network for the node classification task, or aggregate all the nodes to represent the graph and then perform graph classification.

In [24], graph convolution neural network was adopted for bike flow prediction at the station-level. For depicting relation between bike stations, three different weighted graphs were constructed, where the node was station and edge was the degree of a certain type of relations (i.e., distance, ride record correlation) between stations. The weight of distance graph was the reciprocal of Euclidean distance between stations. The weight of interaction graph was the number of ride records between stations. The weight of correlation graph

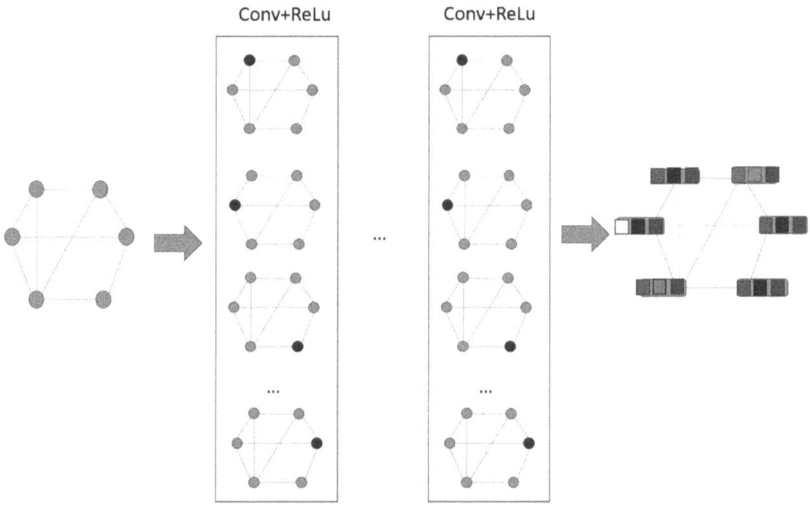

FIG. 10 Basic structure of spatial-based ConvGNN.

was the Pearson coefficient of historical usage records between stations. The authors proposed a graph fusion method to mix these three graphs together, where the weight of edge in the fusion graph was a weighted average of edges from corresponding edges of the three graphs. Then, graph convolutional was adopted to extract latent features of nodes from in–out-flow information of bike usage. The output $[H_1^0, H_1^1, \ldots, H_1^{t-1}]$ of graph convolutional operations at each time step t were put into an encoder–decoder network with LSTM cell. The encoder part encoded the sequence and the code (with temporal pattern) was put into decoder part. The output of the decoder was H_0^t which was the prediction target.

In [17], authors suggested that regions with similar functionality but far apart may have similar ride-hailing demand. Then for describing correlations among regions neighborhood graph, functional similarity graph, and transportation connectivity graph were constructed. In the neighborhood graph, a region was linked with its eight adjacent regions; in the functional similarity graph, edge represented the POI similarity of the surrounding environment of the two nodes; in the transportation connectivity graph, edge was generated between two regions connected by highway, subway, or motorway. With these graphs, the authors proposed multigraph convolution to model the spatial dependency as $X_{l+1} = \sigma\left(\coprod_{A \in \mathbb{A}} f(A; \theta_i) X_l W_l\right)$, where $X_l \in \mathbb{R}^{|V| \times P_l}$, $X_{l+1} \in \mathbb{R}^{|V| \times P_{l+1}}$ were the feature vectors of $|V|$ regions in layer l and $l+1$, respectively. σ denoted the activation function, and $\coprod$ denoted the aggregation function, e.g., sum, max, average, etc. $\mathbb{A}$ denoted the set of graphs, and $f(A; \theta_i) \in \mathbb{R}^{|V| \times |V|}$ represented the aggregation matrix of different samples based on graph $A \in \mathbb{A}$ parameterized by θ_i. $W_l \in \mathbb{R}^{P_l \times P}{}_{l+1}$ denoted the feature transformation matrix.

The abovementioned graph convolutional process can be performed at each time step t, and then the outputs were concatenated together for further time dependency processing.

4.2 Fusion strategy

Through feature extraction, feature vector/tensor can be obtained from data. Then they are fused together for further processing. The general fusion strategies include concatenation, sum, and Hadamard product (element-wise product). Tables 1 and 2 provide a summary of feature fusion strategies used in recent work. Representative examples of each fusion strategy are introduced in the following.

TABLE 1 Summary of feature fusion from multiple datasets.

Ref.	Problem	Data	Feature extraction method	Fusion strategy
[14]	Destination prediction	Taxi trajectory Date, Client info	CNN	Concatenation
[12]		Taxi trajectory Client ID, Taxi ID Date & time	MLP, RNN, bidirectional RNN	Concatenation
[15]	Demand prediction	Taxi trip data Date & time, day of week, weather	LSTM, MDN	Concatenation
[63]		Taxi trajectory POI, weather, date & time	FCN, LSTM, CNN Residual connection	Concatenation
[20]		Taxi trip data Weather	CNN, LSTM LINE embedding	Concatenation
[19]		Online ride-sourcing order data Data & time, weather	LSTM, ConvLSTM	Hadamard product
[17]		Online ride-hailing trip data POI, road network	Graph CNN, RNN Channel-wise attention	Sum

TABLE 1 Summary of feature fusion from multiple datasets—cont'd

Ref.	Problem	Data	Feature extraction method	Fusion strategy
[53]	Travel time estimation	Float-car pick-up data Float-car trip data	Wide & deep learning [64] LSTM	Concatenation
[65]	Bike-sharing demand forecasting	Mobike and of bike trip data Time	ConvLSTM	Hadamard product
[24]		NYC bike-sharing data/Chicago bike-sharing data Weather	Spatial-based ConvGNN, LSTM, AE	Graph fusion
[27]		Mobike trip data Weather, air quality	LSTM, ConvLSTM	Concatenation
[54]	Traffic forecasting	Speed data Accident data	LSTM, AE	Concatenation
[66]	Crowd flow prediction	Taxi trajectory/NYC bike trajectory Weather	CNN Residual connection	Sum Hadamard product
[28]		Beijing/NYC taxi trajectory Weather	FCN, Embedding Residual connection	Sum, Product Concatenation
[67]	Supply-demand gap prediction	Didi order data Weather, traffic data, date & time	FC Residual connection	Hadamard product
[18]		Didi order data Trajectory data of vacant ride-sourcing cars Weather, traffic condition	Hexagon-based CNN	Concatenation

TABLE 2 Summary of feature fusion from one dataset.

Ref.	Problem	Data	Feature extraction method	Fusion strategy
[49]	Speed estimation	Taxi GPS trajectory	CNN, RNN	Concatenation
[68]		GPS trajectory of floating cars	CNN, LSTM	Output of CNN was the input of LSTM
[69]	Travel time estimation	Taxi trajectory	CNN, BiLSTM	Output of CNN was the input of BiLSTM
[25]		Taxi trajectory	LSTM, Attention Residual connection	Concatenation
[56]		Loop detectors	LSTM, DNN	Output of LSTM was the input of DNN
[26]	Origin-destination demand prediction	NYC taxi OD trip	CNN, LSTM	Concatenation
[58]	Demand prediction	NYC taxi OD trip/ NYC bike trip	CNN, ConvLSTM	Attention mechanism
[70]		NYC taxi OD trip/ NYC bike trip	CNN, LSTM, Attention	Concatenation
[71]		Didi online car-hailing orders/ Chicago taxi orders	Spatial-based ConvGNN Variational auto-encoders, GRU, Attention	Concatenation

4.2.1 Concatenation

In [20], at each time interval t, one location i with its surrounding neighborhood was treated as one $S \times S$ image, in which the value of each cell was ride demand. The image was represented as a tensor $Y_t^i \in \mathbb{R}^{S \times S \times 1}$ for each location i and time interval t. Then CNN network takes Y_t^i as input $Y_t^{i,0}$ and feed it into K convolutional layers. After that, a flatten layer was used to transform the result from a convolutional layer to a feature vector s_t^i. For reducing the dimension of s_t^i, a fully connected layer was used to output $\hat{s}_t^i \in \mathbb{R}^d$.Weather information was encoded as a vector $e_t^i \in \mathbb{R}^r$ and it was concatenated with $\hat{s}_t^i$ as $g_t^i \in \mathbb{R}^{d+r}$. Then g_t^i was put into LSTM and hidden state h_t^i can be obtained. Meanwhile, considering the time sequence of demand change trend between regions, a weighted graph of regions was constructed, in which weights represented demand similarity). Based on the graph embedding method (LINE [72]), each node was encoded as a vector m^i, and then was transformed to a vector $\hat{m}^i$ by fully connected layer. Finally, three feature vectors were joined together by concatenating operation as $q_t^i = [h_t^i, \hat{m}^i]$. After that, q_t^i was feed to a fully connected network to get the prediction value for each region.

4.2.2 Sum & Hadamard product

In [28], in/out-flow at a region and transition flow between regions were predicted by multitasking deep learning model. Let $\mathcal{X}_t \in \mathbb{R}^{2\times I\times J}$ represent inflow/out-flow at node at time t, in which I and J were the length and width of the city grid. Let $\mathcal{M}_t \in \mathbb{R}^{2N\times I\times J}$ represents transition flow between nodes, in which $N = I \times J$ and the former N dimensions indicated out-flow to $I \times J$ nodes from the node and the later N dimensions indicated in-flow from $I \times J$ nodes to the node. The goal was to predict $\mathcal{X}_T$ and $\mathcal{M}_T$ based on historical data $\{\mathcal{X}_t, \mathcal{M}_t | t = 1, 2, \ldots, T-1\}$. Authors proposed two models to perform the prediction. These two models were connected by coupling their latent representations of middle layer and trained together. One model was named EDGENET and the other one was named NODENET. In EDGENET, $\mathcal{M}_t$ was first transformed into a low-dimension tensor Z_t. Three different length of time sequences $\mathcal{M}_c, \mathcal{M}_p, \mathcal{M}_t$ were constructed as tensor according to parameters l_c, l_p, l_t ($l_c < l_p < l_t$), representing closeness, period and trend dependents. Then they were feed into fully convolutional network (FCN), in which residual connections were adopted to help train. Finally, the three different outputs of FCN were fused by $\mathcal{M}_{fcn} = \mathrm{W}_c \otimes \mathcal{M}_c + \mathrm{W}_p \otimes \mathcal{M}_p + \mathrm{W}_t \otimes \mathcal{M}_t$, in which $\otimes$ was Hadamard product and W_c, W_p, W_t were the learnable parameters. In NODENET, similar with EDGENET, three time dependents were constructed and feed into FCN and then fused as $\mathcal{X}_{fcn}$. Authors proposed two methods to fuse $\mathcal{X}_{fcn}$ and $\mathcal{M}_{fcn}$: sum fusion and concatenation fusion. The former one was $\mathcal{K} = \mathcal{X}_{fcn} + \mathcal{M}_{fcn}$; the later one was $\mathcal{K} = [\mathcal{X}_{fcn}, \mathcal{M}_{fcn}]$. After fusion step, $\mathcal{K}$ was feed to convolutional layer to obtain output $\mathcal{X}_{Res}$ and $\mathcal{M}_{Res}$, respectively. In addition, weather feature was denoted as $\mathcal{E}_t \in \mathbb{R}^{l_e \times I \times J}$. After two nonlinear transformations, F_m and F_x were obtained. Then the weather information was fused by product operation as $\hat{\mathcal{M}}_{fcn} = \tanh(F_m \otimes \mathcal{M}_{Res})$ and $\hat{\mathcal{X}}_{fcn} = \tanh(F_m \otimes \mathcal{X}_{Res})$. Finally, $\hat{\mathcal{M}}_{fcn}$ and $\hat{\mathcal{X}}_{fcn}$ were used to approximate true value.

4.2.3 Attention mechanism

In [58], OD trip data was processed as a 3D demand tensor $M_t \in \mathbb{R}^{n \times m \times 2}$ over the citywide grid map, in which $(M_t)_{i,\ j,\ 0}$ and $(M_t)_{i,\ j,\ 1}$ represented pick-up demands and drop-off demands in time interval t at location (i,j), respectively. The authors adopted an encoder–decoder architecture to predict future ride demand. The encoder consisted of convolutional units in the lower layers and ConvLSTM in the high layers. The convolutional layers extracted the spatial features I_t for each input tensor M_t. The ConvLSTM layers processed the sequence of $\{I_t\}_{t=1,\ \ldots,\ N}$ and generated the sequence representations as the output. The final sequence representations were stored in the last states of ConvLSTM cell. And then authors copied the hidden states H_t and cell states of last step in encoder as initial hidden state and cell state in decoder. For enhancing the prediction capacity of decoder, K-means++ clustering was performed on historical demand tensors. The K representative tensors $\{(A_k)_{ij}\}_{k=1,\ \ldots,\ K}$ were referred to as annotation tensors. These annotation tensors were used to fuse with input of the decoder by an attention model. Firstly, a convolutional neural network was used to extract spatial features from each $(A_k)_{ij}$ and K annotations $\{(a_k)_{ij}\}_{k=1,\ \ldots,\ K}$ can be obtained. Secondly, $(a_k)_{ij}$ and H_t were flattened and concatenated together for the input of two-layer MLP, followed by a softmax operation. Then a weight vector $\langle \alpha_1, \ldots, \alpha_K \rangle$ can be obtained. Finally, a summation over weighted annotations was performed to get a representation $z = \sum_{k=1}^{K} \alpha_k a_k$. Then, z was feed into ConvLSTM as input for decoder part. The decoder part leveraged the attention information from the attention model and decoded the encoded representations from encoder part to generate future passenger demand.

4.2.4 Graph fusion

In [24], the authors constructed three types of inter-station graphs to represent bike-sharing system, in which nodes were bike stations and edges reflected the spatial relationship, and weights indicated the relationship strength. The first one $G_d(V,E)$ was distance graph, in which weight of edge was the reciprocal of Euclidean distance between nodes. The second one $G_i(V,E)$ was interaction graph, in which weight indicated the number of ride records between nodes. The third one $G_c(V,E)$ was correlation graph, in which weight indicated the Pearson coefficient of historical usages between nodes. Then the three graphs were fused and performed graph convolution operation. The graph fusion step merged different graphs into one graph. Mathematically, the adjacency matrix A of each graph was normalized by $\mathrm{A}' = D^{-1}A + I$, where D was the degree matrix and I was identity matrix. For weight w_{ij}^1, w_{ij}^2, w_{ij}^3 of edge e_{ij} in three graph, a softmax operation was performed to obtain a distribution w'^1_{ij}, w'^2_{ij}, w'^3_{ij}. Then the three graphs can be fused as a new graph G, in which weight w_{ij} of edge e_{ij} can be calculated by $w_{ij} = \sum_{m=1}^{3} \mathrm{w}_{ij}^m \mathrm{A}'_{ij}$.

4.2.5 Output–input structure

Output–input structure refers to using output of one model as the input for another model. This fusion strategy is often adopted by feature fusion from one dataset. For example, in [68] authors developed a hybrid model to combine CNN and LSTM to predict network-wide traffic speeds. Firstly, speed data was transformed into an image and then feed into CNN model. The output of CNN was feed into LSTM for further feature extraction.

5 Decomposition-based methods

Main idea of decomposition-based methods is to extract latent features by leveraging matrix factorization. For example, when forecasting in-flow trips or out-flow trips of an area during timeslot t, it is necessary to extract the geographical features and temporal features that affect the generation of trips in this area. However, we cannot accurately obtain these latent features by human prior knowledge. Through matrix factorization, latent features that are difficult to find can be obtained to explain and predict the occurrence of OD trips

Definition 1 Singular value decomposition (SVD) factorizes an $m \times n$ matrix A into the production of three matrices $A = U\Sigma V^T$, where U is a $m \times m$ orthogonal matrix $UU^T = I$, V is $n \times n$ orthogonal matrix $VV^T = I$, Σ is a $m \times n$ rectangular diagonal matrix composed of nonnegative diagonal elements arranged in descending order $\Sigma = \text{diag}(\sigma_1, \sigma_2, \ldots, \sigma_p)$, $p = \min(m, n)$, $\sigma_1 \geq \sigma_2 \geq \ldots \geq \sigma_p$.

Definition 2 A is a $m \times n$ matrix with rank rank(A) $= r$, truncated singular decomposition means $A \approx U_k \Sigma_k V_k^T$, where U_k is a $m \times k$ matrix consisting of the first $k$columns of U, V_k is a $n \times k$ matrix consisting of the first k columns of V, Σ_k is a $k \times k$ diagonal matrix consisting of the first k diagonal elements of Σ, $0 < k < r$. In practice, matrix $\Sigma_k V_k^T$ can be regarded as one matrix and A can be rewritten as $A \approx UV^T$, where $U \in \mathbb{R}^{m \times k}$, $V \in \mathbb{R}^{n \times k}$.

When all entries of A are nonnegative ($A \geq 0$), with constraint U, $V \geq 0$, this method is called nonnegative matrix factorization (NMF). k is a parameter of NMF.

Based on Definition 2, in/out-flow matrix (mentioned in Section 2.3) can be decomposed into two matrix U and V, where U indicates a geographical correlation between location and hidden features; V indicates the temporal correlation between time and hidden features. Formally the problem comes down to minimize the objective function:

$$f(U, V) = \frac{1}{2} \|A - UV\|_F$$

where $\|\bullet\|_F$ is the Frobenius norm. To avoid over fitting, regularization is usually introduced to the function:

$$f(U,V) = \frac{1}{2}\|A - UV\|_F + \alpha_U\|U\|_F + \alpha_V\|V\|_F$$

where α_U and α_V are parameters for regularization.

There are many algorithms for matrix factorization, and gradient decent is one of the common methods [40]. Firstly, initialize matrix U and V randomly. Secondly, calculate their dot product and modify the initial matrix along the downward direction of gradient of the objective function. This step is performed iteratively until the objective function converges to its minimum. After obtaining matrices U and V, we can use them to reconstruct A.

On one hand, based on NMF, locations/time with similar trip characteristics can be found. In [73], evolution of temporal patterns of usage in the bicycle-sharing systems in Lyon, France was studied based on NMF. Six different usage patterns, including commercial activity, leisure activity, residential/restaurants/bars activity, train stations, universities, business district, were discovered. In [74], authors adopted NMF to compare travel patterns between taxi trips and ride-sharing trips, concluded that ride-sharing was an effective supplement to taxi service, and principally served as an approach for commuting. On the other hand, OD trips can be predicted based on latent features found by NMF. In [75], pattern of OD flows was captured by the basis matrix depicting the trip patterns of the urban residents and a corresponding coefficient matrix indicating the weights of the basis matrix during various periods based on NMF method. An AR model was used for predicting the coefficient matrix and then OD flows can be predicted based on the estimated coefficient matrix and the basis matrix based on NMF.

Previous work adopted collaborative decomposition to solve the poor performance of NMF caused by sparse data [76, 77]. By collaboratively decomposing matrices from different datasets, latent features extracted from multiple datasets can be supplemented with each other, thereby obtaining enhanced features. In [77] a travel speed estimation model based on couple matrix factorization approach was proposed, in which latent features were learned from multiple datasets, including map data and historical taxi trajectories data. Through collective factorization based on common dimensions among geospatial, temporal, and historical datasets, a real-time model for estimating the travel time of a given path was proposed in [78].

There are some problems when applying NMF for extracting latent features. Firstly, the choice of parameter k is a key problem in NMF, as increasing k will decrease the reconstruction error, but at the cost of higher complexity. It is hard to decide the value of k. Secondly, when dealing with a large-size matrix, traditional NMF cannot perform efficiently. Previous work proposed an improved algorithm to solve this problem. For example, Schatten p-norm algorithm was adopted for estimating citywide traffic speed based on ride-sourcing GPS trajectories [79]. However, calculation efficiency is still an obstacle for the wide application of this method. Thirdly, NMF performs linear transform to extract data features. It is difficult to deal with complex nonlinear relationships in datasets.

6 Challenging problems of data fusion in MaaS

Beyond the research efforts surveyed in the previous section, there are several challenges and opportunities that arise in the data fusion in MaaS. In what follows we highlight these challenges and future research directions, and then conclude the chapter.

6.1 Data quality

Data preprocessing is essential before data fusion. For example, GPS trajectory data records cars' moving activities and cannot be replaced or supplemented by other data. However, data quality of GPS trajectory is easily affected by many factors, such as communication failure, battery constraints, and device settings. One common problem is sampling points alternating between sparse and dense episodes due to inconsistent sample rates. A simple solution for this problem is to recognize this kind of trajectory and delete it. From another perspective, this kind of trajectory can be recovered based on similar matching with other trajectories. Note that some researchers have made their efforts towards this direction [80, 81]. However, computation efficiency and accuracy are still challenging problems, and require more attention in the future.

6.2 Model complexity

Researchers have adopted many different deep learning architectures to perform data fusion and achieved promising results. The high accuracy may mostly depend on deep and complex network architectures, rather than fusion strategies (such as concatenation, sum, and Hadamard product). Although deeper and complex network architecture improves the performance of model, it leads to high calculation complexity. Exploring more effective fusion strategies and meanwhile reducing model complexity may be a better choice. Thus, balancing the model complexity and effectiveness of fusion strategies need to be studied in the future. Neural Architecture Search (NAS) is an emerging deep learning paradigm that aims to automatically find or design effective deep learning architectures for mining tasks. Through reinforcement learning, simple and refined model architecture can be learned [82, 83]. How to collaboratively design model architecture and data fusion strategies based on reinforcement learning is an interesting study work, worthy of further discussion.

6.3 Data fusion in comparative analysis

Most of the studies reviewed in the previous sections adapt supervised learning methods that use label data to learn a model for prediction, such as travel time estimation and demand prediction. There are many other important research issues in MaaS, such as the relationship between online ride-hailing and public transits [84], competition with the traditional taxi industry [85], and effects on traffic congestion [86]. The solution of these issues also depends on multiple data sources. For example, when comparing the emission characteristics between online ride-hailing cars and traditional taxis [87], we need GPS trajectory data from online ride-hailing cars and traditional taxis; ridership data and GPS trajectory are necessary when discussing the substitution role and complement role of online ride-hailing for public transit. These data come from different areas, times, devices. How to deal with these data so that they can be compared in statistical domains is a question that needs to be discussed.

7 Conclusions

In this chapter, we conduct a comprehensive review of recent advances in data fusion in the MaaS system. First, we summarize the data types in MaaS system and their input formula for data fusion methods. Second, we discuss the necessary capabilities of data fusion methods used in the MaaS system and then classify data fusion methods into two categories: deep learning-based methods and decomposition-based methods. For deep learning-based methods, basic model units and widely used data fusion strategies are summarized. For decomposition-based methods, we introduce the basic mathematics tools and their applications. Finally, we list some open problems and suggest future study trends.

Acknowledgments

This work was supported by the Young Scientists Fund of the National Natural Science Foundation of China (41706198).

References

[1] Jittrapirom P, et al. Mobility as a service: a critical review of definitions, assessments of schemes, and key challenges. Urban Planning 2017;2(2):13–25.

[2] Goodall W, et al. The rise of mobility as a service. Deloitte Rev 2017;20:112–29.

[3] Smith G, Sochor J, Karlsson IM. Mobility as a Service: development scenarios and implications for public transport. Research in Transportation Economics 2018;69:592–9.

[4] Tirachini A. Ride-hailing, travel behaviour and sustainable mobility: an international review. Transportation 2019;1–37.

[5] Wang H, Yang H. Ridesourcing systems: a framework and review. Transportation Research Part B: Methodological 2019;129:122–55.

[6] Zheng Y. Methodologies for cross-domain data fusion: an overview. IEEE transactions on big data 2015;1(1):16–34.
[7] Xiao Y. Analysis of the influencing factors of the unsafe driving behaviors of online car-hailing drivers in china. PLoS One 2020;15(4), e0231175.
[8] Shi J, Hussain M, Peng D. Aberrant driving behaviours among ride-hailing drivers: direct and indirect effects of personal attributes and work-condition factors. International Journal of Crashworthiness 2020;1–11.
[9] Clewlow RR, Mishra GS. Disruptive transportation: The adoption, utilization, and impacts of ride-hailing in the United States. Davis: Institute of Transportation Studies, University of California; 2017. Research Report UCD-ITS-RR-17-07.
[10] Conway MW, Salon D, King DA. Trends in taxi use and the advent of ridehailing, 1995–2017: evidence from the US National Household Travel Survey. Urban Science 2018;2(3):79.
[11] Dias FF, et al. A behavioral choice model of the use of car-sharing and ride-sourcing services. Transportation 2017;44(6):1307–23.
[12] De Brébisson A, et al. Artificial neural networks applied to taxi destination prediction. arXiv 2015. preprint arXiv:1508.00021.
[13] Endo Y, et al. Predicting destinations from partial trajectories using recurrent neural network. In: Pacific-Asia conference on knowledge discovery and data mining. Springer; 2017.
[14] Lv J, et al. T-CONV: a convolutional neural network for multi-scale taxi trajectory prediction. In: 2018 IEEE international conference on big data and smart computing (bigcomp). IEEE; 2018.
[15] Xu J, et al. Real-time prediction of taxi demand using recurrent neural networks. IEEE Transactions on Intelligent Transportation Systems 2017;19(8):2572–81.
[16] Xu J, et al. Taxi dispatch planning via demand and destination modeling. In: 2018 IEEE 43rd Conference on Local Computer Networks (LCN). IEEE; 2018.
[17] Geng X, et al. Spatiotemporal multi-graph convolution network for ride-hailing demand forecasting. In: Proceedings of the AAAI conference on artificial intelligence; 2019.
[18] Ke J, et al. Hexagon-based convolutional neural network for supply-demand forecasting of ride-sourcing services. IEEE Transactions on Intelligent Transportation Systems 2018;20 (11):4160–73.
[19] Ke J, et al. Short-term forecasting of passenger demand under on-demand ride services: a spatio-temporal deep learning approach. Transportation Research Part C: Emerging Technologies 2017;85:591–608.
[20] Yao H, et al. Deep multi-view spatial-temporal network for taxi demand prediction. In: Proceedings of the AAAI conference on artificial intelligence; 2018.
[21] Pan G, et al. Land-use classification using taxi GPS traces. IEEE Transactions on Intelligent Transportation Systems 2012;14(1):113–23.
[22] Qi G, et al. Measuring social functions of city regions from large-scale taxi behaviors. In: 2011 IEEE International Conference on Pervasive Computing and Communications Workshops (PERCOM Workshops). IEEE; 2011.
[23] Tang J, et al. Uncovering urban human mobility from large scale taxi GPS data. Physica A: Statistical Mechanics and Its Applications 2015;438:140–53.
[24] Chai D, Wang L, Yang Q. Bike flow prediction with multi-graph convolutional networks. In: Proceedings of the 26th ACM SIGSPATIAL international conference on advances in geographic information systems; 2018.
[25] Wang D, et al. When will you arrive? Estimating travel time based on deep neural networks. In: Proceedings of the AAAI conference on artificial intelligence; 2018.

[26] Liu L, et al. Contextualized spatial–temporal network for taxi origin-destination demand prediction. IEEE Transactions on Intelligent Transportation Systems 2019;20(10):3875–87.
[27] Bao J, Yu H, Wu J. Short-term FFBS demand prediction with multi-source data in a hybrid deep learning framework. IET Intelligent Transport Systems 2019;13(9):1340–7.
[28] Zhang J, et al. Flow prediction in spatio-temporal networks based on multitask deep learning. IEEE Transactions on Knowledge and Data Engineering 2019;32(3):468–78.
[29] Li J, et al. Graph CNNs for urban traffic passenger flows prediction. In: 2018 IEEE smartworld, ubiquitous intelligence & computing, advanced & trusted computing, scalable computing & communications, cloud & big data computing, internet of people and smart city innovation (SmartWorld/SCALCOM/UIC/ATC/CBDCom/IOP/SCI). IEEE; 2018.
[30] White FE. Data fusion lexicon. Washington, DC: Joint Directors of Labs; 1991.
[31] Wald L. Some terms of reference in data fusion. IEEE Transactions on Geoscience and Remote Sensing 1999;37(3):1190–3.
[32] El Faouzi N-E, Leung H, Kurian A. Data fusion in intelligent transportation systems: progress and challenges—a survey. Information Fusion 2011;12(1):4–10.
[33] Liu J, et al. Urban big data fusion based on deep learning: an overview. Information Fusion 2020;53:123–33.
[34] Dong J, et al. Advances in multi-sensor data fusion: algorithms and applications. Sensors 2009;9(10):7771–84.
[35] Alam F, et al. Data fusion and IoT for smart ubiquitous environments: a survey. IEEE Access 2017;5:9533–54.
[36] Castanedo F. A review of data fusion techniques. The Scientific World Journal 2013;2013:704504.
[37] Khaleghi B, et al. Multisensor data fusion: a review of the state-of-the-art. Information fusion 2013;14(1):28–44.
[38] Meng T, et al. A survey on machine learning for data fusion. Information Fusion 2020;57: 115–29.
[39] Koks D, Challa S. An introduction to Bayesian and Dempster-Shafer data fusion. Defence Science and Technology Organisation Salisbury (Australia) Systems; 2003.
[40] Wright W. A Markov random field approach to data fusion and colour segmentation. Image and Vision Computing 1989;7(2):144–50.
[41] Arrieta AB, et al. Explainable artificial intelligence (XAI): concepts, taxonomies, opportunities and challenges toward responsible AI. Information Fusion 2020;58:82–115.
[42] LeCun Y, et al. Gradient-based learning applied to document recognition. Proceedings of the IEEE 1998;86(11):2278–324.
[43] Krizhevsky A, Sutskever I, Hinton GE. Imagenet classification with deep convolutional neural networks. Advances in Neural Information Processing Systems 2012;25:1097–105.
[44] Simonyan K, Zisserman A. Very deep convolutional networks for large-scale image recognition. arXiv 2014. preprint arXiv:1409.1556.
[45] Szegedy C, et al. Going deeper with convolutions. In: Proceedings of the IEEE conference on computer vision and pattern recognition; 2015.
[46] Long J, Shelhamer E, Darrell T. Fully convolutional networks for semantic segmentation. In: Proceedings of the IEEE conference on computer vision and pattern recognition; 2015.
[47] He K, et al. Deep residual learning for image recognition. In: Proceedings of the IEEE conference on computer vision and pattern recognition; 2016.
[48] Ma X, et al. Learning traffic as images: a deep convolutional neural network for large-scale transportation network speed prediction. Sensors 2017;17(4):818.

[49] Wang J, et al. Traffic speed prediction and congestion source exploration: a deep learning method. In: 2016 IEEE 16th international conference on data mining (ICDM). IEEE; 2016.
[50] Wang L, et al. Ridesourcing car detection by transfer learning. arXiv 2017. preprint arXiv:1705.08409.
[51] Hochreiter S, Schmidhuber J. Long short-term memory. Neural Computation 1997;9(8): 1735–80.
[52] Cho K, et al. Learning phrase representations using RNN encoder-decoder for statistical machine translation. arXiv 2014. preprint arXiv:1406.1078.
[53] Wang Z, Fu K, Ye J. Learning to estimate the travel time. In: Proceedings of the 24th ACM SIGKDD international conference on knowledge discovery & data mining; 2018.
[54] Yu R, et al. Deep learning: a generic approach for extreme condition traffic forecasting. In: Proceedings of the 2017 SIAM international conference on data mining. SIAM; 2017.
[55] Zhang H, et al. DEEPTRAVEL: a neural network based travel time estimation model with auxiliary supervision. In: Proceedings of the 27th international joint conference on artificial intelligence: IJCAI 2018, Stockholm, Sweden, July 13; 2018.
[56] Liu Y, et al. Short-term travel time prediction by deep learning: a comparison of different LSTM-DNN models. In: 2017 IEEE 20th international conference on intelligent transportation systems (ITSC). IEEE; 2017.
[57] Shi X, et al. Convolutional LSTM network: a machine learning approach for precipitation nowcasting. arXiv 2015. preprint arXiv:1506.04214.
[58] Zhou X, et al. Predicting multi-step citywide passenger demands using attention-based neural networks. In: Proceedings of the eleventh ACM international conference on web search and data mining; 2018.
[59] Bruna J, et al. Spectral networks and locally connected networks on graphs. arXiv 2013. preprint arXiv:1312.6203.
[60] Veličković P, et al. Graph attention networks. arXiv 2017. preprint arXiv:1710.10903.
[61] Hamilton WL, Ying R, Leskovec J. Inductive representation learning on large graphs. arXiv 2017. preprint arXiv:1706.02216.
[62] Wu Z, et al. A comprehensive survey on graph neural networks. IEEE Transactions on Neural Networks and Learning Systems; 2020.
[63] Guo G, Zhang T. A residual spatio-temporal architecture for travel demand forecasting. Transportation Research Part C: Emerging Technologies 2020;115:102639.
[64] Cheng H-T, et al. Wide & deep learning for recommender systems. In: Proceedings of the 1st workshop on deep learning for recommender systems; 2016.
[65] Ai Y, et al. A deep learning approach on short-term spatiotemporal distribution forecasting of dockless bike-sharing system. Neural Computing and Applications 2019;31(5):1665–77.
[66] Zhang J, Zheng Y, Qi D. Deep spatio-temporal residual networks for citywide crowd flows prediction. In: Proceedings of the AAAI conference on artificial intelligence; 2017.
[67] Wang D, et al. DeepSD: Supply-demand prediction for online car-hailing services using deep neural networks. In: 2017 IEEE 33rd international conference on data engineering (ICDE). IEEE; 2017.
[68] Yu H, et al. Spatiotemporal recurrent convolutional networks for traffic prediction in transportation networks. Sensors 2017;17(7):1501.
[69] Zhang H, et al. Deeptravel: a neural network based travel time estimation model with auxiliary supervision. arXiv 2018. preprint arXiv:1802.02147.
[70] Yao H, et al. Revisiting spatial-temporal similarity: a deep learning framework for traffic prediction. In: Proceedings of the AAAI conference on artificial intelligence; 2019.

[71] Jin G, et al. Urban ride-hailing demand prediction with multiple spatio-temporal information fusion network. Transportation Research Part C: Emerging Technologies 2020;117:102665.
[72] Tang J, et al. Line: large-scale information network embedding. In: Proceedings of the 24th international conference on world wide web; 2015.
[73] Cazabet R, Jensen P, Borgnat P. Tracking the evolution of temporal patterns of usage in bicycle-sharing systems using nonnegative matrix factorization on multiple sliding windows. International Journal of Urban Sciences 2018;22(2):147–61.
[74] Dong Y, et al. An empirical study on travel patterns of internet based ride-sharing. Transportation Research Part C: Emerging Technologies 2018;86:1–22.
[75] Li X, et al. A hybrid algorithm for estimating origin-destination flows. IEEE Access 2017;6:677–87.
[76] Bhargava P, et al. Who, what, when, and where: multi-dimensional collaborative recommendations using tensor factorization on sparse user-generated data. In: Proceedings of the 24th international conference on world wide web; 2015.
[77] Singh AP, Gordon GJ. Relational learning via collective matrix factorization. In: Proceedings of the 14th ACM SIGKDD international conference on knowledge discovery and data mining; 2008.
[78] Wang Y, Zheng Y, Xue Y. Travel time estimation of a path using sparse trajectories. In: Proceedings of the 20th ACM SIGKDD international conference on knowledge discovery and data mining; 2014.
[79] Yu J, et al. Urban network-wide traffic speed estimation with massive ride-sourcing GPS traces. Transportation Research Part C: Emerging Technologies 2020;112:136–52.
[80] Li X, et al. Deep representation learning for trajectory similarity computation. In: 2018 IEEE 34th international conference on data engineering (ICDE). IEEE; 2018.
[81] Yao D, et al. Computing trajectory similarity in linear time: a generic seed-guided neural metric learning approach. In: 2019 IEEE 35th international conference on data engineering (ICDE). IEEE; 2019.
[82] Zoph B, Le QV. Neural architecture search with reinforcement learning. arXiv 2016. preprint arXiv:1611.01578.
[83] Liu C, et al. Progressive neural architecture search. In: Proceedings of the European conference on computer vision (ECCV); 2018.
[84] Kong H, Zhang X, Zhao J. How does ridesourcing substitute for public transit? A geospatial perspective in Chengdu,Chine. Journal of Transport Geography 2020;86:102769.
[85] Jiang W, Zhang L. The impact of the transportation network companies on the taxi industry: evidence from Beijing's GPS taxi trajectory data. IEEE Access 2018;6:12438–50.
[86] Erhardt GD, et al. Do transportation network companies decrease or increase congestion? Science Advances 2019;5(5), eaau2670.
[87] Sui Y, et al. GPS data in urban online ride-hailing: a comparative analysis on fuel consumption and emissions. Journal of Cleaner Production 2019;227:495–505.

Chapter 5

Data-driven optimization technologies for MaaS

Haoning Xi[a,b]
[a]*School of Civil and Environmental Engineering, University of New South Wales, Sydney, Australia,*
[b]*Data61, CSIRO, Canberra, Australia*

1 Overview of data-driven optimization for the urban mobility system

This section provides data-driven optimization models and algorithms used in the existing urban mobility systems, proving their use in MaaS systems.

1.1 Data-driven dispatching methods for on-demand ridesharing

Driven by the trend of "Everything-as-a-Service," car ownership will be converted into car usage, thus on-demand mobility systems, which has proven to be a viable solution for providing more timely and personalized transportation services, led by such companies as Uber, Lyft, and DiDi. By connecting potential mobility requests with eligible drivers, these platforms can also utilize idle vehicle vacancies more efficiently to meet the growing demand for on-demand transportation. This section aims to disseminate the latest findings, research achievements, and ideas through the lens of data-driven optimization, to balance theoretical research ideas and their practicability as well as industrial applicability.

As is shown in Fig. 1, the passenger submits the ride request, including information related to the passenger's origin and destination, which is translated into coordinates for the global positioning system (GPS). After entering the trip details, the platform provides a quote. As soon as the order is submitted, it enters the dispatching system, which assigns it to a driver according to the dispatching policy. In most cases, the assignment time is determined by the dispatching policy and cannot be earlier than when the order is submitted. The order will be placed on hold in the system if no driver matching the order is available at that moment. The driver leaves his location after accepting the assignment to pick up the passenger. Passengers receive an estimated time of pick-up and drop-off.

Big Data and Mobility as a Service. https://doi.org/10.1016/B978-0-323-90169-7.00006-3

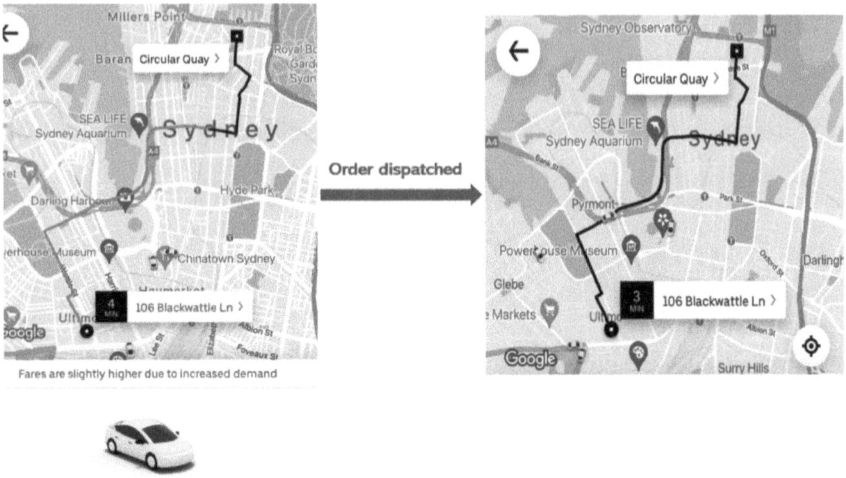

FIG. 1 Ridesharing order dispatched.

Drivers earn a predetermined percentage of the actual price. In Fig. 1, the screenshot on the left shows a trip request, including the origin and destination information. The screenshot on the right shows a driver matched with a rider.

Increasing urban populations poses several challenges in transportation, resource consumption, and environmental concerns [92]. As a means to reduce traffic congestion, gasoline consumption, and pollution, ridesharing has been proposed as a practical approach. The data-driven optimization technologies have been widely applied to the ridesharing system [1–3]. The order-dispatching problem in the ridesharing domain is related to several classical NP-hard combinatorial optimization problems in the operations research literature [4, 5]. Wang and Wu [6] proposed and analyzed a driver dispatching system to reduce the waiting times of passengers at a control center. In order to manage operations, the system allocates a minimum number of drivers across different regions' risks. The data-driven system is based on the Rolling Time Horizon approach and utilizes knowledge learned from historical data. During testing, the authors simulated the performance of the system using actual past data for taxi orders. A significant decrease in average waiting time has been observed with the new system. Comparatively, it saves planning time. Furthermore, the results demonstrate that the dispatching system finds a solution that approximately satisfies allocation constraints while requiring an abrupt increase in waiting time in non-extreme cases.

Alisoltani et al. [7] proposed a new taxi-dispatching system called DTaD, which can allocate fleet taxis for trips according to a 10-min rolling horizon corresponding to the requested trips. This method evaluates all the feasible assignments while considering the maximum waiting time for passengers. To speed up the process, the authors propose a heuristic method based on clustering to group trips with the greatest potential for being served sequentially by the same taxi. Ota et al. [8] proposed a data-driven, real-time simulation framework that can be used to analyze taxi ridesharing efficiently. Since the proposed framework treats taxis and trips as separate entities, a wide range of realistic scenarios can be simulated. In addition, a comprehensive set of parameters allows one to examine the taxi ridesharing issue from various angles while considering the interests and constraints of different stakeholders. This approach can be scaled up due to the use of an efficient indexing scheme and a linear optimization algorithm, which mitigates the computational complexity of the model. Further, a real-world evaluation of the framework and algorithm is conducted by examining 360 million trips taken by 13,000 taxis in New York City between 2011 and 2012. The results of the study indicate that the proposed framework is effective and provides insights into the implementation of urban ridesharing systems. The average passenger demand usually follows a complicated probability distribution in data-driven environments, which cannot be characterized exactly by parametric methods. Therefore, Li and Zhao [9] proposed a new data-driven optimization framework that integrates robust kernel density estimation and two-stage SP modeling. The SP model is then reformulated and solved using a Monte Carlo method known as sample average approximation.

Huo et al. [10] proposed a data-driven optimization model that considers the uncertainty of demand to increase efficiency and overall profitability. In order to characterize the dynamics of the vehicles and the behavioral features of users, historical order data from an electric car-sharing systems (ECS) company are analyzed. It is noteworthy that the daily demand for pick-ups by users follows a Poisson distribution, and that the arrival rates vary throughout the day at four major time intervals. As a consequence of this observation, the present research proposes a data-driven optimization model for the ECS reallocation problem, which is a combination of a probabilistic expectation model and a linear programming problem that utilizes real-time input data. In addition, this study formulates the profit as the mathematical expectation of a discrete random variable together with uncertain consumer demand.

In the context of improving the average system performance, understanding how to deal with uncertainties in the predicted demand probability distribution is still a challenging and unresolved issue. Miao et al. [11] developed a distributionally robust vehicle balancing method to minimize the worst-case expected cost based on data. This approach exploits the quad-tree dynamic region partitioning technique to capture the spatial-temporal dynamics of uncertain demand and develops efficient algorithms for constructing uncertainty sets of probability distributions for different predictions methods. The distributionally robust problem is then reformulated into an equivalent computationally tractable form that can be solved numerically. A 4-year taxi trip dataset of New York City (NYC) is being used to evaluate how the data-driven vehicle balancing algorithm performs under different demand prediction and region partitioning methods. In contrast to vehicle balancing techniques based on static region partitions without considering unknowable demand conditions, the authors demonstrate that the average total idle driving distance can be reduced by 30% using distributionally robust vehicle balancing methods using quad-tree dynamic region partitions.

In order to reach the customer in the shortest time, the taxi with the shortest route is assigned to the booking job. In this manner, customers are served within the shortest possible time and are more likely to be satisfied. Most of the time, when there is no customer on board, taxi drivers will cruise the streets in search of customers either directly or indirectly. Verma et al. [12] developed a Reinforcement Learning (RL) model for cruising taxis to understand real-time driver behavior logs and advise them to find customers at the best places for maximum revenue. In order to build this RL system on the basis of real data, there are multiple translational challenges, such as monitoring activities observed in the trajectory logs, identifying relevant features, action space, and measuring it against actual drivers' behaviors. A dynamic abstraction mechanism is also provided to improve the basic learning mechanism. The above research focused on online bipartite matching markets, where a predetermined set of available resources are matched. Dickerson et al. [13] proposed a new model, online matching with (offline) reusable resources under known adversarial distributions, considering offline resources as reusable as opposed to disposable. The model is shown to be robust enough to capture the application of taxi dispatching services and ridesharing systems using data-driven analysis on a massive, openly available dataset.

A ridesharing order-dispatching problem is challenging to solve due to the dynamic and stochastic nature of supply and demand within this context. The complexity is further increased by the fact that there are multiple objectives, such as system response time and reliability. Qin et al.[14] presented how the proposed optimization problem evolved from a traditional combinatorial approach to incorporating semi-Markov decision-process models along with deep reinforcement learning. During dispatching, a separate module must be used to predict the estimated arrival time of the candidate driver, as well as a separate module that estimates the charge based on the historical data. Verma et al. [12] developed a Reinforcement Learning (RL) based system which can learn from real vehicle trajectory logs to assist drivers in finding customers in the most effective manner. This RL system based on real-world data presents several translational challenges, such as annotation of user movements observed via logs, identifying the right features for a state, action space, and comparing these features with real driver performance observed in the dataset. A dynamic abstraction mechanism is also provided to enhance the basic learning mechanism. Shou et al. [15] integrated Markov Decision Process (MDP) to model the optimal sequential decisions that idle e-hailing drivers make in seeking passengers. The reward function can be discovered by applying the inverse reinforcement learning technique to the MDP model. Then the model is trained using 44,160 three-day trajectory data of Didi drivers.

1.2 Data-driven scheduling methods for public transit

In recent years, bus companies have collected a huge amount of travel data from passengers. These data offer great potential for optimizing bus schedules once been fully utilized. Wang et al. [16] proposed an economical and sustainable strategy so that the profitability of bus companies can be increased. By optimizing how the scheduling algorithm works from blindfold to data-driven, the authors aim to maintain a comparable level of service with fewer buses while maintaining an expected average waiting time. The proposed approach utilizes historical bus operating data to develop a time-dependent bus network correlated with traffic and demand data during different hours of one day. Throughout the dataset, there are hundreds of millions of records for people living in the city, which can be used to infer historical patterns of traffic and demand. Specifically, each day is divided into a fixed number of time intervals. The average travel time between each pair of neighboring bus stations is calculated for each interval, along with the number of waiting passengers at each bus station. A time-dependent bus network is used to develop an optimal model for scheduling departure times of each bus service into the network to minimize the average waiting time. Enrique et al. [17] presented a methodology for dealing with the Public Transport Network Design Problem (PTNDP) and extended its application to a Design Study developed for recommending the development of a new structure for the transit system of Santiago, Chile. The first step is to briefly describe the PTNDP as a multilevel programming problem and discuss the method used for solving it. Following this, the proposed methodology is applied to the Santiago transit system. Considering the stochastic nature of travel demand as well as variability throughout the day, Shariat and Amirpour [18] developed a model to minimize waiting times concerning the stochastic nature of transit demand during the day, so that the transit system could also be prevented from being over or underutilized. Additionally, the proposed model considers the fleet size constraint, making it more applicable and flexible for the urban public transit systems.

As opposed to bus systems, metro systems provide better service reliability in terms of operating speed, dwell time at stations, and service regularity, simplifying the process of service scheduling. The metro service reliability is dependent on the design of the timetable. In daily operations, designing a schedule based on peak and off-peak periods is a common strategy. The advent of smart card-based automated fare collection systems allows us to quantify spatial-temporal demand based on the microscopic level. Smartcard data allows us to identify the demand of spatial and temporal patterns. Sun et al. [19] described a method for analyzing smart card data collected in Singapore to identify characteristics of dynamic demand of rapid mass transit (MRT) services. Passengers enter and leave MRT stations by using a smart card reader, and they either walk to and from the platforms as they board and alight from trains or traveling in the trains between tapping into and leaving MRT stations. As a way to determine the effective position of the passengers, a regression model based on data from the fastest passengers has been developed for each origin-destination pair. As a result of this model, all the other observations can be divided into two groups: passengers boarding on the trains and passengers waiting at the stations. In the estimation model, the spatial and temporal density of passengers is calculated. Using the density plots, trains' paths can be determined, and passengers can be assigned to single trains in accordance with their estimated locations. Further, Sun et al. [20] proposed three optimization models to develop demand-sensitive timetables by using the equivalent time. The first model can capture the dynamic nature of a timetable; the second model can optimize the timetable with a capacity limitation; the third model can be used to devise an off-peak/peak timetable to capacitate concerning demand. The performance of the three models were evaluated by conducting sensitivity analyses on different parameters of the Singapore metro. Finally, the authors concluded that schedule built with capacity constraints is the most advantageous.

1.3 Data-driven rebalancing methods for bicycle-sharing

Over the past 10 years, bicycle-sharing systems have developed from being interesting experiments to mainstream public transport options in urban mobility. An important aspect of bicycle sharing is that individuals can use bicycles as needed without being saddled by the costs and responsibilities of owning a bicycle. As bicycle-sharing systems gradually expand throughout the urban areas, the availability of publicly available data has led researchers address the challenges faced by these systems. One of the key challenges involves rebalancing the system to ensure that more bicycles are available to users. Many collaborative efforts have been made to develop and implement real data-driven optimization methods to guide the rebalancing of bicycle-sharing systems.

Most of the approaches are based on integer programming, e.g., Refs. [21–23]. Considering the dynamic case in which the demand occurs during/after the rebalancing period, Kloimüllner et al. [21] proposed an efficient way to model and simulate these dynamics as well as adapt a greedy and PILOT construction heuristic, Variable Neighborhood Search and Greedy Randomized Adaptive Search accordingly. Contardo et al. [24] presented a decomposition technique based on Dantzig-Wolfe and Benders to obtain upper and lower bounds for instances up to 100 stations. Schuijbroek et al. [25] combined two aspects treated separately in the literature: determining the service level requirements at each bicycle-sharing station and establishing (near-)optimal vehicle routes to rebalance the inventory. Sun et al. [19] proposed a new cluster-first route-second heuristic, in which a polynomial-size clustering problem considers the approximate routing costs and feasibility of service level simultaneously. It is showed that the proposed heuristic outperforms mixed-integer programming and constraint programming formulations while testing on the real-world data from Hubway (Boston, MA) and Capital Bikeshare (Washington, DC). Pfrommer et al. [26] demonstrated the efficiency of shared mobility systems via the application of intelligent routing decisions to redistribute vehicles based on a staff-based allocation method and real-time pricing incentives for customers. Based on the historical data, the authors simulated the Barclays Cycle Hire scheme in London. The mechanics of model-based predictive control are used to calculate and offer dynamically varying rewards to customers throughout their journey, according to the current and anticipated condition of the bicycle-sharing system.

2 Overview of the general concept in MaaS System

The term "Mobility as a Service (MaaS)" refers to the integration of various forms of transportation into a single mobility service that can be accessed on-demand. MaaS operators facilitate a variety of travel options to meet customers' needs, including public transportation, ride-, car-, or bicycle-sharing, taxi or car rental/lease, or a combination of these options. With MaaS, users can benefit from using one application to access mobility with a single payment channel rather than multiple ticketing and payment channels. It should be the best value proposition for its users, helping them meet their mobility needs and removing the inconvenient aspects of individual journeys as well as the entire mobility service system.

2.1 Overview of the MaaS systems

Ho et al. [27] highlighted the subscription in MaaS in accordance with booking and buying the mobility packages, e.g., a traveler can either choose "Pay-as-You-Go" for immediate traveler "Pay-as-a-Package" for long-term usage, e.g., weekly subscription and monthly subscription. Jittrapirom et al. [28] indicated that the mobility package can offer bundles of various transport modes and the amount of km/minutes/points for a monthly usage with a lump-sum payment. This customized subscription can tailor the bundles to satisfy the demand of heterogeneous travelers [29–31]. Cottrill [32] defined MaaS as a multi-modal ecosystem, which can be predicted based on the data from transportation providers, travels, financial companies, and social platforms, etc. The characteristics of MaaS emphasized in different literature are summarized in Table 1.

In sum, the urban transportation planning sector will contribute to a conducive environment for MaaS development as well as maximizing social welfare. The existing literature suggested that a new business model that would benefit all participants and ensure the awareness of these benefits should be pursued to encourage the adoption of MaaS and the mobility resources in the MaaS systems should be allocated efficiently.

TABLE 1 Summary of the characteristics of MaaS.

Summary	Description	Literature
Integration of transport modes	Bundling different modes such as public transport, taxi, car-sharing, ridesharing, bicycle-sharing, car-rental, and on-demand bus services.	[31, 33–35]
Customized subscription	User-centric MaaS bundles customized based on users' preference and willingness to pay (WTP).	[29,31,36,37]
Operation mode	"Pay-as-a-Package" (PAAP) and "Pay-as-You-Go" (PAYG).	[37–39]
Third-Platform	The third platform is allowing users to book and request services in real-time.	[28, 31]
MaaS ecosystem	The participants of MaaS ecosystem include travelers, suppliers, financial and telecommunication companies.	[40–42]
Internet of Things (IOT)	GPS, wireless telecommunications networks, e-ticketing e-payment systems, and autonomous vehicles.	[43–46]
Preference and WTP	MaaS aims to provide the service maximizing travelers' utilities based on preference and WTP.	[47, 48]
Transition of mobility	Car ownership will be transformed into car usage, and public and private companies will collaborate with each other.	[47, 49–51]

2.2 Overview of data in MaaS systems

MaaS platform plays the role of a standardized layer by integrating and cooperating data from various transportation service providers, and mainly delivers four types of data [33]:

- Basic data of the transport system
  MaaS platforms provide both dynamic and static data to travelers. Static data includes public transport schedules, location data such as points of interest (POI), public transport stations and lines, parking information, charging stations, etc. Dynamic data includes dynamic pricing, real-time road traffic conditions, real-time idle vehicle information, estimated waiting time, estimated distance, and estimated service time from origin to the destination, etc.
- Routing data of travelers
  The MaaS platform tailors intermodal and multi-modal travel routes with the most logical combination of travel modes based on travelers' preferences and willingness to pay. Each traveler's routing data includes the

specific route from origin to destination, the travel time on each mode, and transfer time among different modes.

- Booking, ticketing, and payment data
  The platform employs a variety of booking workflows, including route-related and product-related booking. Furthermore, the MaaS platform collects a single charge from each traveler by using the services of multiple vendors.
- Feedback data from travelers
  The MaaS platform collects comprehensive feedback from travelers and service providers. Its dashboard can track, e.g., mobility service data, processing and enrichment of mobility information, transactions and booking information, etc.

Moreover, the data of different partners (e.g., investors, public authorities) can be used for further processing, analysis, and mining or integration into existing services.

After exploring the data in MaaS, the next step is to understand what kind of activities can be applied to extract business value from data. These activities translate into either customer services or internal cost optimization with direct and indirect monetization options in MaaS ecosystem (Table 2): they can provide additional features (e.g., parking), increase convenience (e.g., in-car payment) or reduce cost (e.g., usage-based-insurance), and improve customer experience (e.g., personal travel planner).

In MaaS ecosystem, data-driven services can be categorized into four categories:

- *Mobility services*: Additional services which are out of the scope of the current mobility services.

TABLE 2 Data sof different services in MaaS ecosystem [42].

Mobility services	Connected car services	IOT	Digital life
• Car pooling • Vehicle sharing • E-hailing • Crowded-sourced • MaaS	• Personalized settings • Remote services • Predictive maintenance • Driving style • Online booking • Maintenance service • Charing service	• Health services • Smart logistics • Road sign recognition • Traffic flow monitoring	Personal travel planner In-car entertainment Smartphone APPs Digital identity Mobile payment

- *Connected car services*: Smart and connected services which can improve the car with additional features.
- *Internet of Things (IOT)*: Growing connectivity of devices in the physical world will drive business opportunities along the intersection of different industries, like smart grids, smart cities, logistics, health, and financial services.
- *Digital life*: The mobility services are expected to expand beyond transportation and develop other relevant fields; services are only gradually emerging; and these technologies will be used for digital assistants, location-based services, and augmented reality.

3 Mobility resource allocation in MaaS system

MaaS would become the mainstream in bringing people and activities together by converting the ownership of a vehicle to the right to use the sharing vehicles under a sharing culture of society, such as the shared use of different resources. There is a common challenge to smart mobility: limited resources to be optimized. In smart mobility, resource allocation relies on optimization problems that generate social interests and unsolved challenges. This is the main reason for their popularity and importance. Typically, transportation problems used in routing problems are based on real-world transportation scenarios. Hence, resolving them efficiently will be both socially and commercially beneficial. Nevertheless, most of the issues appearing in this field have a significant computational complexity. For this reason, the efficient resolution of the resource allocation in MaaS system persists as a significant challenge. Thus, we introduce the optimization technologies in MaaS system in the length of mobility resource allocation problems. In Section 3.1, we introduce a deterministic mobility resource allocation framework for MaaS systems; In Section 3.2, we introduce different stochastic mobility resource allocation methods that can be applied in MaaS systems.

3.1 Mobility resource allocation framework in MaaS

In the government regulated markets, Xi et al. [39] proposed an innovative and alternative representation for MaaS system in which continuous and unified mobility resources integrated from both public and private multimodal transportation providers can be effectively and efficiently allocated to travelers based on their preference and willingness to pay (WTP) through a single payment on a digital platform. The authors take the perspective of a MaaS regulator, which aims to maximize social welfare by allocating mobility resources to user requests.

Further, under economic deregulation, Xi et al. [52] proposed a tractable MaaS framework in the two-sided markets by using the single-leader multi-follower approach where the MaaS operator is the leader and travelers, as well

FIG. 2 Keywords of the proposed MaaS framework.

as TSPs, represent two groups of agents in the two-sided MaaS markets. In the proposed single-leader multi-follower game, the MaaS operator (leader) aims to maximize its profits, travelers aim to minimize their travel cost and TSPs aims to maximize their profits, respectively. The keywords of the proposed MaaS frameworks are summarized in Fig. 2.

Xi et al. [39] presented a framework for online auctions, where travelers can place bids on continuous mobility resources according to their needs and willingness to pay. Pay-as-You-Go (PAYG) and Pay-as-a-Package (PAAP) are two different MaaS mechanisms that allow travelers to either pay for immediate use of mobility services or subscribe to services packages for more prolonged use. To implement the MaaS mechanism, the authors develop efficient online primal-dual algorithms. Furthermore, a rolling horizon framework is proposed to improve social welfare by using online algorithms and incorporating booking flexibility. The main optimization methodologies presented in the proposed MaaS framework are shown in Fig. 3.

Xi et al. [39] presented two examples in Fig. 4 to illustrate PAYG and PAAP mechanisms. For example, a traveler with a fixed inconvenience tolerance of

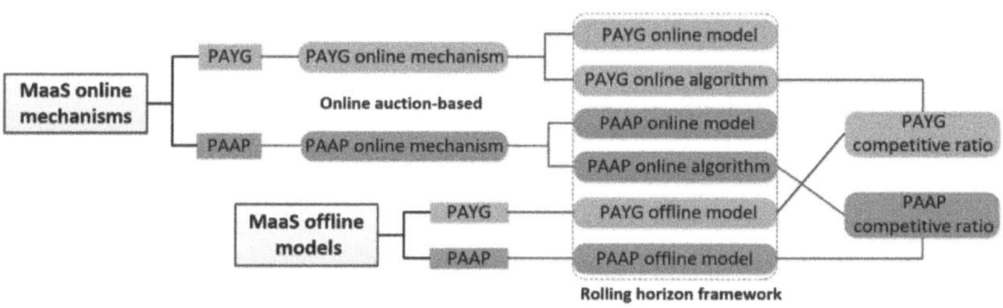

FIG. 3 Overview of methodologies in the proposed MaaS framework.

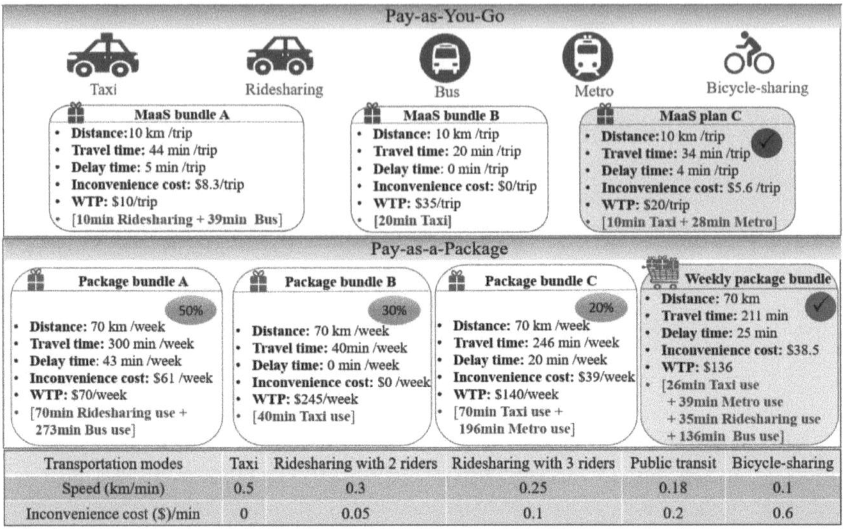

FIG. 4 Illustration of MaaS bundles [39].

100 for a particular trip may purchase a 10km-mobility service for immediate use or subscribe to a 70km-mobility service for regular use. Through the PAYG mechanism, travelers may bid on different speeds for different prices to obtain a "10km-service." Once the traveler submits three bids, the MaaS supplier will combine the most practical combination of transportation modes for each bid, such as the three bundle plans shown in Fig. 2. Although travelers have the option of choosing any of the three bundle plans, PAYG's online system assigns the traveler with the bundle plan that maximizes their convenience, e.g., "Bundle plan C." Assume that the traveler submits three bids in the PAAP mechanism, and the MaaS operator arranges package bundle plans A, B, and C with a fraction, such as 50%, 30%, and 20%, and the package plan associated with the traveler will be weighted by the average distance, speed, and inconvenience cost in three bundle packages.

The procedure of the online auction mechanism is shown in Fig. 5. After travelers and transportation service providers join the MaaS platform, the MaaS operator updates the information on the allocated resources and available mobility resources in real-time. When travelers arrive at the system and request mobility resources in terms of bids, the MaaS operator determines acceptance/rejection based on travelers' bidding price on the unit quantity of mobility resources. If a traveler is accepted by the MaaS operator, he/she will be arranged with the most efficient combinations of travel modes, referred to as MaaS bundles. Among all the feasible MaaS bundles, the MaaS operator will select the one which can maximize the traveler's utility. Moreover, when the requested mobility resources reach the capacity, the MaaS operator will preferentially allocate the mobility resources to travelers with a higher WTP (bidding price) for per unit of mobility resources.

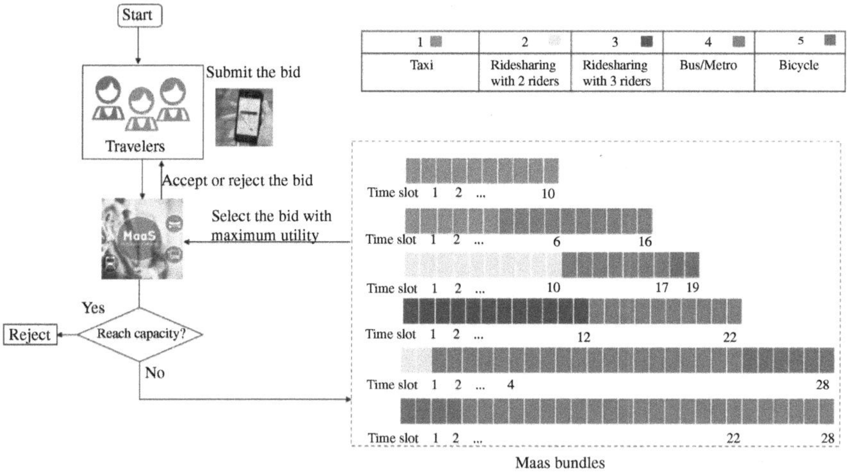

FIG. 5 Online auctions in the proposed MaaS system.

The authors cast the proposed auction-based mechanisms as online resource allocation problems where users compete for MaaS resources and bid for travel time per trip. They propose (integer-) linear programming formulations to accommodate user bids based on available mobility resources in an online optimization approach. Specifically, the resource allocation problem with the PAYG option can be formulated as a mixed-integer linear programming (MILP) and the resource allocation problem with the PAAP option can be formulated as linear programming (LP), respectively. A summary of mathematical notations is given in Table 3.

In the proposed MaaS framework [39], each traveler can submit the service requests in terms of bids. Each traveler can submit multi-bids, including the bidding price (WTP) for Traveler's MaaS bundle and the quantity of requested motility resources for Traveler's MaaS bundle. A traveling is presumed to

TABLE 3 Mathematical notations.

Notation	Description
A_t	The quantity of available mobility resources at time t
$b_{i,s}$	Traveler i's bidding price for service s
$Q_{i,s}$	Traveler i's requested quantity of mobility resources.
$I(t)$	The set of travelers requesting mobility service at time t
S_i	The set of feasible MaaS bundles for traveler
Ω	The set of time slots

bid for a mobility service for immediate use in the PAYG mechanism. The traveler's requests will be either accepted or refused, i.e., this decision is represented as binary allocation. In exchange, under the PAAP mechanism, a trip is assumed to commit to a mobility plan to be used over a predefined span of time, such as a monthly mobile phone subscription. In this case, travelers are assumed to bid for a mobility bundle but can be assigned with a fraction of their requested mobility facilities. The fraction of a package MaaS bundle denotes the requested distance and travel time proportion while maintaining the requested speed constant. The resulting mixed-integer linear programming (MILP) formulation for the PAYG online mobility resource allocation problem at time slot t, $t \in \Omega$, is summarized as follows:

(PAYG online resource allocation problem)

$$\max \sum_{i \in I(t)} \sum_{s \in S_i} b_{i,s} x_{is}, \tag{1}$$

subject to:

$$\sum_{i \in I(t)} \sum_{s \in S_i} Q_{is} x_{is} \leq A_t, \tag{2}$$

$$\sum_{s \in S_i} x_{is} \leq 1, \quad \forall i \in I(t), \tag{3}$$

$$x_{is} \in \{0, 1\}, \ \forall i \in I(t), \ s \in S_i, \tag{4}$$

where x_{is} is a binary variable indicating whether MaaS bundle s is allocated to Traveler i, Eq. (1) aims to maximize the social welfare of the MaaS system, Eq. (2) indicates that the quantity of allocated mobility resources cannot exceed that of available resources at time t. Eqs. (3), (4) indicate that at most one bid is allocated to traveler i, $\forall i \in I(t)$.

Moreover, the resulting linear programming (LP) formulation for the online Pay-as-a-Package (PAAP) mobility resource allocation problem is summarized as follows.

(PAAP online resource allocation problem)

$$\max \sum_{i \in I(t)} \sum_{s \in S_i} b_{i,s} x_{is}, \tag{5}$$

subject to:

$$\sum_{i \in I(t)} \sum_{s \in S_i} Q_{is} x_{is} \leq A_t, \tag{6}$$

$$\sum_{s \in S_i} x_{is} \leq 1, \quad \forall i \in I(t), \tag{7}$$

$$x_{is} \geq 0, \ \forall i \in I(t), \ s \in S_i, \tag{8}$$

where x_{is} is a continuous variable indicating the proportion of mobility resources in MaaS bundle s allocated to traveler i. Note that PAYG online resource allocation problem is a binary resource allocation problem while PAAP online resource allocation problem is a fractional resource allocation problem.

3.2 Data-driven online stochastic resource allocation problems

In the context of online stochastic resource allocation, Funaro et al. [53] presented a mechanism for generating fixed availability (traditional) resource allocations in conjunction with stochastic resource allocations. The authors first proposed an innovative stochastic allocation mechanism that allows the provider to sell reserved resources along with an additional stochastic allocation. Additionally, the authors presented a validated infrastructure for cloud simulation known as the Stochastic Allocation Simulator. With this infrastructure, it is possible to create a large dataset of realistic clients with different behaviors and simulate their rational set of bundle selections based on the known resource distributions. Li and Ye [54] studied the online linear programming problem with a provable regret bound and empirically observed that bounded regret can be achieved on some problem instances through an action-history-dependent design. Li and Ye [54] proposed an action-history-dependent learning algorithm for an online linear programming problem, which can improve the performance of the previous algorithms by considering the past input data and decisions. Asadpour, Wang, and Zhang [55] derived bounded regret for a resource allocation problem based on an online linear programming problem with a random input model, in which the columns of the constraint matrix, as well as the corresponding coefficients in the objective function, are generated from an unknown distribution and revealed sequentially over time. Furthermore, Chen et al. [56] investigated an online stochastic resource allocation problem with linear programming as its general basic form and then analyzed an adaptive allocation algorithm with a constant regret bound dependent on the number of time periods.

4 Data-driven optimization technologies for resource allocation in MaaS

Considering the response time, reliability, different objectives of multiple stockholders, the dynamic and stochastic supply and demand in the MaaS system, it is challenging to solve the mobility resource allocation problem. Data-driven optimization has become omnipresent and well-studied in various systems. However, different applications of data-driven optimization diverge into the concept of uncertainty through data analysis, e.g., defining a set with uncertainty as a random variable with a probability distribution. A fundamental challenge in data-driven decision problems is constructing estimators for the optimal solutions of stochastic optimization problems based on the limited training data. Thus data-driven optimization focuses on defining the probability distribution through analyzing the pertinent data. In several real-world applications, data-driven decision-making is based on the combination of both machine learning and optimization technologies. The optimization model is typically used to make decisions, while a machine learning tool is used to build a prediction model that predicts key unknown parameters of the optimization model. Then we introduce sample average approximation in Section 4.1, robust optimization in Section 4.2, predictive analysis and prescriptive analysis in Section 4.3, machine learning and case study in Section 4.4.

4.1 Sample average approximation

The sample average approximation (SAA) method is a Monte Carlo simulation-based approach to solve stochastic optimization problems. The estimated objective function of the stochastic problem is approximated using this technique by a sample average approximation obtained from a random sample. The procedure is replicated for separate samples to produce candidate solutions as well as statistical estimates of their optimality gaps [57].

Consider the following optimization problem in Eq. (9):

$$\max_{x \in D} h(\boldsymbol{x}), \tag{9}$$

where D is a convex set of feasible solutions, and the real-valued function $h(\cdot)$ cannot be calculated exactly but can be estimated through a stochastic simulation. Sample average approximation (SAA) is a method for dealing with such problems that employ sampling and optimization methods for deterministic problems [58].

It is assumed that the function f cannot be observed or computed directly, but we know $f(x) = \mathbb{E}[Y(x, \xi)]$, where ξ is a random element with a distribution that does not depend on x, and $Y(\cdot)$ is a real-valued function. Namely, $\forall x \in D, \mathbb{E}[Y(x, \xi)] < \infty$.

In SAA, we select and fix $\xi_1, \xi_2, \ldots, \xi_n$, all having the same distribution as ξ, and set in Eq. (10):

$$f_n(x) = \frac{1}{n} \sum_{i=1}^{n} Y(x, \xi_i). \tag{10}$$

Given the fixed sample, $\xi_1, \xi_2, \ldots, \xi_n$, the function $h_n(\cdot)$ is deterministic, and then we can apply deterministic optimization algorithms to solve the problem (11):

$$\min_{x \in D} h_n(x). \tag{11}$$

Assume that $\xi_1,\xi_2, \ldots, \xi_n$, form an independent and identically distributed (i.i.d.) sample. The independence assumption is sometimes relaxed, mostly in variance reduction schemes or randomized quasi-Monte Carlo schemes where dependence is deliberately introduced.

Then we use the MaaS framework proposed by Xi, Liu, Rey, Waller and Kilby [39] to illustrate SAA. Assume that travelers buy x units of mobility resources with the per unit of salary paid to transportation service providers cost $c>0$, and observe travelers' demand ξ, the MaaS operator allocates as many units as possible mobility resources to travelers at a unit price $p, p>c$. The goal of the MaaS operator is to determine the allocation of mobility resources for each traveler x, so asso as to maximize its profits. The profits of the MaaS platform for a given realization ξ is denoted as Y, $Y(x,\xi)=p\min\{x,\xi\}-cx$, this function is concave in x and has a slope $s-c>0$ for sufficiently small x and a sufficiently large x. It is simple to demonstrate that an optimizer X_n^* of $h_n(\cdot)$ occurs at $1-c/p$ quantile of the empirical distribution associated with the observed travel demands $\xi_1,\xi_2, \ldots, \xi_n$. If we assume that the distribution of ξ is continuous at its $1-c/p$ quantile, which is optimal for the true problem, then X_n^* converges to this value as $n\rightarrow\infty$ almost surely. In this case, SAA is successful in that the sequence of optimizers $\{X_n^*\}$ converges to a true optimizer [59, 60].

Overall, SAA is a practical approach for data-driven decision-making in the MaaS system. Moreover, Lagoudakis and Parr [61], Mahadevan [62], and Michels et al. [63] introduced several types of problems such that SAA works in the sense that X_n^* can be expected to converge to the set of optimizers of Eq. (9) as $n\rightarrow\infty$, introduced several types of problems such that optimistic algorithms can be applied in Eq. (11) with comparable computational quality.

4.2 Robust optimization

Ye [64] presented several optimization models and computational algorithms dealing with uncertain, dynamic, online, structured, and massively distributed data. Many decision-making problems in unpredictable situations can be efficiently described using the stochastic optimization programming. For example, given that one is interested in solving a convex optimization problem in Eq. (12):

$$\max_{x\in D} h(\boldsymbol{x}, \boldsymbol{\xi}), \tag{12}$$

where $h(\boldsymbol{x},\boldsymbol{\zeta})$ is a convex cost function in terms of $\boldsymbol{x}$ that depends on some vector of parameters, it is often the case that the parameters have not yet been fully resolved. For example, a MaaS operator cannot know the exact future travel demand if MaaS operator chooses to represent his uncertainty through a distribution F to maximize its expected profits, leading to a stochastic program in Eq. (13):

$$\max_{x\in D}\mathbb{E}_{F_\xi}[h(x, \xi)], \tag{13}$$

where $\boldsymbol{x}$ is the decision variable vector with the feasible region D, ξ is a random parameter vector with density or distribution F_ξ. This type of stochastic optimization problem has a wide range of applications and has been shown that the predicted value is an accurate indicator of performance. Thus, this model can also serve as a decision-making tool for the MaaS operators. Based on a well-formulated stochastic model, the MaaS operator can choose to take an early decision that leads to the highest expected profits. However, a more challenging difficulty in practice is the need to commit to the distribution of ξ given only limited information about the stochastic parameter. Deviating from the assumed distribution may cause sub-optimal solutions. In order to resolve the distribution's lack of information, Arrow et al. [65] propose the following robust optimization problem:

$$\max_{\boldsymbol{x}\in D} \min_{\xi\in\Omega} h(\boldsymbol{x}, \xi) \tag{14}$$

where Ω is the support region of ξ. Although this approach has the advantage of only requiring uncertain parameters, the decision that maximizes the worst-case pay-off may perform poorly in real applications.

In reality, people commonly know such moments based on rich observational evidence, even though the precise distribution of the random variables is unknown. Furthermore, where there is uncertainty in choosing a distribution for the random parameters, their approach can be misleading. We suggest a model that defines uncertainty in terms of distribution structure (discrete, Gaussian, exponential, and so on) and moments (mean and covariance matrix). In order to choose an intermediate approach between stochastic optimization, which has no robustness to the error of distribution, and robust optimization. Thus, the distributionally robust optimization approach is introduced as follows:

$$\max_{x\in D} \min_{\xi\in\Gamma} \mathbb{E}_{F_\xi}[h(x, \xi)], \tag{15}$$

where Γ is a set of density functions or distributions and maximize the worst-case expected cost value among those distributions in Γ. The real meanings, tractability, and potential loss, compared with the robust approach, need to be considered while choosing Γ. For example, Γ is given as follows:

$$\Gamma = f_\xi \geq 0 \left| \begin{array}{l} \mathbb{E}[I(\xi\in\Omega) = 1] \\ (\mathbb{E}[\xi] - \mu_0)^T \sum_0^{-1} (\mathbb{E}[\xi] - \mu_0) \leq \gamma_1, \\ \mathbb{E}\left[(\xi - \mu_0)(\xi - \mu_0)^T\right] \preccurlyeq \gamma_2 \sum_0 \end{array} \right. \tag{16}$$

where μ_0 and $\sum_0$ are mean vector and covariance matrix of ξ. Namely, the density function or distribution set is defined based on the support, first- and second-order moment constraints [65–70].

Van Parys et al. [69] investigated stochastic programming in which the decision-maker is unable to track the distribution of exogenous uncertainty but has access to a finite number of independent samples from that distribution. In this context, the aim is to find a procedure that converts the data into an approximation of the expected cost function under the uncertain data generation distribution, a predictor, and an optimizer of the predicted cost function that serves as a near-optimal candidate decision, a prescription. Delage and Ye [71] showed that the related distributionally stable stochastic programming can be solved efficiently for various cost functions. Furthermore, the authors offer probabilistic reasons for using the proposed model in problems that depend heavily on historical data by deriving a new confidence area for a random variable's mean and covariance matrix.

4.3 Predictive analysis and prescriptive analysis

Let x denote the feature or covariate, and y denote the information such as weather, demographics, and economic factors. We often have contextual information associated with y; thus, the data consists of $\{(x_1,y_1),(x_2,y_2),\ldots,(x_n,y_n)\}$. The goal is to solve the following conditional (contextual) stochastic optimization problem:

$$\max_{z\in S}\mathbb{E}_{Y\sim D_x}[h(z,Y)|x]. \tag{17}$$

Supervised learning is mainly interested in discovering the relationship between x and y. Assume $y\approx f(x)$ we can learn the best prediction model by minimizing a certain loss function:

$$\min_{f\in\mathcal{F}}\sum_{i=1}^{n} l(f(x_i),y_i), \tag{18}$$

where the loss function can be such as linear regression, e.g., $(y_i-f(x_i))^2$.

Bertsimas and Kallus [72] considered the data that primarily consists observations of correlated auxiliary quantities rather than observations of quantities with direct impacts on costs/revenues, such as demand or returns. The key problem of interest is a conditional stochastic optimization problem with incomplete observations and undefined joint probability distributions that define the problem. Punia et al. [73] proposed a distribution-free, data-driven solution approach to the suggested system that finds order quantities using sample demand data as input and computer (and deep) learning methods with the analytical risk minimization theorem. We are interested in the relationship between x and $f(x)$, which may not be continuous and thus is challenging to learn. However, it is doable in some cases, e.g., the newsvendor problem (NVP).

A more recent approach for solving NVP advocates integrating the demand estimation and optimization steps into a single step in data-driven methods. Ban and Rudin [74] proposed a data-driven NVP in which the machine learning algorithms along with empirical risk minimization and kernel weights optimization are used to solve a multi-feature NVP. The authors showed that the empirical risk minimization approach is equivalent to a high-dimensional quantile regression and can be solved by convex optimization methods. Oroojlooyjadid et al. [75] and Punia et al. [93] used deep neural networks to solve the NVP using an integrated method. However, they have not used quantile regression for the same. A similar approach is proposed by Huber et al. [76], in which the authors used nonlinear machine learning methods like a decision tree and single-layer neural network in quantile regression to solve the single item NVP.

Machine learning (ML) methods gained popularity in recent years for demand forecasting [95]. The ML methods model patterns in demand based on correlations with independent variables. In ML methods, artificial neural networks (ANN) are widely used for demand prediction and reported improvements in the accuracy of forecasts in various cases [77].

As a part of the newsvendor model, it is needed to estimate the travelers' demand. We first tried to find out an accurate demand estimation method used in the proposed data-driven inventory optimization model. The advanced machine learning method (e.g., random forest, feed-forward neural network, etc.) and deep learning method (e.g., deep neural network) are used to estimate the demand for future periods [95]. The predictions from these methods are compared with widely used time-series and regression methods to investigate the performance of the proposed method.

Ban and Rudin [74] investigated the data-driven newsvendor problem when there are different number of measurements of demand features and historical demand data. Instead of a two-step method of first calculating a demand distribution and then optimizing the optimum order quantity, the authors suggest solving the data-driven NVP using single-step machine-learning algorithms. The authors suggest algorithms based on the scientific risk minimization theorem, with and without regularization, as well as a kernel-weights optimization algorithm.

4.4 Machine learning-based robust optimization

In this section, Bertsimas and Kallus [72] presented approaches that incorporate ideas from machine learning, operations research, and management science to provide a system, as well as basic methods for using data to recommend optimal decisions in operations research or management science.

Typically, the optimization model is used to make decisions, while a machine learning tool is used to generate a simulation model that forecasts the optimization model's primary unknown parameters [94]. Because of the inherent complexity of both tasks, a broad-purpose approach that is often employed in analytics practice is the predict-then-optimize paradigm [88].

Deep neural network machine learning methods have been used as new approaches within an encoding–decoding framework [78, 79]. To improve a driver's chances of receiving a request, reinforcement learning has been applied to the driver routing prospect. Verma et al. [12] used a Monte Carlo learning approach, and Shou et al. [15] used dynamic programming to learn the state-action value function. Oda and Joe-Wong [80] employed a deep Q-network in a distributed setting to account for the multiagent nature of the fleet-management problem. In value iteration or Q-learning, the use of a neural network, a nonlinear value function approximator, raises established convergence problems [81]. Qin et al. [14] explained how to progress from a myopic, combinatorial optimization approach to incorporate deep reinforcement learning for long-term optimization.

Consider the mobility resource allocation problem in the MaaS system. To begin, a previously trained prediction model predicts travel time on all edges of a road network based on current traffic, temperature, holidays, time, and so on. The estimated travel times are then fed into an optimization solver, which generates near-optimal routes. The authors stress that most real-world analytics solution systems have a combination of estimation and optimization [82–87].

Overall, Machine learning tools are designed to reduce prediction error but do not consider how the predictions can be used in the downstream optimization problem. In contrast, Elmachtoub and Grigas [88] proposed Smart "Predict, then Optimize" (SPO), a modern and very general paradigm that explicitly leverages the optimization problem structure: its objective and constraints, for designing better prediction models. In order to better solve the problems, Zhu et al. [89] proposed a joint estimation and robustness optimization framework to mitigate estimation uncertainty in optimization problems by seamlessly incorporating both the parameter estimation procedure and the optimization problem. They specifically built an uncertainty set that includes all the data, and the size of the uncertainty set is determined by how well the parameters are estimated from that data when using a particular estimation procedure: regressions, the least absolute shrinkage, and selection operator, and maximum probability estimation.

Overall, the data-driven optimization technologies for the MaaS resource allocation problem introduced in this section can be summarized in Fig. 6.

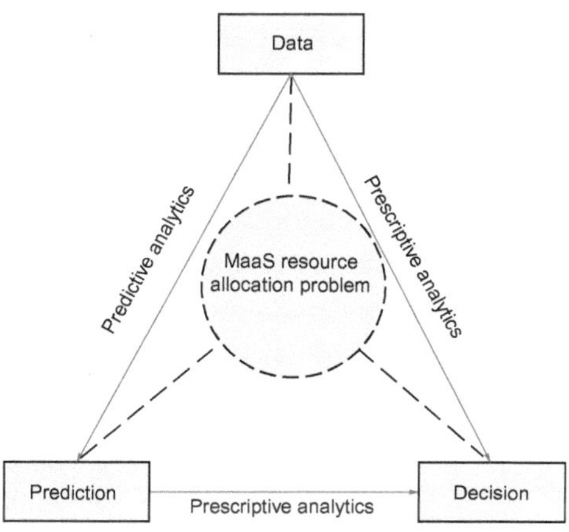

FIG. 6 Data-driven optimization technologies in this chapter.

5 Real-world application and case study

In this section, we introduce a real-world application of Customized Bus (CB), which is a significant component of MaaS systems. We introduce a Data-driven approach for bus lines design in CB system. Yu et al. [90] proposed a method to generate planning suggestions for CB lines and stops from massive demand data. From the demand input, a link network is constructed to represent the sharing route of the demand. Community detection is applied on the link network to segment the link network into network communities with similar travel routes. By examining the core-peripheral structure and match the core part of communities into the road network, the CB lines are generated. Analyzing the potential demand of CB lines, boarding and alighting hotspots are identified as the suggestion for CB bus stops. A case study is conducted using mobile phone data in Tokyo, in which 29 CB lines are extracted. According to the shape and spatial location of the CB lines, three types of CB lines serving different travel patterns are classified, including radiation type lines, ring-type lines, and suburban lines.

5.1 Problem description

A dynamic CB line planning system is a key element for a CB service, Fig. 7 shows an overview of the system. First of all, passengers will submit their demands via the online platform. After each certain period of time, the demand will be upload into the cloud server and aggregated into origin (O) and destination (D) as the input for the dynamic CB line planning system. Then, the system will search for the best route choice for each OD pair. Based on the routes, the system will CB lines in the city in order to meet the demand. The stops of the CB lines will then be deployed so that the CB system can publish CB riding information to passengers and telling the location and time to get on the bus.

In this system, the most difficult part is the CB lines planning. The aim is to generate CB lines and their directions base on the massive volume of travel routes. The system has to decide how many CB lines to deploy and how is the direction of each CB line. Naturally, the more CB lines deployed as possible, the more demands will be served, but at the same time, the more of the CB lines will be uneconomical with less demand for each CB line. So the solution to this problem involves the balance between two objectives: (i) The lesser number of

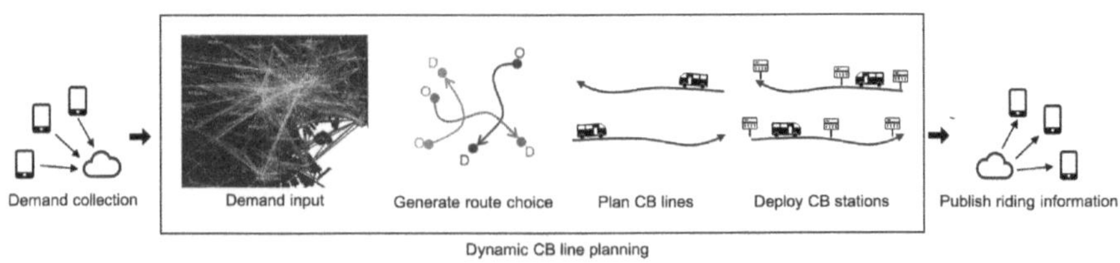

FIG. 7 Overview of a dynamic customized bus line planning system.

CB lines with higher demand assembled for each CB line yielding higher profits and (ii) more demand can be effectively served.

Due to the constant change of the demand, the algorithm of generating CB lines also requires high computing speed. Recent research on CB line planning has been mostly restricted to establishing complicated optimization models and can hardly be applied to the city-wide range of areas with massive demand input. Instead of treating the CB routing problem as an optimization problem, the algorithm in this paper reconsiders the problem in another way: To discover the traveling demand clusters sharing with similar routes and extract the set of routes to generate a suggestion for the bus lines.

5.2 Methodology

The framework of this paper design is shown in Fig. 8. Mobile phone data is used here to offer suggestions for CB lines design. The input of our methodology is the demand routing trajectories of CB service. In this study, the basic assumption is that urban car travel demand shares the same distribution with CB demand. As a sampling from all travel, the car trips extracted from mobile phone data can reflect the distribution of CB travel demand.

The methodology is as follows: In the beginning, the trajectory of car mode is extracted from the mobile data by using the data mining method introduced in the previous work [91]. After matching trajectories into mesh grids, a link network is constructed from the route sharing of trajectories. The network community detection method is applied to segment the link network into communities which reflects the travel demand clusters with a similar traveling pattern. By detecting the core-peripheral structure of communities, the core part of communities can be extracted to suggest the direction of CB lines. Identifying the boarding and alighting hotspots also offer the suggestion for choosing the location of CB stops. Finally, the potential travel demand of the CB lines is identified to analyze the travel pattern.

5.3 Results and discussion

From the car trajectory dataset, a link network with 189,829 nodes and 2,336,738 edges is generated. Applying the fast unfolding algorithm on the link network, the algorithm produces a result of community segmentation with the modularity of 0.73, indicating that there is a community structure in the link network. As a result of community segmentation, there are 10,313 communities in total, 851 communities are with over 10 links, and only 152 communities are with over 100 links. The complementary cumulative distribution function (CCDF) of the number of links in link communities is shown in Fig. 9. The CCDF curve decreases quickly in the small number of links, indicating that a relatively small number of communities are containing a significantly large number of links and play an essential role in the link network. Examine the

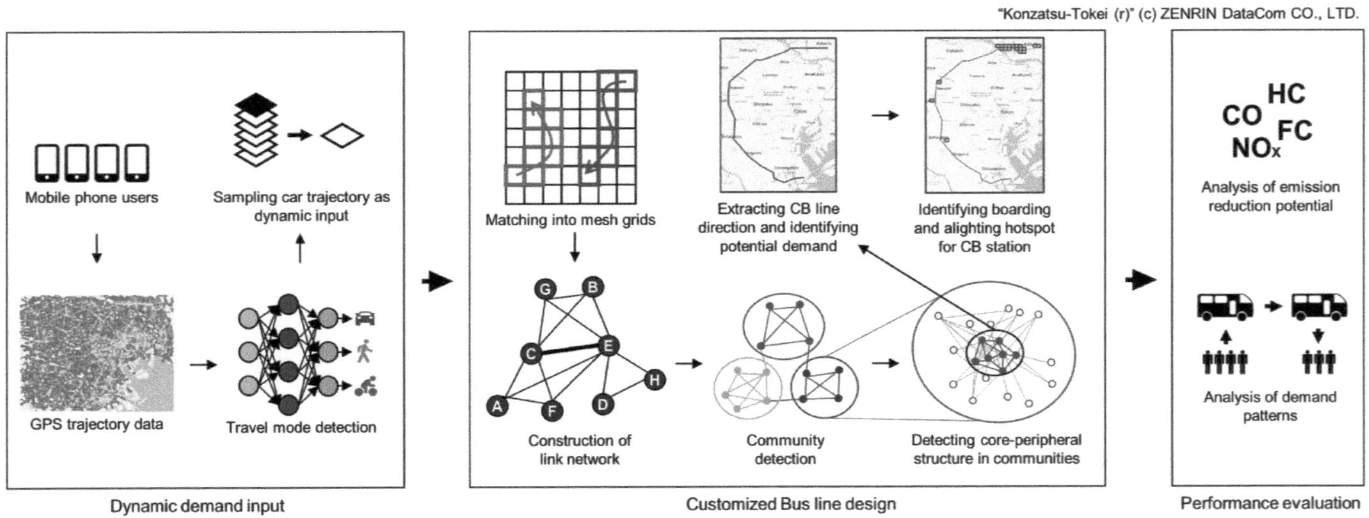

FIG. 8 Framework of CB lines design.

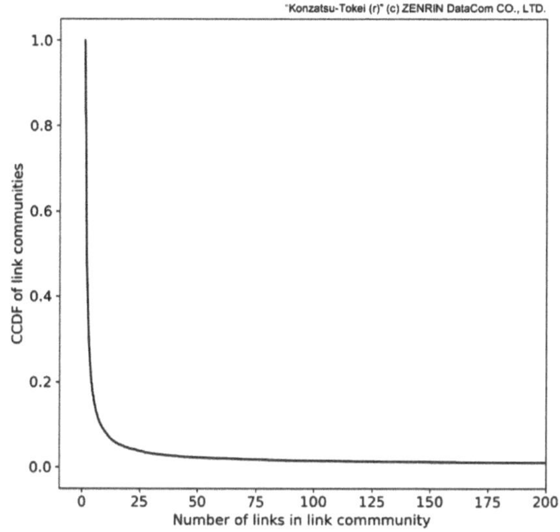

FIG. 9 Complementary cumulative distribution function of the number of links for link communities.

core-peripheral structure of the large communities, most of the core links of link communities are in a line shape and suitable to extract bus lines. Fig. 10 shows two examples of the core-peripheral structure of link communities and the bus line extracted.

From the core-peripheral structure of 152 link communities with over 100 links, 29 CB lines are extracted in the area of Tokyo. The average length of the CB lines is 18.7 km, with the longest of 38.1 km and the shortest of 11.6 km. The number of CB lines is less than the number of link communities, and the reason is that the CB lines extracted from some of the small communities are much shorter, most of them are in the same direction as the larger communities. The travel demand from these small link communities can be satisfied by the CB lines extracted from large communities if the location of bus stops is properly arranged. The core-peripheral structure of link communities that extract the 29 CB lines are shown in Supplementary Material A.

After extracting CB lines, the potential demand for each line is identified. The CB lines are named in descending order from 1 to 29 according to the number of potential demands. Fig. 11A shows the spatial distribution of CB lines' direction. According to the shape and spatial location of the CB lines, they can be classified into three types:

- *Radiation type lines*: This type of CB line include line 2, 3, 4, 5, 7, 12, 13, 14, 15, 16, 17, 19, 20, 21, 23, 25 and 29. These CB lines are extending from the city center to suburban areas in all directions and offering an efficient transportation method for passengers from some specific suburban residential

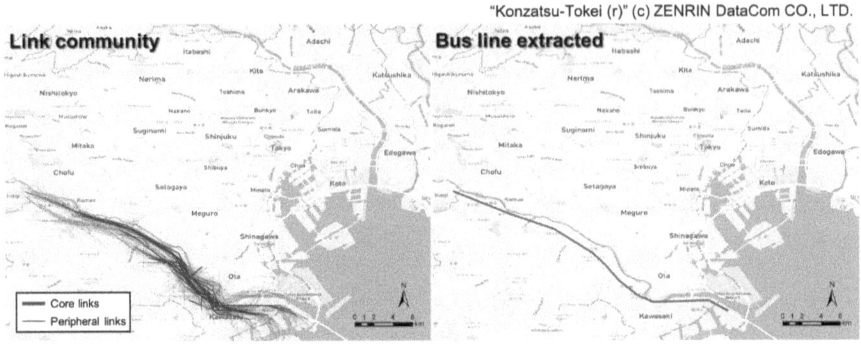

(A)

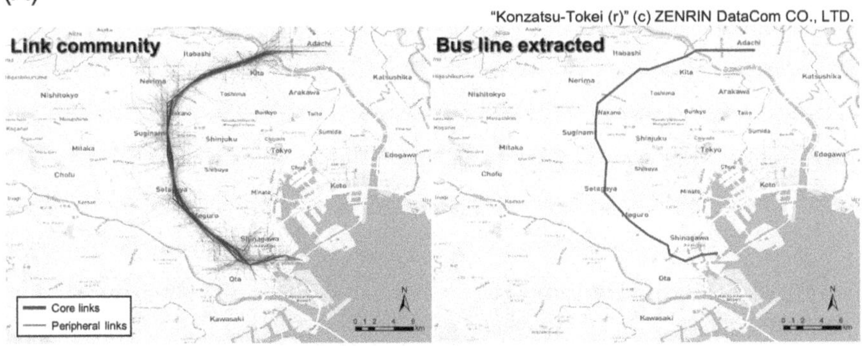

(B)

FIG. 10 Example of Link communities and CB line extracted.

area to enter the city center or subcenter rapidly. These CB lines are mostly in the straight-line shape as the shortest path from the origin to the destination, which can ensure the efficiency of the CB lines.

- *Ring-type lines*: This type of CB line include line 1, 6, 8, 9, 11, 22, 26, and 28. This type of CB line is on the edge of the central urban district connecting subcentres of the city.
- *Suburban lines*: This type of CB line include line 10, 18, 24, and 27. These CB lines are connecting multiple suburban centers.

These three types of CB lines are reasonable and in line with the position of CB lines as a supplement in the urban transportation system. Notice that the presenting CB lines network does not connect multiple Central Business Districts (CBDs) in one line. The reason is that the travel demand from one CBD to another can be well served by urban subway lines.

Fig. 11B shows the regional hotspots and individual hotspots identified from boarding and alighting of potential demand from each CB line. Regional hotspots indicate the cluster of high-demand locations along the CB line, which

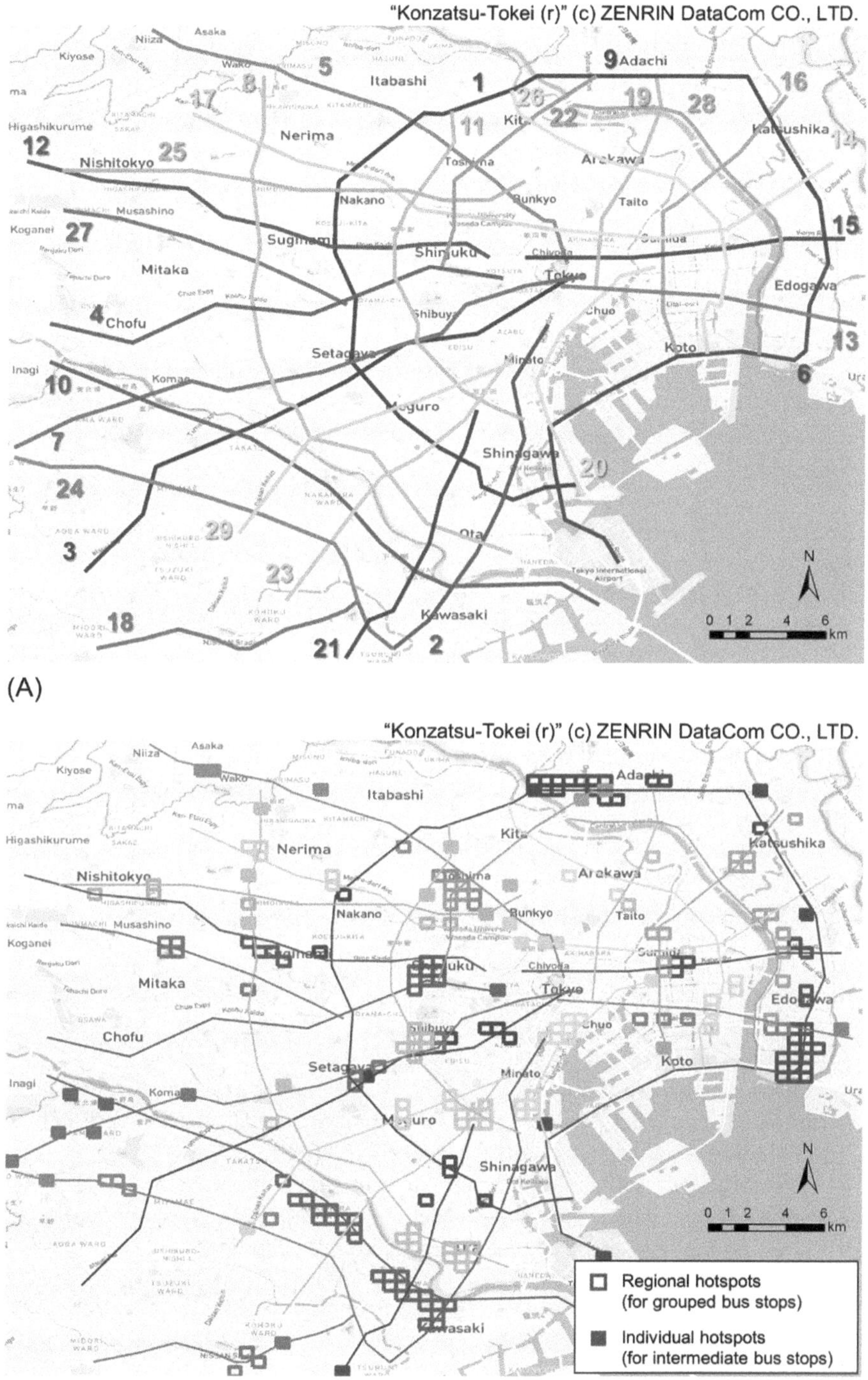

(A)

(B)

FIG. 11 CB lines planning suggestions from link communities. (A) The direction of 29 CB lines extracted and (B) Hotspot identified for bus stop.

can be the location for grouped bus stops. Grouped stops in a range of areas can assemble passengers and reduce their walk distance to the bus stops; Individual hotspots indicate a high-demand location in a low-demand neighborhood. Intermediate bus stops can be arranged in these hotspots. Arranging these two types of stops according to the hotspots can ensure the satisfaction of demand without devastating the efficiency and directness of the CB line.

As a new type of Internet-based transportation mode in the public transportation system, the CB system has a high potential of replacing the private car in the urban area. However, designing a demand-oriented and efficient system is the key element and highly related to the public acceptance and operation performance of the CB system. Under this background, this paper proposes a methodology to generate planning suggestions for CB lines and stops based on massive demand data with high computing speed that is capable of integrating into a dynamic CB planning system. The proposed method involved the following steps:

The car trajectories are extracted from mobile phone data as the input of potential CB travel demand. From the demand input, a link network is constructed to represent the sharing route of the demand. Community detection is applied on the link network to segment the link network into network communities with similar travel routes, and the core-peripheral structure of each community is examined. By extracting the core part of the link network communities and match into the road network, the CB lines are generated. Potential demand of CB lines are identified, and boarding and alighting hotspots are extracted as the suggestion for CB bus stops. A case study is conducted using mobile phone data in Tokyo, 29 CB lines are extracted. According to the shape and spatial location of the CB lines, three types of CB lines serving different travel patterns are classified, including radiation type lines, ring-type lines, and suburban lines.

The proposed method has several advantages:

- The method can automatically classify similar travel paths into clusters and generate bus lines direction and stops location suggestions, which is more demand-oriented and efficient than the existing traditional CB line planning method.
- Potential demand reduction of each CB line can be examined from human mobility data, which can help the CB operating company and government to evaluate the potential performance of the CB system.
- The algorithm is capable of processing massive demand data at a fast computing speed, which has a high potential of integrating into a dynamic CB line planning system to generate real-time CB planning.

6 Conclusions

In the transition to smart cities and the development of technology, a huge amount of data is collected in real-time. Several on-demand mobility services are available, such as taxis, ride-hailing, on-demand public transit, and bicycle-sharing services. This chapter provides a detailed overview of data-driven optimization technologies used in the existing urban mobility system, such as data-driven dispatching methods for ridesharing services, data-driven scheduling methods for public transit, and data-driven rebalancing methods for bicycle-sharing systems. Then we analyzed and summarized the characteristics of MaaS systems in the existing literature. It is concluded that the mobility resource allocation that relies on optimization problems is a big challenge in smart mobility systems. Hence, we provide a general MaaS framework by resorting to the resource allocation and introduce both deterministic mobility resource allocation problems as well as stochastic mobility resource allocation problems; Further, we introduce more advanced data-driven optimization technologies which are applicable for the mobility resource allocation problems in MaaS system from the following four perspectives: data uncertainty (robust optimization), data dynamics (online mobility resource allocation), data structure (sample average approximation) and data prediction (predictive analysis and prescriptive analysis). Finally, a real-world case study conducted in Tokyo based on massive demand data is presented. Here the proposed data-driven methodology can generate planning suggestions for CB lines and stops with high computing speed that are capable of integrating into a dynamic CB planning system. Overall, in the era of big data, data-driven optimization technologies can serve as a more accurate tool to facilitate the MaaS operator to manage the MaaS system and make better decisions to maximize its profits and improve the customer experience in contrast with the traditional optimization methods.

References

[1] Liu S. Data analytics and smart city operations. Berkeley: UC; 2019.

[2] Liu S, He L, Max Shen Z-J. On-time last-mile delivery: order assignment with travel-time predictors. Management Science 2020. https://doi.org/10.1287/mnsc.2020.3741.

[3] Qi W, Li L, Liu S, Shen ZJM. Shared mobility for last-mile delivery: design, operational prescriptions, and environmental impact. Manufacturing and Service Operations Management 2018;20(4):731–51. https://doi.org/10.1287/msom.2017.0683.

[4] Miao F, Han S, Lin S, Stankovic JA, Zhang D, Munir S, Huang H, He T, Pappas GJ. Taxi dispatch with real-time sensing data in metropolitan areas: a receding horizon control approach. IEEE Transactions on Automation Science and Engineering 2016;13(2): 463–78.

[5] Özkan E, Ward AR. Dynamic matching for real-time ride sharing. Stochastic Systems 2020;10 (1):29–70. https://doi.org/10.1287/stsy.2019.0037.

[6] Wang T, Wu D. Data-driven driver dispatching system with allocation constraints and operational risk management for a ride-sharing platform. Decision Sciences 2020;51(6):1490–520. https://doi.org/10.1111/deci.12433.

[7] Alisoltani N, Zargayouna M, Leclercq L. A sequential clustering method for the taxi-dispatching problem considering traffic dynamics. IEEE Intelligent Transportation Systems Magazine 2020;12(4):169–81.

[8] Ota M, Vo H, Silva C, Freire J. A scalable approach for data-driven taxi ride-sharing simulation. In: Proceedings—2015 IEEE international conference on big data, IEEE big data 2015; 2015. p. 888–97. https://doi.org/10.1109/BigData.2015.7363837.

[9] Li X, Zhao L. Towards smart transportation: a learning-based data-driven optimization approach for electric taxi dispatch problem. Internet Technology Letters 2021;4(1). https://doi.org/10.1002/itl2.164, e164.

[10] Huo X, Wu X, Li M, Zheng N, Yu G. The allocation problem of electric car-sharing system: a data-driven approach. Transportation Research Part D: Transport and Environment 2020;78:102192. https://doi.org/10.1016/j.trd.2019.11.021.

[11] Miao F, He S, Pepin L, Han S, Hendawi A, Khalefa ME, Stankovic JA, Pappas G. Data-driven distributionally robust optimization for vehicle balancing of mobility-on-demand systems. ACM Transactions on Cyber-Physical Systems 2021;5(2):261–71. https://doi.org/10.1145/3418287.

[12] Verma T, Varakantham P, Kraus S, Lau HC. Augmenting decisions of taxi drivers through reinforcement learning for improving revenues. In: Proceedings international conference on automated planning and scheduling, ICAPS; 2017.

[13] Dickerson J, Sankararaman K, Srinivasan A, Xu P. Allocation problems in ride-sharing platforms: Online matching with offline reusable resources. In: Proceedings of the AAAI conference on artificial intelligence (Vol. 32, No. 1); 2018, April.

[14] Qin Z, Tang X, Jiao Y, Zhang F, Xu Z, Zhu H, Ye J. Ride-hailing order dispatching at DiDi via reinforcement learning. INFORMS Journal on Applied Analytics 2020;50(5):272–86.

[15] Shou Z, Di X, Ye J, Zhu H, Zhang H, Hampshire R. Optimal passenger-seeking policies on E-hailing platforms using Markov decision process and imitation learning. *Transportation Research Part C: Emerging Technologies* 2020. https://doi.org/10.1016/j.trc.2019.12.005.

[16] Wang Y, Zhang D, Hu L, Yang Y, Lee LH. A data-driven and optimal bus scheduling model with time-dependent traffic and demand. IEEE Transactions on Intelligent Transportation Systems 2017;18(9):2443–52. https://doi.org/10.1109/TITS.2016.2644725.

[17] Enrique FL, de Cea Ch J, Malbran RH. Demand responsive urban public transport system design: methodology and application. Transportation Research Part A: Policy and Practice 2008;42(7):951–72. https://doi.org/10.1016/j.tra.2007.12.008.

[18] Shariat M, Amirpour S. Creating bus timetables under stochastic demand. International Journal of Industrial Engineering & Production Research 2009;2009:83–91.

[19] Sun L, Lee DH, Erath A, Huang X. Using smart card data to extract passenger's spatio-temporal density and train's trajectory of MRT system. In: Proceedings of the ACM SIGKDD international conference on knowledge discovery and data mining, 12; 2012. p. 142–8. https://doi.org/10.1145/2346496.2346519.

[20] Sun L, Jin JG, Lee DH, Axhausen KW, Erath A. Demand-driven timetable design for metro services. Transportation Research Part C: Emerging Technologies 2014;46:284–99. https://doi.org/10.1016/j.trc.2014.06.003.

[21] Kloimüllner C, Papazek P, Raidl G, Hu B, Raidl GR. Balancing bicycle sharing systems: an approach for the dynamic case planning bicycle sharing systems view project. In: Balancing bicycle sharing systems: an approach for the dynamic case, 8600. Springer; 2014. p. 73–84.

[22] Raidl G, Papazek P, Raidl GR, Hu B, Rainer-Harbach M. Balancing bicycle sharing systems: improving a VNS by efficiently determining optimal loading operations NEEMO networking for excellence in electric mobility operations view project. In: Balancing bicycle sharing systems: improving a VNS by efficiently determining optimal loading operations. Springer, 7919 LNCS; 2013. p. 130–43. https://doi.org/10.1007/978-3-642-38516-2_11.

[23] Rainer-Harbach M, Papazek P, Hu B, Raidl GR. Balancing bicycle sharing systems: a variable neighborhood search approach. In: Lecture notes in computer science (including subseries

lecture notes in artificial intelligence and lecture notes in bioinformatics), 7832 LNCS; 2013. p. 121–32. https://doi.org/10.1007/978-3-642-37198-1_11.

[24] Contardo C, Morency C, Rousseau L-M. Balancing a dynamic public bike-sharing system; 2012. Bureaux de Montréal/Bureaux de Québec. www.cirrelt.ca.

[25] Schuijbroek J, Hampshire RC, van Hoeve WJ. Inventory rebalancing and vehicle routing in bike sharing systems. European Journal of Operational Research 2017;257(3):992–1004. https://doi.org/10.1016/j.ejor.2016.08.029.

[26] Pfrommer J, Warrington J, Schildbach G, Morari M. Dynamic vehicle redistribution and online price incentives in shared mobility systems. IEEE Transactions on Intelligent Transportation Systems 2014;15(4):1567–78. https://doi.org/10.1109/TITS.2014.2303986.

[27] Ho CQ, Mulley C, Hensher DA. Public preferences for mobility as a service: insights from stated preference surveys. Transportation Research Part A: Policy and Practice 2020. https://doi.org/10.1016/j.tra.2019.09.031.

[28] Jittrapirom P, Caiati V, Feneri AM, Ebrahimigharehbaghi S, Alonso González MJ, Narayan J. Mobility as a service: a critical review of definitions, assessments of schemes, and key challenges; 2017.

[29] Hensher DA, Ho CQ, Reck DJ. Mobility as a service and private car use: evidence from the Sydney MaaS trial. Transportation Research Part A: Policy and Practice 2021;145:17–33.

[30] Ho CQ, Hensher DA, Reck DJ. Drivers of participant's choices of monthly mobility bundles: key behavioural findings from the Sydney Mobility as a Service (MaaS) trial. Transportation Research Part C: Emerging Technologies 2021. https://doi.org/10.1016/j.trc.2020.102932.

[31] Matyas M, Kamargianni M. The potential of mobility as a service bundles as a mobility management tool. Transportation 2019;46(5):1951–68.

[32] Cottrill CD. MaaS surveillance: privacy considerations in mobility as a service. Transportation Research Part A: Policy and Practice 2020;131:50–7. https://doi.org/10.1016/j.tra.2019.09.026.

[33] Expósito-Izquierdo C, Expósito-Márquez A, Brito-Santana J. Mobility as a Service. In: Smart cities: foundations, principles, and applications; 2017. https://doi.org/10.1002/9781119226444.ch15.

[34] Giesecke R, Surakka T, Hakonen M. Conceptualising mobility as a service. In: 2016 eleventh international conference on ecological vehicles and renewable energies (EVER). IEEE; 2016, April. p. 1–11.

[35] Goodall W, Bornstein J, Fishman TD, Bonthron B. The rise of mobility as a service; 2017. Deloitte Review.

[36] Ho CQ, Hensher DA, Mulley C, Wong YZ. Potential uptake and willingness-to-pay for Mobility as a Service (MaaS): a stated choice study. Transportation Research Part A: Policy and Practice 2018.

[37] Jittrapirom P, Marchau V, van der Heijden R, Meurs H. Future implementation of mobility as a service (MaaS): RESULTS of an international Delphi study. Travel Behaviour and Society 2018. https://doi.org/10.1016/j.tbs.2018.12.004.

[38] Wong YZ, Hensher DA, Mulley C. Mobility as a service (MaaS): charting a future context. Transportation Research Part A: Policy and Practice 2020;131:5–19. https://doi.org/10.1016/j.tra.2019.09.030.

[39] Xi H, Liu W, Rey D, Waller ST, Kilby P. Incentive-compatible mechanisms for continuous resource allocation in mobility-as-a-service: Pay-as-You-Go and Pay-as-a-Package. ArXiv 2020. Preprint ArXiv:2009.06806.

[40] Merkert R, Bushell J, Beck MJ. Collaboration as a service (CaaS) to fully integrate public transportation—lessons from long distance travel to reimagine mobility as a service. *Transportation Research Part A: Policy and Practice* 2020. https://doi.org/10.1016/j.tra.2019.09.025.

[41] Mulley C, Ho C, Balbontin C, Hensher D, Stevens L, Nelson JD, Wright S. Mobility as a service in community transport in Australia: can it provide a sustainable future? Transportation Research Part A: Policy and Practice 2020;131:107–22.

[42] Seiberth G, Gruendinger W. Data-driven business models in connected cars. In: Mobility services and beyond (Vol. 18, Issue 01). BVDW Research; 2018.

[43] Barreto L, Amaral A, Baltazar S. Urban mobility digitalization: towards mobility as a service (MaaS). In: 9th international conference on intelligent systems 2018: theory, research and innovation in applications, IS 2018—proceedings; 2018. https://doi.org/10.1109/IS.2018.8710457.

[44] Cooper P, Tryfonas T, Crick T, Marsh A. Electric vehicle mobility-as-a-service: exploring the "Tri-Opt" of novel private transport business models. Journal of Urban Technology 2019;26(1):35–56.

[45] Franco P, Johnston R, McCormick E. Demand responsive transport: generation of activity patterns from mobile phone network data to support the operation of new mobility services. Transportation Research Part A: Policy and Practice 2020. https://doi.org/10.1016/j.tra.2019.09.038.

[46] Schikofsky J, Dannewald T, Kowald M. Exploring motivational mechanisms behind the intention to adopt mobility as a service (MaaS): insights from Germany. Transportation Research Part A: Policy and Practice 2020. https://doi.org/10.1016/j.tra.2019.09.022.

[47] Polydoropoulou A, Pagoni I, Tsirimpa A, Roumboutsos A, Kamargianni M, Tsouros I. Prototype business models for Mobility-as-a-Service. Transportation Research Part A: Policy and Practice 2020;131:149–62. https://doi.org/10.1016/j.tra.2019.09.035.

[48] Reck DJ, Axhausen KW. How much of which mode? Using revealed preference data to design mobility as a service plans. Transportation Research Record 2020;2674(7):494–503. https://doi.org/10.1177/0361198120923667.

[49] Ambrosino G, Nelson JD, Boero M, Pettinelli I. Enabling intermodal urban transport through complementary services: from flexible mobility services to the shared use mobility agency: workshop 4. Developing inter-modal transport systems. Research in Transportation Economics 2016;59:179–84.

[50] Gould E, Wehrmeyer W, Leach M. Transition pathways of e-mobility services. WIT Transactions on Ecology and the Environment 2015;194:349–59.

[51] Hensher DA. Future bus transport contracts under a mobility as a service (MaaS) regime in the digital age: Are they likely to change? Transportation Research Part A: Policy and Practice 2017. https://doi.org/10.1016/j.tra.2017.02.006.

[52] Xi H, Aussel D, Liu W, Waller ST, Rey D. Single-leader multi-follower games for the regulationof two-sided Mobility-as-a-Service markets. ArXiv 2021. Preprint ArXiv:2106.02151.

[53] Funaro L, Ben-Yehuda OA, Schuster A. Stochastic resource allocation. In: VEE 2019—Proceedings of the 15th ACM SIGPLAN/SIGOPS international conference on virtual execution environments; 2019. p. 122–36.

[54] Li X, Ye Y. Online linear programming: dual convergence, new algorithms, and regret bounds. arXiv 2019;preprint arXiv:1909.05499.

[55] Asadpour A, Wang X, Zhang J. Online resource allocation with limited flexibility. Management Science 2020;66(2):642–66.

[56] Chen G, Li X, Ye Y. How linear reward helps in online resource allocation. ArXiv 2021. Preprint ArXiv 2101.11092.

[57] Verweij B, Ahmed S, Kleywegt AJ, Nemhauser G, Shapiro A. The sample average approximation method applied to stochastic routing problems: a computational study. Computational Optimization and Applications 2003;24(2):289–333. https://doi.org/10.1023/A:1021814225969.

[58] Kim S, Pasupathy R, Henderson SG. A guide to sample average approximation. Handbook of Simulation Optimization 2015;207–43.
[59] Bertsimas D, Gupta V, Kallus N. Robust sample average approximation. Mathematical Programming 2018;171(1):217–82.
[60] Kleywegt AJ, Shapiro A, Homem-de-Mello T. The sample average approximation method for stochastic discrete optimization. SIAM Journal on Optimization 2002;12(2):479–502.
[61] Lagoudakis MG, Parr R. Least-squares policy iteration. Journal of Machine Learning Research 2004;4:1107–49.
[62] Mahadevan S. Learning representation and control in markov decision processes. Now Publishers Inc.; 2009.
[63] Michels J, Saxena A, Ng AY. High speed obstacle avoidance using monocular vision and reinforcement learning. In: ICML 2005—proceedings of the 22nd international conference on machine learning; 2005. p. 593–600.
[64] Ye Y. Data-driven optimization. In: Powerpoint slides. Stanford University; 2014.
[65] Arrow K, Karlin S, Scarf H. Studies in the mathematical theory of inventory and production. Stanford University Press; 1958.
[66] Bertsimas D, Popescu I. Optimal inequalities in probability theory: a convex optimization approach. SIAM Journal on Optimization 2005;15(3):780–804. https://doi.org/10.1137/S1052623401399903.
[67] Dupačová J. The minimax approach to stochastic programming and an illustrative application. Stochastics 1987;20(1):73–88.
[68] Prékopa A. Programming under probabilistic constraint and maximizing probabilities under constraintsss. In: Stochastic programming; 1995. p. 319–71. https://doi.org/10.1007/978-94-017-3087-7_11.
[69] Van Parys BPG, Esfahani PM, Kuhn D. From data to decisions: distributionally robust optimization is optimal. Management Science 2020.
[70] Wang Z, Glynn PW, Ye Y. Likelihood robust optimization for data-driven problems. Computational Management Science 2016;13(2):241–61. https://doi.org/10.1007/s10287-015-0240-3.
[71] Delage E, Ye Y. Distributionally robust optimization under moment uncertainty with application to data-driven problems. Operations Research 2010;58(3):596–612. https://doi.org/10.1287/opre.1090.0741.
[72] Bertsimas D, Kallus N. From predictive to prescriptive analytics. Management Science 2020;66(3):1025–44. https://doi.org/10.1287/mnsc.2018.3253.
[73] Punia S, Singh SP, Madaan JK. From predictive to prescriptive analytics: a data-driven multi-item newsvendor model. Decision Support Systems 2020;136:113340. https://doi.org/10.1016/j.dss.2020.113340.
[74] Ban GY, Rudin C. The big data newsvendor: practical insights from machine learning. Operations Research 2019;67(1):90–108.
[75] Oroojlooyjadid A, Snyder LV, Takáč M. Applying deep learning to the newsvendor problem. IISE Transactions 2020;52(4):444–63. https://doi.org/10.1080/24725854.2019.1632502.
[76] Huber J, Müller S, Fleischmann M, Stuckenschmidt H. A data-driven newsvendor problem: from data to decision. European Journal of Operational Research 2019;278(3):904–15. https://doi.org/10.1016/j.ejor.2019.04.043.
[77] Alon I, Qi M, Sadowski RJ. Forecasting aggregate retail sales: a comparison of artifcial neural networks and traditional methods. Journal of Retailing and Consumer Services 2001;8(3):147–56.

[78] Bello I, Pham H, Le QV, Norouzi M, Bengio S. Neural combinatorial optimization with reinforcement learning. In: 5th international conference on learning representations, ICLR 2017—workshop track proceedings; 2019.

[79] Nazari M, Oroojlooy A, Snyder LV, Takáč M. Reinforcement learning for solving the vehicle routing problem. arXiv 2018. preprint arXiv:1802.04240.

[80] Oda T, Joe-Wong C. Movi: a model-free approach to dynamic fleet management. In: IEEE INFOCOM 2018-IEEE conference on computer communications. IEEE; 2018, April. p. 2708–16.

[81] Sutton RS, Barto AG. Reinforcement learning: an introduction. MIT Press; 2018.

[82] Angalakudati M, Balwani S, Calzada J, Chatterjee B, Perakis G, Raad N, Uichanco J. Business analytics for flexible resource allocation under random emergencies. Management Science 2014;60(6):1552–73.

[83] Besbes O, Gur Y, Zeevi A. Optimization in online content recommendation services: beyond click-through rates. Manufacturing and Service Operations Management 2016;18(1):15–33.

[84] Chan CW, Green LV, Lu Y, Leahy N, Yurt R. Prioritizing burn-injured patients during a disaster. Manufacturing & Service Operations Management 2013;15(2):170–90.

[85] Cohen MC, Leung NHZ, Panchamgam K, Perakis G, Smith A. The impact of linear optimization on promotion planning. Operations Research 2017;65(2):446–68. https://doi.org/10.1287/opre.2016.1573.

[86] Mehrotra M, Dawande M, Gavirneni S, Demirci M, Tayur S. Or practice—production planning with patterns: a problem from processed food manufacturing. Operations Research 2011;59(2):267–82.

[87] Mišić VV, Perakis G. Data analytics in operations management: a review. Manufacturing and Service Operations Management 2020;22(1):158–69. https://doi.org/10.1287/msom.2019.0805.

[88] Elmachtoub AN, Grigas P. Smart "predict, then optimize". Management Science 2021. https://doi.org/10.1287/mnsc.2020.3922. In press.

[89] Zhu T, Xie J, Sim M. Joint estimation and robustness optimization. Management Science 2021;1–19. https://doi.org/10.1287/mnsc.2020.3898.

[90] Yu Q, Zhang H, Li W, Song X, Yang D, Shibasaki R. Mobile phone GPS data in urban customized bus: dynamic line design and emission reduction potentials analysis. Journal of Cleaner Production 2020;272:122471. https://doi.org/10.1016/j.jclepro.2020.122471.

[91] Zhang H, Song X, Xia T, Yuan M, Fan Z, Shibasaki R, Liang Y. Battery electric vehicles in Japan: human mobile behavior based adoption potential analysis and policy target response. Applied Energy 2018;220:527–35. https://doi.org/10.1016/j.apenergy.2018.03.105.

[92] Xi H, He L, Zhang Y, Wang Z. Differentiable road pricing for environment-oriented electric vehicle and gasoline vehicle users in the bi-objective transportation network. Transp Lett 2021;1–15. https://doi.org/10.1080/19427867.2021.1919468.

[93] Punia S, Singh SP, Madaan JK. From predictive to prescriptive analytics: a data-driven multi-item newsvendor model. Decis Support Syst 2020;113340. https://doi.org/10.1016/j.dss.2020.113340.

[94] Lu W, Yu Y, Chang Y, Wang Z, Li C, Yuan B. A dual input-aware factorization machine for CTR prediction. IJCAI 2020;3139–45.

[95] Yu Y, Wang Z, Yuan B. An Input-aware Factorization Machine for Sparse Prediction. IJCAI 2019;1466–72.

Chapter 6

Data-driven estimation for urban travel shareability

Qing Yu, Weifeng Li, and Dongyuan Yang
Key Laboratory of Road and Traffic Engineering of the Ministry of Education, Tongji University, Shanghai, China

1 Introduction

1.1 The emergence of sharing transportation

With the continuous planning, development, and construction, the area of the city is growing and gradually expands to the surrounding areas. In the past decade, new types of "Urban Disease" appear in most of the large cities all over the world, including population expansion, traffic congestion, environmental deterioration, housing shortage, employment difficulties, and so on. Urban transportation is an important part of the urban system. On one hand, it is closely related to the liveability of the city and the wellbeing of residents; on the other hand, it produces "Urban Diseases" including emissions and traffic congestions. How to effectively utilize the limited traffic resources to provide convenient and efficient public transportation for urban residents has become an important issue that earns great attention from policymakers, transportation modelers, researchers, and a wide range of stakeholders.

In human society, the limited access to resources forces us to share the resource with others. Sharing behavior has been proved to be the history of several thousand years and is called "the most universal form of human economic behavior" [1]. Transportation can also be shared. In the modern era, the Internet and big data technology give rise to the new form of Internet-based sharing transportation modes. Only through a simple operation on their mobile phone applications can users easily access transportation services and find an appropriate match for sharing in realtime. In the development of the next-generation smart cities, shared transportation is considered to be the mode with great potential to relieve urban disease and, currently, it presents a trend of rapid growth.

Big Data and Mobility as a Service. https://doi.org/10.1016/B978-0-323-90169-7.00007-5

1.2 The significance of shareability estimation

However, although sharing modes seem to be promising in decreasing the urban congestion, fuel consumption, and emissions generated by transportation, some researchers are skeptical about it [2]. The system layout can be a crucial part of designing a sharing transportation mode and it can directly affect the performance and public acceptance of this mode. The inappropriate design of the sharing transportation system may lead to new urban diseases [3, 4]. For example, in the promotion of bicycle-sharing mode, serious problems have emerged including abandoned bicycles, streets blocked by random parking, unbalance between demand and supply that causing undersupply and oversupply. Another example is the carsharing mode; although it can undoubtedly reduce the number of cars on the street, it may force the drivers to detour in order to pick up extra passengers causing new congestions.

Urban travel shareability that can be interpreted as the capability of the urban sharing transportation system has become a significant concept. The estimation of shareability not only includes predicting the future demand but also includes the task of the assessment of the potential performance of the sharing transportation mode. How to estimate the shareability in a city? How to design the layout of the sharing system and the corresponding operation strategies in order to improve shareability? When shareability is improved, what impact will it have on the urban system? These are crucial questions to the planning and designing of the sharing travel system in the next-generation smart city.

In recent years, data generated by human beings began to grow explosively. The new types of geospatial-based big data involving human activity have been generated and collected. These data are called "geospatial big data" or "track and trace data" [5], which generally contain three levels of information: who, when, and where. These high-quality, pervasive, and detailedly recorded geospatial data allow us to fully capture the daily activity and travel for each individual in the city in both temporal and spatial dimensions. The emergence of big data also offers a new way and has triggered the opportunity to estimate the potential demand and performance of the sharing transportation mode.

1.3 Chapter organization

In this chapter, we seek to address-specific research questions regarding data-driven estimation for urban travel shareability:

- (Q1) How can data-driven research overcome the limitation of traditional data and methodologies in shareability estimation?
- (Q2) What are the key methodologies and technologies? How to properly apply them?
- (Q3) What are the opportunities and challenges in the data-driven shareability estimation approaches?

The rest of this chapter is organized as follows: Section 2 introduces several existing types of sharing transportation mode and their operation strategies. Sections 3 and 4 address the Q1. In Section 3, we describe the traditional types of data in sharing transportation estimation and summarize their limitations. In Section 4, we introduce the new types of data and their advantages with the comparison to the traditional data. Sections 5 and 6 address the Q2. In Section 5, we establish the potential of the Agent-Based Model (ABM) towards enhancing sharing transportation modeling, data-driven estimation of shareability, and policy making. In Section 6, we proposed two examples about the shareability estimation for bike sharing and taxi sharing, the two examples utilize ABM in two different ways. Section 7 addresses the Q3 by proposing the opportunities and challenges in this field. Finally, we draw our conclusions and summarized our response to all research questions in light of our discussions in Section 8.

2 Emerging sharing transportation mode

In this section, we will describe the operation mechanism of three types of emerging sharing transportation (Fig. 1): bicycle sharing, ride sharing, and taxi

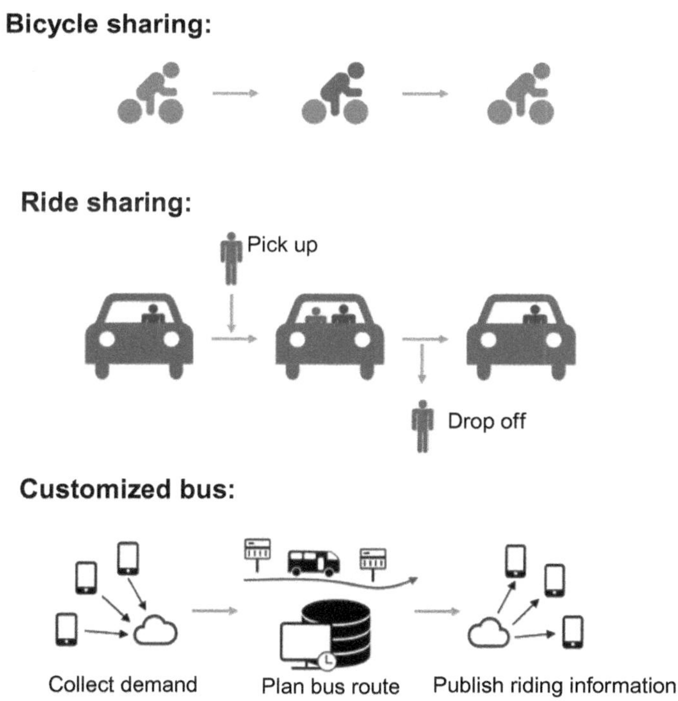

FIG. 1 Emerging sharing transportation modes.

sharing, and customized bus. In recent years, these three modes have appeared to become popular rapidly; they all have successful operation cases that won public acceptance.

2.1 Bicycle sharing

Bicycle sharing is one of the Internet-supported transportation modes that emerged and gradually become warmly welcome for a great number of cities in recent years. In some cities, they are electric scooters sharing with a similar operation mode. As an alternative to motorized public transportation for short-distance trips, bicycle sharing is widely adopted as a green travel mode to mitigate urban emissions [6]. Depending on whether the bicycle docking stations are needed, bicycle-sharing system is divided into a dock-based bicycle-sharing system and a dockless bicycle-sharing system. In the past decade, the number of cities operating a bicycle sharing system has increased from 13 in 2004 to 855 as of 2014, the global bicycle-sharing fleet is estimated at 946,000 bicycles, of which 750,500 are in China [7]. Mobike is one of the most popular bicycle-sharing companies in China. Using Mobike's bicycle smartphone application, users can check the bicycle location, make an appointment and find the bicycle with their mobile phone. Mobike abandons fixed bicycle-sharing dock stations and allows users to park their bikes at any side of the road where there is a government-drawn parking area [8].

In the recent study of bicycle-sharing systems, big data analysis methodologies are widely considered as powerful tools for improving the bicycle-sharing system. Based on massive human mobility data, dozens of researches focus on bicycle-sharing demand prediction, system layout design, and optimization, and improve system attractiveness based on user preference.

Bicycle-sharing demand prediction focuses on efficiently expressing and extracting people's movement from massive and large-scale data, and predicts bicycle-sharing potential. The existing bicycle-sharing demand predicting researches are focusing on three levels [9]: city level, cluster level, and station level. Researches dedicate to forecasting the hourly or daily bicycle-sharing demand. More recently, deep learning models have been widely used to solve spatiotemporal transportation demand prediction. Such methods include ridge regression, AdaBoost regression, support vector regression, random forecast tree and gradient boosting regression tree, convolutional neural network (CNN), graph CNN (GCNN), etc. [10–15].

For bicycle-sharing system layout, the design of bicycle-sharing system layout includes the siting of bicycle stations, the management policy, the subdivision of coverage area, and dispatching and rationing strategy [3, 4, 16–20]. The existing methods can be divided into two types: mathematical programming models and GIS-based methods. The mathematical model approach usually develops mixed-integer linear programming (MILP) or mixed-integer nonlinear programming (MINLP) models to plan a new bicycle-sharing system.

The GIS-based approach makes use of GIS as a tool to measure potential demand and optimize bicycle facilities in bicycle-sharing systems.

By examining the bicycle-sharing behavior with different user frequency and trip purposes, researchers also seek for key elements of users choosing bicycle sharing as their transportation mode [21–24]. The most frequently mentioned factors include convenience, cycling environment, location of the docking stations, money-saving, etc. Such research focuses on improving the attractiveness of urban bicycle-sharing systems to enhance system performance.

2.2 Ride sharing and taxi sharing

As another popular sharing transportation concept, ride sharing is a mode that combining two or more rider groups to travel together in the same vehicle riding trip. And if the vehicle is a taxi, it becomes the concept of taxi sharing [25, 26]. For ride sharing, to match up the sharing trips, the car owner will pick up and drop-off sharing passengers on his travel route. But for taxi sharing, taxis do not have to start from a certain origin or destination, drivers can schedule the pick-up and drop-off sequence of the passengers to better utilize their taxi resources. This novel form of taxi service is widely regarded as a solution to limited taxi resources in urban transportation and a way to relieve pressure on the transportation system [27].

With the recent emerging of Internet-based smartphone application platforms, the exchange of realtime user demand information enables to promote the taxi ride sharing. Ride-sharing trips mainly refer to two types of traveling modes: end-to-end ride sharing (ride sharing of users with similar origins and destinations) and en-route ride sharing (ride sharing of users sharing portions of their paths between dissimilar origins and/or destinations) [2]. In the emerging Internet-based taxi sharing service, users who want to share trips will allow the taxi sharing system to match the other travelers with similar travel demands, and to pick up or drop-off travelers along the travel trip, without restricting the sharing trips to have the same origins or destinations [28].

The matching algorithm of taxi sharing trips is an important topic in this field. Current methodologies can be divided into two aspects: static taxi sharing and dynamic taxi sharing. The static method is to match sharing trips on the preknow trips and the dynamic method is to match on realtime with running taxis [25, 29, 30]. However, the existing matching systems are facing the following week's points including high computing costs with slow matching speed, long detour distance against driver's behavior [31], and lack of economical consideration [32].

There are also several concerns about taxi sharing. Firstly, to what extent the taxi sharing service will relieve the traffic pressure and mitigate traffic emissions in the urban area is still a question. To answer this, the most common way is to establish the model of travel demand based on big data, to match the taxi ride-sharing trips on temporal and spatial aspects, and to evaluate the impact of taxi sharing service on urban congestion and emission [33, 34].

2.3 Customized bus

Customized bus (CB) is a new type of Internet-supported transportation mode that emerged and was warmly applauded by the public in recent years. Comparing with traditional bus lines, a CB system requires passengers to submit their origin, destination, and departure time via the Internet to reserve seats. The design of CB lines emphasizes efficiency and directness with few or no intermediate stops along the route [35].

E-drive is a CB company committed to exploring and implementing CB service in Shanghai. E-drive offers several types of CB service including long-term CB service called micro railway, CB service for parking areas, CB service for temporary leases, special customized line, etc. The operation procedures of E-drive are Internet based. Customers can submit their demands including the information of their origins, destinations, and departure times via online platforms such as websites and mobile phone apps. CB company will match the demand and plan bus lines for customers with similar travel demands. In their micro railway service, E-drive divides the area of Shanghai into 11 subareas and offering 45 CB lines to connect between subareas. Customers can check the existing CB lines and stations, bus timetables, realtime bus GPS location, submitted orders, and reserved bus seats via online platforms.

2.4 Characteristics of sharing transportation modes

The sharing transportation modes mentioned above have their common characteristics. The opportunity for their birth and development is the Internet-based mobile application and big data technology. In urban transportation systems, they are all playing different roles and are designed as the supplement of the existing public transportation system: For example, bicycle sharing is mainly designed to solve the "last mile problem" connecting people with the transit system; Ride sharing is designed to solve the travel mode of low price and high comfort and taxi sharing is the supplement of urban taxi resources; customized bus is a supplement to the existing public transit network. The recent fast development of these sharing transportation modes largely relies on Internet-based mobile applications. All these sharing modes require operators to provide and allocate transportation resources and matching services for them to facilitate the share trips. Therefore, the design and operation strategy of the service mechanism will directly affect the performance of the system.

Sharing travel involves the interaction behavior among multiple agent groups, including operators, traffic resources, and users. Therefore, in data-driven shareability estimation, the interaction between multiple agents is the most important point and should be fully considered.

3 Background to traditional data and their limitations

The data used in traditional traffic demand estimation are usually aggregated and statistical data, which are collected by government institutes or private researchers. These data include census, travel survey, safety accident data, traffic volume data, etc.

Travel survey data is one of the most frequently used data in the transportation field. These data are usually actively collected with the respondents agreeing to take the survey, which can collect detailed personal information and user preference. Some cities have regular census and resident travel surveys, which are time and manpower consuming. Due to the cost, government-dominated surveys are usually conducted with an interval of years. It is difficult to produce long-term observation data when collecting individual mobility through survey data, and it is also difficult for individuals to fully record the specific details of each trip in the questionnaire.

The goal of transportation planning has changed in recent years. Under the new concept of shared transportation, transportation planning of the next-generation smart cities is no longer focused on predicting the traffic volume or capturing the relationship between traffic OD and urban infrastructure, but in pursuit of livability, sustainability, shareability, social equity, public health, environmental quality, carbon-neutral, etc.

However, the traditional transportation model and data can hardly support these new development goals. For traditional data, they are highly aggregated without the precise travel path of the users. The data collection costs a great amount of manpower and material resources, which decides that it will be difficult to obtain large-scale and long-time observation data. For traditional models, they are mostly established on traditional aggregated data. The travel behavior of individual travelers is often considered separately and independently. The traditional analysis methods offer limited information to have a clear grasp of the travel characteristics and internal mechanism of individual travel behavior.

When considering sharing transportation, the sharing behavior is always generated by the interaction between multiple travelers. In the face of the change of demand, the traditional traffic system analysis theory has begun to expose its limitations in practical work. Traditional analysis models are difficult to adapt into sharing transportation scenarios and can hardly provide support for the designing and planning of next-generation traffic systems.

4 New and emerging source of data

In recent years, with the popularity and development of powerful hardware, an increasing number of smart mobile devices are used to collect and process data. New types of data are produced and collected, which have great potential to be utilized in data-driven researches in the transportation field. These emerging

FIG. 2 Multisource geospatial big data.

sources of data, together with the traditional data, have constructed the multi-source database to offer detailed travel behavior information of travelers (Fig. 2). Here in this section, we will describe the two types of newly emerged data form: Track and trace data and geographic information data.

4.1 Track and trace data

The concept of track and trace (T&T) refers to spatiotemporal data sets of individual travel [5]. T&T data consists of three important elements: Who? When? And where? Through continuous observation of individuals, T&T data can capture the travel chain of individuals. Here, we will introduce several types of commonly used T&T data and their characteristics.

4.1.1 Mobile phone data

There are generally two types of mobile data providers: mobile signal operators and mobile application operators. According to different data providers, mobile phone data have different characteristics. The data coverage of mobile signaling data provided by signal operators is usually more extensive. The data usually contains the basic information of the wireless communication between mobile stations and base transceiver stations (BTS), including the encrypted mobile phone identifier, the service time, the service type, the geographic location of the connected BTS, and the location area (LA). For this type of data, the position of users will be represented by the location of the BTS they are connecting, which will lead to distance error. The data provided by mobile application operators are limited to mobile phone users using a specific application. Compared with signaling data, this kind of data can provide more abundant information and a more accurate geographical location.

4.1.2 Smart card data

Smart card data is provided by public transportation system operators, which often includes the time and place of ticketing transactions (the place is represented by bus or metro line and station), card type (ordinary card, student card or staff card, etc.), transaction number, driver number, and vehicle number, etc. In some cities, the card data includes both metro and bus transit transactions. For bus transit, the smart card data needs to fuse with the bus GPS data to obtain the boarding and alighting station, which highly depend on the data quality and the sampling interval of bus GPS.

4.1.3 Taxi GPS data

Taxi GPS trajectories data are collected and provided by the taxi company. Apart from taxi vehicle ID, time, and coordinates, taxi GPS data often includes occupancy status (with or without passengers), speed, direction angle, and other information. The pick-up/drop-off time and location of the taxi trips can be inferred from the occupancy status. Taxi data can not only reflect the urban taxi resources and travel demand but also contain the traffic status of the city.

4.1.4 Bicycle-sharing data

The bicycle-sharing data can be divided into two types: bicycle order data and bicycle trajectory data. The data quality of bicycle-sharing data varies from the different bicycle-sharing companies. Bicycle order data only generate records when locking or unlocking the bicycle. Bicycle trajectory data will contain the full trajectory of each bicycle trip. The bicycle-sharing data is usually analyzed from two aspects: parking and traveling. However, bicycles not in use may not be recorded in the data, which may potentially lead to the underestimation of parking demand.

4.2 Geographic information data

In recent years, Internet-based, location-based services (LBS) generate a new form of geographic information data. LBS is a new concept that denotes applications integrating geographic location with the general notion of services [36, 37]. Such applications are mostly offered by mapping companies like Google Maps, Apple Maps, Baidu Maps, etc. We categorize these geographic information data into four types: "transportation network," "vector data," "point of interest" and "navigation data." In the research of spatial–temporal big data, the most important element of track and trace data is spatial information. The data fusion of spatial information in track and trace data and geographic information data can provide us with more detailed information about human activities.

4.2.1 Transportation network

Transportation network data is a type of link model abstracted from the real-world transportation system, which includes the road network, metro network, bus transit network, etc. The basic structure of a transportation network consists of points (intersections, stations, terminals, etc.) and edges (road sections, bus routes, metro lines, etc.). With the corresponding attributes and connections between elements, transportation network data can show the detailed topology and connectivity of the transportation system.

4.2.2 Vector data

Geographic vector data is the most commonly used data type in studying geographic areas. These data are usually collected and measured by remote sensing images obtained by remote sensing [38]. Geographic vector data provides information on the land cover including the urban area, forest, mountain, water, administration division, traffic analysis zone (TAZ), etc.

4.2.3 Point of interest data

Point of interest (POI) refers to the point data in the Internet-based map service, which usually includes four attributes: name, address, coordinate, and category. In this type of data, the urban facilities in the real world are abstracted as point elements for the convenience of analysis. However, in the process of abstracting 3D objects into data points, the POI data will inevitably lose some of the spatial information. To address the weak point of POI, the region of interest (ROI) or area of interest (AOI) data are also introduced, which are the POI data with spatial polygon geometry information.

4.2.4 Navigation data

Navigation data is also a data type provided by the LBS service, which is built on the transportation network of map suppliers. By submitting the information of origin and destination, transportation mode, and other related pieces of information, LBS service can return the corresponding navigation path planning information, which often includes: geographic coordination of travel path, travel time, travel cost, etc. Some LBS services also provide the isochrone, also known as the reachable circle [39]. The isochrone tool provides polygon or line features that show areas reachable within a specified amount of time from a given location.

4.3 Advantages and disadvantages of new data sources

Comparing with the traditional types of data, the advantages of the new data are obvious. New sources of data can obtain samples with a high sampling rate at a low cost. The data set can completely capture the individual movement with a larger and more comprehensive range of time and space. Multisource data fusion can also provide a comprehensive understanding of human mobility.

The new sources of data also have the following weaknesses:

- Sampling problem: Most of these data sources are passively collected. The data contains only the user group of the data provider, which hardly guaranteed uniform sampling and may lead to deviation. The result generated from new data needs to be checked and validate with traditional data.
- Privacy issue: These data contain massive private information from users. The abuse of the data may raise public consideration of the privacy policy.
- Lack of travel purposes: The capture of user mobility by new data often focuses on time and space. However, the new data cannot capture the reasons and purposes of travel behavior. In this aspect, the traditional survey data can be a supplement.

5 Emerging form of key technologies

5.1 Agent-based modeling

Traditional traffic models are based on the hypothesis that the travel between individuals is independent of each other. However, there are certain drawbacks associated with the use of traditional models. In the context of urban travel shareability, users claim no possession of vehicles offered by operators. In opposite, users only accept the transport services provided by operators. When operators implement a specific operation strategy, users make decisions under the rules of operation strategy. Consequently, the nonlinear superposition of a large number of individual behaviors will lead to the high complexity of the system. So the interaction between individuals should be taken into account during the shareability modeling (Fig. 3). With the increasing prominence of the large-scale dataset, ABMs, which enable to simulate the individual actions of diverse agents and measure the resulting system tend to be the effective approach in urban simulation [40].

ABMs, also known as multiagent systems (MAS), have been used in many fields, such as pollution exposure estimation [41, 42], residential demand

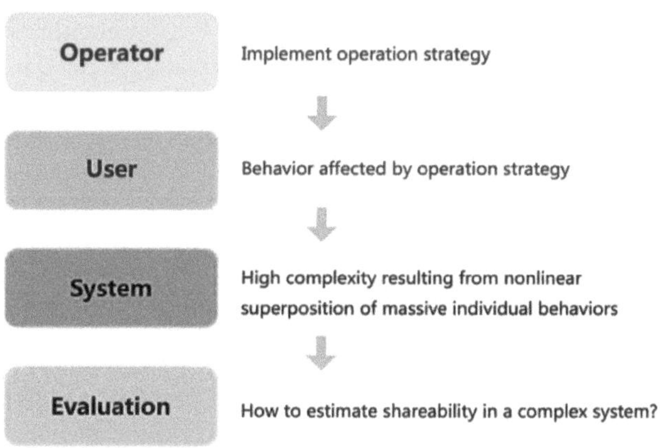

FIG. 3 Motivation of agent-based modeling.

modeling [43], and infrastructure investment decision [44]. There is no universal definition of the term 'agent'[45], and to the best of our knowledge, the word **agent** usually refers to an individual (humans, cities, and other abstract representations) who acts according to specific rules in a system. Several features are common to most agents, such as autonomy, heterogeneity, and activity. Especially, if agents are heterogeneous, they are called the heterogeneous agent model (HAM).

In the macrolevel model, both supply and demand sides are considered. If both sides are represented by a typical function or can be divided into a limited class of economic participants who can be discussed simply, then this model can be considered as a typical representative participant's model. In other words, such models aggregate individual entities into representative entities. In contrast, HAMs expand the representative participants of both supply and demand sides from limited to unlimited. For example, in the most basic economic model, it is assumed that there is a representative manufacturer for production and a representative consumer for consumption and savings. When HAM is introduced, the assumption will be changed accordingly. The producer could be composed of different manufacturers and the consumers could have different characteristics.

As microlevel models, ABMs simulate the simultaneous action and interaction of multiple agents to reproduce and predict complex phenomena from microlevel to macrolevel. Therefore, the key of this model is that simple behavior rules can produce complex system performance, which is the KISS principle (keep it simple, stupid) widely adopted in the modeling field. Generally speaking, independent individuals are bounded rationally. Assuming that they act for personal interests, such as reproduction, positive interests, or social status, and can only make decisions through exploratory or simple decision-making rules, agents may experience the process of learning, adaptation, and reproduction in ABM.

The vast majority of ABMs include: (1) a large number of agents distinguished by various indicators (individual granularity); (2) exploratory decision-making; (3) learning rules or adaptation process; (4) a communication topology network and (5) a nonagent environment. ABM is mainly applied in a computer simulation, usually through special software or ABM toolkits. At the same time, the software can also be used to test how individual behavior changes affect the overall state of the whole system.

5.2 How ABM can be applied in shareability estimation

ABMs are widely utilized in transportation simulations. According to the role of ABM in the research projects, the mainstream ABMs can be divided into three levels: (1) macroscopic policy assessment, (2) microscopic strategy evaluation, and (3) ABM in both macroscopic and microscopic strategy optimization. The difference between the three levels is shown in Table 1.

TABLE 1 Different levels of applying ABM.

	Operation strategy	Agent behavior
Level 1	Preset by ABM designer	Homogeneous rules decided by designer
Level 2	Preset by ABM designer	Heterogeneous rules decided by designer
Level 3	Optimized by ABM	Heterogeneous rules optimized by ABM

5.2.1 Level 1: ABM in macroscopic policy assessment

For level 1 macroscopic policy assessment, ABMs provide effective tools for policymakers to simulate various scenarios and obtain the potential effect of a specific policy. Numerous studies have attempted to use ABMs in their fields to assess policy implement [46–48]. In Lu's work [46], using the GAMA platform, a high-resolution (spatial and temporal) ABM was developed based on the collected bike-sharing travel information. Three agents were defined in this model: bike users, operators, and bikes. To optimize the location and capacity of parking lots, two scenarios (parking incentives and bike lane extension) were simulated, and user cost and system cost were extracted from simulation results. For the case study in Hong Kong, results showed that for most target people to be willing to change the parking location, the minimum fare discount rate for doing so should be set to 30%. In addition, the research revealed that the average trip time can be reduced by 3.8%, and per-user cost can be reduced by 2.4% with an expected investment of 0.12 million USD to build new cycle tracks and connect existing cycle ways. The study indicates that ABM serves as an advanced tool to assess the impact of macroscopic policy.

5.2.2 Level 2: ABM in microscopic strategy evaluation

In level 2 microscopic strategy evaluation, the agents in ABM are divided into several groups and each group has different characteristics or obeys different rules.

Previous studies [44, 49] have explored the application of ABM in microscopic strategy evaluation. In Benenson's research, PARKAGENT, an ABM was developed to study the effect of additional parking supply. In this model, four groups of drivers (residents, guests, employees, and customers) were distinguished, which may differ in the parameters of their behavior. The four groups of agents showed differences in their destination, arrival time, and duration of parking. To estimate the effects of a new parking facility, four scenarios with various additional parking places were compared with the base case. The ABM makes it possible to record the lifepath of every model driver. Three key distributions were constructed: one of parking search time, one of the air distance to destination, and one of the paid parking fees. For the case study, the results showed that additional parking supply linearly affects the occurrence of extreme values, but has only a weak impact on the average search time for a parking place or the average walking distance between the parking place and the destination. A more recent study [44] developed a high-resolution ABM that utilized data from a synthetic population simulator that generates agents for a geographic region along with socio-demographic attributes. And active transportation mode choice for each agent was simulated based on socio-demographic attributes, built environment, walk and bike transportation network attributes, and interaction with other agents. Results from New York City case study indicate that infrastructure investments such as widening the sidewalk and increasing the bike lane network can positively influence the active transportation mode choices.

5.2.3 Level 3: ABM in both macroscopic and microscopic strategy optimization

In level3 integration of microscopic strategy optimization and ABM, multiagent reinforcement learning (MARL) is introduced to cope with large-scale systems in realtime with partial or full data availability. Specifically, MARL addresses the sequential decision-making problem of multiple autonomous agents that operate in a common environment, each of which aims to optimize its own long-term return by interacting with the environment and other agents [50]. Largely, MARL algorithms can be placed into three groups, fully cooperative, fully competitive, and a mix of the two [51, 52].

In Wen's work [53], a reinforcement learning approach was proposed, which adopts a deep Q network and adaptively moves idle vehicles to regain balance. This approach was applied to an agent-based simulator and tested it in a London case study. The research concludes that the proposed rebalancing approach is effective under various demand scenarios and will benefit both travelers and operators if implemented in shared mobility-on-demand systems.

6 Case study of ABM in urban shareability estimation

Here in this section, we will introduce an example utilizing ABM to estimate the shareability in the case of dynamic electric fence planning of bicycle sharing. In this case, the new source of data is introduced including the bicycle-sharing trips (track and trace data) and the electric fence data (ROI data). An ABM is used to simulate the user behavior and several indexes are proposed to evaluate the performance of the dynamic electronic fence strategy.

6.1 Dynamic electric fence for bicycle sharing

An electric fence is a virtual parking area, which works by using GPS, Radio Frequency Identification Device (RFID), or Bluetooth signals to determine the spatial relationship between shared bicycles and dedicated parking spaces.

In bicycle-sharing service, if the electric fence strategy is applied, bicycles are only allowed to park inside the area of the electric fence.

The advantage of the electric fence is obvious. Electric fences can regulate the parking behavior under the user's spontaneous behavior. Electric fences can avoid random parking, reduce the occupation of public space as well as increase the utilization rate of bicycles.

Dynamic electric fence is a newly proposed approach to govern and guide the parking behavior of bicycles. The principle is that the state of electric fences can be adjusted in realtime according to the demand. If the parking demand exceeds the parking ability of a fence, it will be closed. Compared with the static electric fence, a dynamic electric fence has a positive impact on rebalancing parking demand and parking ability.

6.2 ABM simulation

How to evaluate the performance of a dynamic electric fence strategy? Planning and designing a dynamic electric fences strategy is the decision-making part. Under this strategy, users are instructed to pick up and return bicycles according to the status of electric fences. The nonlinear superposition of massive individual behavior makes up a system with high complexity. Multiple users share the limited bicycle resources and one's behavior will affect others. For example, a user returns a bicycle in an electric fence, there will be one more bicycle resource in this fence for the next user to pick up. But under the dynamic electric fence strategy, if the fence is closed, the previous user cannot return the bicycle here, which will lead to the change of behavior of the subsequent users.

Here, an ABM is proposed to solve this problem by precisely simulating individuals' pick-up and return behavior, so as to evaluate the performance of dynamic electric fences. The model consists of the following steps:

Step 1: Simulate the bicycle-sharing trips under the dynamic electric fence restriction. For each trip, the pick-up and return location may be changed and the behavior of travelers obeys the following rules:

- Pick-up bicycles: when a trip starts, the user will pick up the nearest bicycle.
- Return bicycles: when a trip ends, the user will return the bicycle to the nearest fence which is to the destination.

After each of the times that pick-up and return behavior happens, the algorithm will immediately update the parking location of the bicycles. The next trip will be processed based on the latest fence position.

Step 2: To estimate the bicycle resource required in the simulation process, the next step of the ABM is to dispatch the matched sharing trips to bicycles. The dispatching algorithm is as follows:

- Generate a new bicycle: When a bicycle is generated, the trip with the earliest end time will be dispatched to this bicycle.

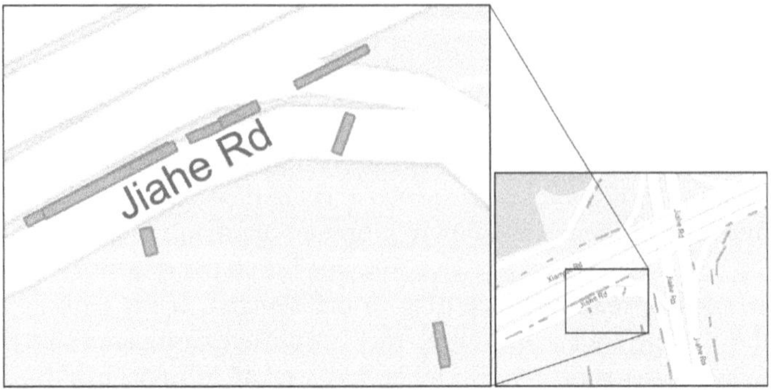

FIG. 4 Distribution of electric fences at an intersection.

- The algorithm will continuously dispatch the next schedulable trip for the bicycle until no more trips can be dispatched.
- The result will then be stored, and the next bicycle is generated until all the trips are dispatched.

6.3 Data and study area

The bicycle-sharing trip data utilized in this case study involves 50,366 bicycle-sharing trips. The data was collected during the morning rush hours (6:00–10:00) on 22 December 2020 in Xiamen, Fujian province of China. At present, the bicycle-sharing service in Xiamen already has a static electric fence strategy. Fig. 4 shows the distribution of electric fences at an intersection. Each of these electric fences is ROI data with a spatial polygon shape. Here in the case study, dynamic electric fence schemes are already generated, and the purpose of the ABM is to evaluate the schemes. As is shown in Fig. 5, the switch status of the electronic fence changes with time.

6.4 Result of simulation

Inputting the travel demand and the scheme of the dynamic electric fences into the ABM, the travel behavior of each traveler is simulated. Fig. 6 shows the visualization of the simulation, Fig. 6A, C, and E are the original distribution of the bicycles in the raw data, and Fig. 6B, D, and F show the ABM simulation result with dynamic electronic fences.

As is shown in Fig. 6A and B, on the city scale, there is basically no difference in the distribution of bicycles before and after the implementation of the dynamic electric fence. The dynamic electric fences will only change the distribution of bicycles in a relatively small area.

Zoom in to observe the parking hotspot areas in Fig. 6C–F, the simulation result shows that during the morning peak hour the bicycles tended to park

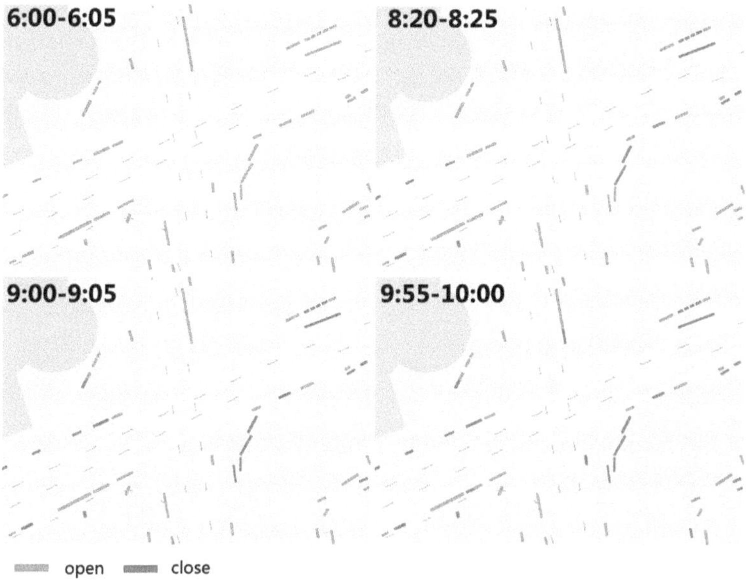

FIG. 5 Dynamic electric fence scheme.

FIG. 6 Visualization of simulation results.

orderly. In raw data, the bicycles are easy to cluster in small areas causing the redundancy of bicycle resources. With the dynamic electric fence strategy, the constant changing characteristic of electric fence can avoid the bicycles concentrating too much on a few locations. In this way, the redundancy of bicycle resources can be alleviated.

6.5 Evaluation of the result

In the case study, four indexes are proposed to evaluate the simulation result in different aspects, including parking demand, spatial entropy, bicycle resources saving, and walk distance.

(1) Parking demand

One of the main purposes of implementing dynamic electric fences is to alleviate the parking pressure in the demand hotspot area. Here, we examine the parking demand of the 40 hotspot areas with the largest gap between parking demand and capacity. These areas are the important areas that affect the overall performance of the bicycle-sharing system in the city.

Fig. 7 shows the change of the parking demand during the morning peak hour in the 40 hotspot areas. In the raw data, the parking demand of the 40 hotspot area gradually increases and exceeds the parking capacity after 8:15. By comparison, in the simulation of dynamic electric fence scenarios, the parking demand in these areas gradually reduced. For the hotspot areas, when the parking demand exceeds the capacity, the electric fence will be closed and the bicycles are only allowed to be picked up. As a result, the number of bicycles will

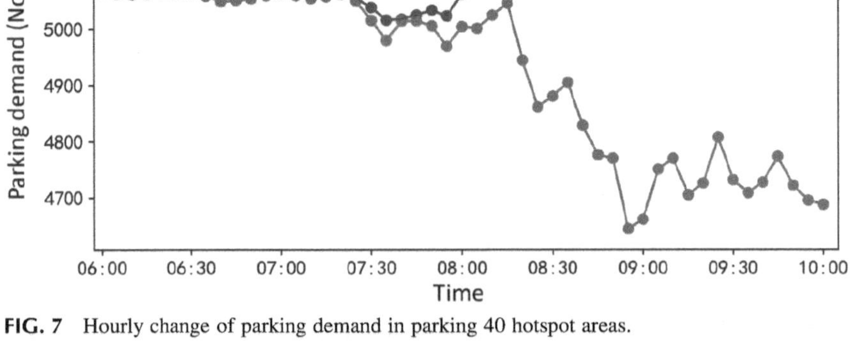

FIG. 7 Hourly change of parking demand in parking 40 hotspot areas.

only reduce rather than increase. The result shows that the electric fence can effectively control the spatial distribution of parking demand to avoid the high concentration of bicycle resources.

(2) Spatial entropy

With the application of electric fences, users will regularly pick up and return bicycles. The parking position of bicycles will gradually change from a chaotic and random state to a centralized and orderly state. To evaluate the tidiness of bicycle parking, spatial entropy is introduced. The higher the spatial entropy is, the more chaotic the parking state is; the lower the spatial entropy is, the more concentrated the parking state is. For each area, the spatial entropy is calculated as follows:

$$H(u) = -\sum_{i} p(g_i) logp(g_i) s.t. g_i \in u$$

where g_i are the smaller grids in area u and $p(g_i)$ represent the probability of bicycles parked in the grid g_i. A better parking status is with the spatial entropy value that is neither too high to cause redundant bicycle resources nor too low to cause random parking.

Fig. 8 shows the change of spatial entropy with and without the dynamic electric fence. After the implementation of the electric fence, the overall spatial entropy decreases while the spatial entropy of key hotspot areas increases. For nonhotspot areas, the parking status of bicycles gradually changed from random parking to orderly parking and lead to the decrease of spatial entropy; for

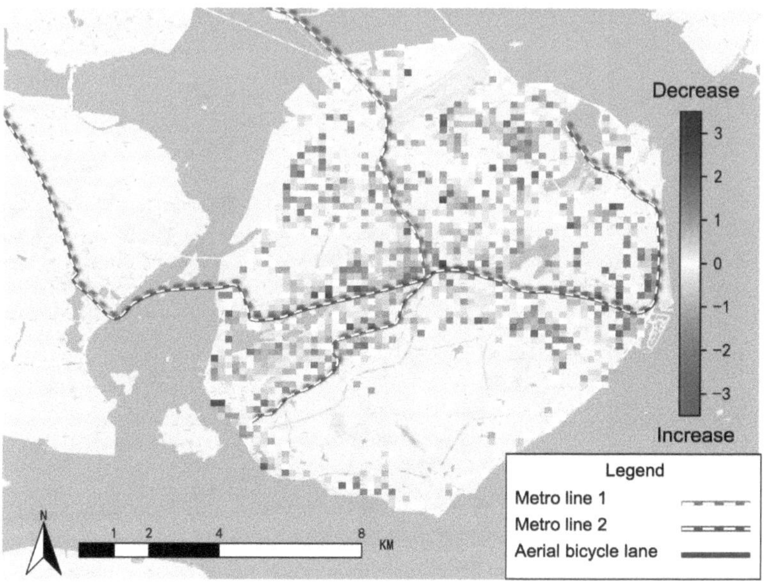

FIG. 8 The change of spatial entropy with and without a dynamic electric fence.

hotspot areas, the parking status of bicycles gradually changed from centralized and concentrated distribution to decentralized distribution and lead to the increase of spatial entropy.

(3) Bicycle resource saving

The dynamic electric fence can also save bicycle resources. With the bicycles parked in fewer concentrated areas, a bicycle will have a higher probability to be found by the next user and utilized. Fig. 9 shows the change in the number of bicycle resources required for each traffic analysis zones (TAZs). From the simulation result, for most of the area, the dynamic electric fences can effectively reduce the number of bicycles resource required.

(4) Walk distance

With the implementation of the dynamic electric fence, users may need to walk for a certain distance to pick up or return bicycles. Walk distance of the electric fence is another important assessment index that directly affects the service quality and public acceptance of the electric fence strategy. Extracted and evaluated from the ABM model, Fig. 10 shows the distribution of walking distance for all the bicycle-sharing trips. The walk distance of 75% of users is less than 40m.

In summary, the ABM used in this case study is in level 1 with the dynamic electric fence strategy preset and all the travelers obeying the same rule.

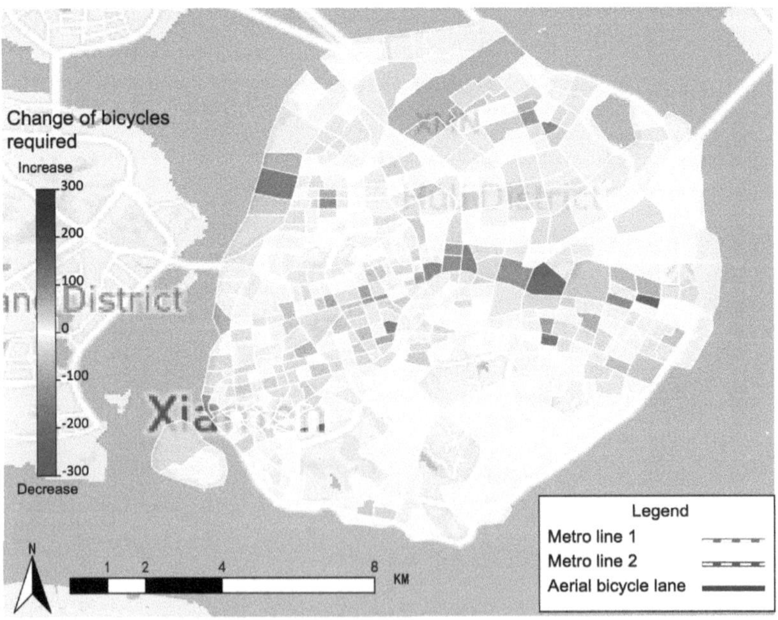

FIG. 9 Distribution of bicycle resource saving.

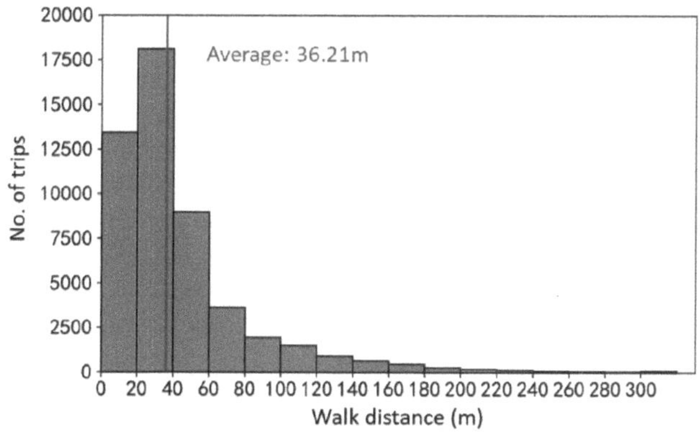

FIG. 10 Distribution of walk distance simulated by ABM.

The purpose of introducing ABM is to evaluate the performance of the electric fence strategy. From the simulation result, several indexes are proposed to describe the shareability of bicycle-sharing services. The simulation results strongly prove the feasibility and effectiveness of the electric fence strategy. The advantage of introducing ABM is that the decision maker and operator can test and evaluate the scheme of the electric fence before putting it forward into real-world practice.

7 Opportunities and challenges

The potential analysis of urban travel shareability is a fast-growing research topic, but it is still in its infancy. In the previous sections, we have introduced the potential of using ABM to estimate urban travel shareability. Here in this section, we will discuss the challenges and possible future directions of this research field.

7.1 Data acquisition

The acquisition of large-scale multisource data with high quality is still a high barrier in the field of big data. Although there is a large amount of open-source data on the network, they are usually limited to a certain area in a certain period. The access of most of the individual trajectory data is limited by the privacy policy or company business secrets. To make a comprehensive and detailed analysis of the urban travel shareability, the ideal situation is that the researchers have multisource data from transportation modes within the same observation period, which is extremely difficult. Most of the academic research projects are limited by the data. And most of the time, data determines what kind of research can be conducted.

7.2 Demand prediction

The geospatial big data acquired reflects the historical activities of users. When estimating the urban travel shareability, the objective is to consider the performance of the sharing transportation mode and the potential impact of the operation strategies in the future prospective. In recent years, the most popular demand prediction models are the machine learning models such as recurrent neural network (RNN) and long short-term memory (LSTM). To predict future demand, these models usually need historical data as input for training. However, for a new sharing transportation mode or an operation strategy, it has not existed or appeared in the historical data. For these demand forecasting tasks, the prediction models can hardly be adopted. Combining the prediction model with ABM may be the solution. Inputting the historical data into the prediction model to predict the future demand, and inputting the predicted demand into the ABM model, the performance of the sharing transportation system can be predicted more accurately.

7.3 Design improvement of ABM

ABM is a flexible simulation system, a promising research direction is to introduce a variety of mechanisms and improve the complexity of this system. ABMs can be improved in the following aspects:

Introducing multiagents: Urban sharing transportation usually involves multiple parties. Correspondingly, more agent types can be introduced in the ABM to simulate the interaction between multiple parties.

Differentiated agent behavior: In real-world situations, individual travel behavior has certain randomness. For ABM, the decision rules of the same type of agents can be differentiated by introducing individual preference or stochastic factors.

Intelligent decision: The operation strategies can introduce more intelligent decision mechanisms such as dynamic pricing and management. Reinforcement learning can also be introduced to support decision making.

7.4 Acceleration of large-scale ABM

The more complex ABM involves more agents that will generate more data with greater computation required. Therefore, find the proper way to accelerate the model can help researchers establish a more powerful ABM to simulate a large-scale sharing travel system. In the ABM, the simulation of the sharing trip is based on the result of the previous trip. The high correlation between the behavior of agents generates difficulties in parallel computing. The possible solution of computing acceleration may start from two aspects: 1. Finding individual behaviors that do not interfere with each other and design parallel computing; 2. Sacrificing part of the computation accuracy by assuming that some behaviors will not interfere with each other. Splitting the computing task into multiple batches and conducting parallel computing. The larger the batch size, the lesser the computation accuracy but higher computing speed.

8 Conclusions

In recent years, the newly emerged sharing mobility service can potentially solve the problems of limited transportation resources and relieve the pressure of transportation, energy, and emission of the urban system. On one hand, conventional data offers limited information to have a clear grasp of the travel characteristics and internal mechanism of individual travel behavior. On the other hand, the lack of transportation modeling analysis technologies on the individual level makes it difficult to accurately estimate the potential of the emerging sharing transportation mode. With the support of pervasively collected geospatial data generated by individuals, this chapter establishes the potential of the ABM towards enhancing sharing transportation modeling, data-driven estimation of shareability, and policy making. We summarized the key technologies, the existing application, future opportunities, and challenges of ABM in travel shareability evaluation.

From the discussions, we answered the following question from our point of view:

(Q1) How can data-driven research overcome the limitation of traditional data and models in shareability estimation?

Comparing with traditional data, new sources of data can capture individual mobility with a more comprehensive range of time and space at a lower cost. With the support of multisource data fusion, we can detailedly model the sharing travel activities of individuals to support the estimation of urban travel shareability.

(Q2) What are the key methodologies and technologies? How to properly apply them?

ABM that allows one to simulate the individual actions of diverse agents and measure the resulting system tends to be the effective approach in data-driven urban shareability estimation. We summarized the existing application of ABM and found that ABM can be integrated into the existing research in the following ways: (1) ABM in macroscopic policy assessment, (2) ABM in microscopic strategy evaluation, and (3) ABM in both macroscopic and microscopic strategy optimization.

(Q3) What are the opportunities and challenges in the data-driven shareability estimation approaches?

For the future prospects of this research field, data acquisition and demand prediction are the fundamental part of the estimation model. Model design and computation acceleration decide the complexity and the scale of the ABM and the task it can handle. Using the more powerful model with multisource data fusion can enhance the data-driven estimation of urban travel shareability. This is of significant importance due to the growing awareness of a need to make decisions based on the monitoring, simulation, and evaluation of shared transportation to deliver the next-generation smart city development.

Acknowledgment

This work is supported by the National Natural Science Foundation of China (Grant No. 71734004).

References

[1] Price JA. Sharing: the integration of intimate economies. Anthropol Forum 1975;3–27.

[2] Alexander LP, González MC. Assessing the impact of real-time ridesharing on urban traffic using mobile phone data. Proc UrbComp 2015;1–9.

[3] Yu Q, Zhang H, Li W, Sui Y, Song X, Yang D, Shibasaki R, Jiang W. Mobile phone data in urban bicycle-sharing: market-oriented sub-area division and spatial analysis on emission reduction potentials. J Clean Prod 2020;254:119974.

[4] Zhang H, Song X, Long Y, Xia T, Fang K, Zheng J, Huang D, Shibasaki R, Liang Y. Mobile phone GPS data in urban bicycle-sharing: layout optimization and emissions reduction analysis. Appl Energy 2019;242:138–47.

[5] Harrison G, Grant-Muller SM, Hodgson FC. New and emerging data forms in transportation planning and policy: opportunities and challenges for "track and trace" data. Transp Res Part C Emerg Technol 2020;117:102672.

[6] Long Y, Yoshida Y, Meng J, Guan D, Yao L, Zhang H. Unequal age-based household emission and its monthly variation embodied in energy consumption–a cases study of Tokyo, Japan. Appl Energy 2019;247:350–62.

[7] Fishman E. Bikeshare: a review of recent literature. Transp Rev 2016;36:92–113.

[8] Lan J, Ma Y, Zhu D, Mangalagiu D, Thornton TF. Enabling value co-creation in the sharing economy: the case of mobike. Sustainability 2017;9:1504.

[9] Lin L, He Z, Peeta S. Predicting station-level hourly demand in a large-scale bike-sharing network: a graph convolutional neural network approach. Transp Res Part C Emerg Technol 2018;97:258–76.

[10] Bao J, Xu C, Liu P, Wang W. Exploring bikesharing travel patterns and trip purposes using smart card data and online point of interests. Netw Spat Econ 2017;17:1231–53.

[11] Chen L, Zhang D, Wang L, Yang D, Ma X, Li S, Wu Z, Pan G, Nguyen T-M-T, Jakubowicz J. Dynamic cluster-based over-demand prediction in bike sharing systems. In: Proceedings of the 2016 ACM international joint conference on pervasive and ubiquitous computing; 2016. p. 841–52.

[12] Faghih-Imani A, Eluru N, El-Geneidy AM, Rabbat M, Haq U. How land-use and urban form impact bicycle flows: evidence from the bicycle-sharing system (BIXI) in Montreal. J Transp Geogr 2014;41:306–14.

[13] Giot R, Cherrier R. Predicting bikeshare system usage up to one day ahead. In: 2014 IEEE symposium on computational intelligence in vehicles and transportation systems (Civts). IEEE; 2014. p. 22–9.

[14] Li Y, Zheng Y, Zhang H, Chen L. Traffic prediction in a bike-sharing system. In: Proceedings of the 23rd SIGSPATIAL international conference on advances in geographic information systems; 2015. p. 1–10.

[15] Zhou X. Understanding spatiotemporal patterns of biking behavior by analyzing massive bike sharing data in Chicago. PLoS One 2015;10, e0137922.
[16] Chen Q, Sun T. A model for the layout of bike stations in public bike-sharing systems. J Adv Transp 2015;49:884–900.
[17] Kabak M, Erbaş M, Çetinkaya C, Özceylan E. A GIS-based MCDM approach for the evaluation of bike-share stations. J Clean Prod 2018;201:49–60.
[18] Landis BW. Bicycle system performance measures. ITE J 1996;66:18–26.
[19] Lopez Gonzalez L. Optimal location for bike sharing stations in downtown Kalamazoo; 2016.
[20] Rybarczyk G, Wu C. Bicycle facility planning using GIS and multi-criteria decision analysis. App Geogr 2010;30:282–93.
[21] Buck D, Buehler R, Happ P, Rawls B, Chung P, Borecki N. Are bikeshare users different from regular cyclists? A first look at short-term users, annual members, and area cyclists in the Washington, DC, region. Transp Res Rec 2013;2387:112–9.
[22] Fishman E, Washington S, Haworth N. Barriers and facilitators to public bicycle scheme use: a qualitative approach. Transp Res F: Traffic Psychol Behav 2012;15:686–98.
[23] Fishman E, Washington S, Haworth N, Watson A. Factors influencing bike share membership: an analysis of Melbourne and Brisbane. Transp Res A Policy Pract 2015;71:17–30.
[24] Ogilvie F, Goodman A. Inequalities in usage of a public bicycle sharing scheme: socio-demographic predictors of uptake and usage of the London (UK) cycle hire scheme. Prev Med 2012;55:40–5.
[25] Santi P, Resta G, Szell M, Sobolevsky S, Strogatz SH, Ratti C. Quantifying the benefits of vehicle pooling with shareability networks. Proc Natl Acad Sci 2014;111:13290–4.
[26] Zhang H, Chen J, Li W, Song X, Shibasaki R. Mobile phone GPS data in urban ride-sharing: An assessment method for emission reduction potential. Appl Energy 2020;269:115038.
[27] Zhang C, Dong M, Ota K, Guo M. A social-network-optimized taxi-sharing service. IT Prof 2016;18:34–40.
[28] Lokhandwala M, Cai H. Dynamic ride sharing using traditional taxis and shared autonomous taxis: a case study of NYC. Transp Res Part C Emerg Technol 2018;97:45–60.
[29] Fagnant DJ, Kockelman KM. Dynamic ride-sharing and fleet sizing for a system of shared autonomous vehicles in Austin, Texas. Transportation 2018;45:143–58.
[30] Wang Y, Zheng B, Lim E-P. Understanding the effects of taxi ride-sharing—a case study of Singapore. Comput Environ Urban Syst 2018;69:124–32.
[31] Ma S, Zheng Y, Wolfson O. Real-time city-scale taxi ridesharing. IEEE Trans Knowl Data Eng 2014;27:1782–95.
[32] Alonso-Mora J, Samaranayake S, Wallar A, Frazzoli E, Rus D. On-demand high-capacity ride-sharing via dynamic trip-vehicle assignment. Proc Natl Acad Sci 2017;114:462–7.
[33] Li Z, Hong Y, Zhang Z. Do ride-sharing services affect traffic congestion? An empirical study of uber entry. Soc Sci Res Netw 2016;2002:1–29.
[34] Sui Y, Zhang H, Song X, Shao F, Yu X, Shibasaki R, Sun R, Yuan M, Wang C, Li S. GPS data in urban online ride-hailing: a comparative analysis on fuel consumption and emissions. J Clean Prod 2019;227:495–505.
[35] Lyu Y, Chow C-Y, Lee VC, Ng JK, Li Y, Zeng J. CB-planner: a bus line planning framework for customized bus systems. Transp Res Part C Emerg Technol 2019;101:233–53.
[36] Schiller J, Voisard A. Location-based services. Elsevier; 2004.
[37] Ratti C, Frenchman D, Pulselli RM, Williams S. Mobile landscapes: using location data from cell phones for urban analysis. Environ Plann B Plann Des 2006;33:727–48.
[38] Körting TS, Fonseca LMG, Câmara G. GeoDMA—geographic data mining analyst. Comput Geosci 2013;57:133–45.

[39] Mapbox. Isochrone | API [WWW document]. URL: Mapbox; 2021. https://docs.mapbox.com/api/navigation/isochrone/. [Accessed 6 April 2021].

[40] Crooks A, Castle C, Batty M. Key challenges in agent-based modelling for geo-spatial simulation. In: Computers, environment and urban systems. GeoComputation: modeling with spatial agents, 32; 2008. p. 417–30. https://doi.org/10.1016/j.compenvurbsys.2008.09.004.

[41] Gurram S, Stuart AL, Pinjari AR. Agent-based modeling to estimate exposures to urban air pollution from transportation: exposure disparities and impacts of high-resolution data. Comput Environ Urban Syst 2019;75:22–34. https://doi.org/10.1016/j.compenvurbsys.2019.01.002.

[42] Lund AM, Gouripeddi R, Facelli JC. STHAM: an agent based model for simulating human exposure across high resolution spatiotemporal domains. J Expo Sci Environ Epidemiol 2020;30:459–68. https://doi.org/10.1038/s41370-020-0216-4.

[43] Fontaine CM, Rounsevell MDA. An agent-based approach to model future residential pressure on a regional landscape. Landsc Ecol 2009;24:1237. https://doi.org/10.1007/s10980-009-9378-0.

[44] Aziz HMA, Park BH, Morton A, Stewart RN, Hilliard M, Maness M. A high resolution agent-based model to support walk-bicycle infrastructure investment decisions: a case study with new York City. Transp Res Part C Emerg Technol 2018;86:280–99. https://doi.org/10.1016/j.trc.2017.11.008.

[45] Crooks AT, Heppenstall AJ. Introduction to agent-based modelling. In: Heppenstall AJ, Crooks AT, See LM, Batty M, editors. Agent-based models of geographical systems. Dordrecht: Springer Netherlands; 2012. p. 85–105. https://doi.org/10.1007/978-90-481-8927-4_5.

[46] Lu M, An K, Hsu S-C, Zhu R. Considering user behavior in free-floating bike sharing system design: a data-informed spatial agent-based model. Sustain Cities Soc 2019;49:101567. https://doi.org/10.1016/j.scs.2019.101567.

[47] Lu M, Hsu S-C, Chen P-C, Lee W-Y. Improving the sustainability of integrated transportation system with bike-sharing: a spatial agent-based approach. Sustain Cities Soc 2018;41:44–51.

[48] Van der Veen RAC, Kisjes KH, Nikolic I. Exploring policy impacts for servicising in product-based markets: a generic agent-based model. J Clean Prod 2017;145:1–13.

[49] Benenson I, Martens K, Birfir S. PARKAGENT: An agent-based model of parking in the city. In: Computers, environment and urban systems. Geocomputation: modeling with spatial agents, 32; 2008. p. 431–9. https://doi.org/10.1016/j.compenvurbsys.2008.09.011.

[50] Busoniu L, Babuska R, De Schutter B. A comprehensive survey of multiagent reinforcement learning. IEEE Trans Syst Man Cybern Part C-Appl Rev 2008;38:156–72. https://doi.org/10.1109/TSMCC.2007.913919.

[51] Buşoniu L, Babuška R, De Schutter B. Multi-agent reinforcement learning: An overview. In: Srinivasan D, Jain LC, editors. Innovations in multi-agent systems and applications—1. Berlin Heidelberg, Berlin, Heidelberg: Studies in Computational Intelligence. Springer; 2010. p. 183–221. https://doi.org/10.1007/978-3-642-14435-6_7.

[52] Zhang K, Yang Z, Başar T. Multi-agent reinforcement learning: a selective overview of theories and algorithms. arXiv 2019. 1911.10635 [cs, stat].

[53] Wen J, Zhao J, Jaillet P. Rebalancing shared mobility-on-demand systems: A reinforcement learning approach. In: 2017 IEEE 20th international conference on intelligent transportation systems (ITSC); 2017. p. 220–5. https://doi.org/10.1109/ITSC.2017.8317908. Presented at the 2017 IEEE 20th international conference on intelligent transportation systems (ITSC).

Chapter 7

Data mining technologies for Mobility-as-a-Service (MaaS)

Wen-Long Shang[a], Haoran Zhang[b], and Yi Sui[b,c,d]
[a]*Beijing Key Laboratory of Traffic Engineering, College of Metropolitan Transportation, Beijing University of Technology, Beijing, China,* [b]*Center for Spatial Information Science, The University of Tokyo, Kashiwa, Chiba, Japan,* [c]*College of Computer Science and Technology, Qingdao University, Qingdao, China,* [d]*Institute of Smart City and Big Data Technology, Qingdao, China*

1 Introduction of data mining technologies in MaaS system

In recent years, mobility-as-a-service (MaaS) has become a hot topic in the field of transportation, and its popularity is increasing. In short, the idealized MaaS breaks through the barriers between trains, subway, buses, taxis, car sharing, bike-sharing, and other modes of transportation, which tries to utilize joint digital channel/platform to complete the traveling trips based on all modes of transportation.

Currently, the definitions of MaaS are diverse and mainly focus on the role of technology, the establishment of the ecosystem, and the integration of transportation modes.

According to the MaaS Alliance, MaaS is regarded as:

> *The integration of various forms of transport services into a single mobility service accessible on demand. To meet a customer's request, a MaaS operator facilitates a diverse menu of transport options, be they public transport, ride-, car- or bike-sharing, taxi or car rental/lease, or a combination thereof. For the user, MaaS can offer added value through the use of a single application to provide access to mobility, with a single payment channel instead of multiple ticketing and payment operations. For its users, MaaS should be the best value proposition, by helping them meet their mobility needs and solve the inconvenient parts of individual journeys as well as the entire system of mobility services.*

The aim of MaaS is to provide an alternative to the use of private vehicles so as to achieve more convenient, more sustainable, and even cheaper services.

How does big data play a role in transportation? Driven by the rapid development of information and communication technology (ICT) and the

Big Data and Mobility as a Service. https://doi.org/10.1016/B978-0-323-90169-7.00008-7

prevalence of data science, intelligent transportation and traffic big data are becoming the new concept and practice of worldwide urbanization for the next generation. Multi-source traffic big data, including GPS data, bus IC card data, subway AFC data, road crossing data, and so on, are being generated at an unprecedented scale and speed, and application scenarios for such data are becoming more widespread than ever before. Through dynamic monitoring, analysis, integration, and utilization of multi-source data of urban transportation, real-time sensing and accurate estimation of traffic state can be realized. For example, mobile location data can be used to analyze the characteristics of urban population distribution and traffic trip distribution, so as to serve traffic planning. The traffic flow of the road networks is able to be analyzed by using the data of car-hailing. Based on this, the signal timing scheme of each intersection can be optimized to improve road efficiency. The multi-source data such as inductive loop, video technology, and floating car data can be used to achieve the recognition and monitor of traffic state in real-time.

In the era of big data, researchers and engineers can obtain multi-source traffic data from many dimensions. However, due to the complexity and huge volume of traffic big data, the traditional methods of data analysis and modeling have limitations in handling the data. Without powerful tools, it is difficult to understand these data for further applications. Furthermore, as a century-old discipline, traffic engineering has accumulated rich knowledge and data. Therefore, based on the emerging data analysis and machine learning technologies, we build a new generation of analysis theories and methods for traffic big data to facilitate relevant decision-makers and planners to extract key information from increasingly complex big data and combine this with MaaS to solve traffic problems (Fig. 1).

MaaS apparently can provide a new business model for transportation, which requires the supports of many technologies and analyses, such as information and communication technologies (ICT), demand analyses and management, estimation of travel trajectory and time. Therefore, data mining technologies will play important role in offering such supports.

2 Data mining technologies in MaaS system

2.1 What is data mining?

Data Mining refers to the process that searches underlying or hidden information from massive data through algorithms. Data mining is the general term of a series of technical tasks, and its main purpose is to find the internal pattern of data from large-scale datasets. During this process, machine learning and statistics are required. The core of data mining is to extract information from datasets and transform the information into a type of data structure that is easy to understand, so as to set up the foundation for further applications [1].

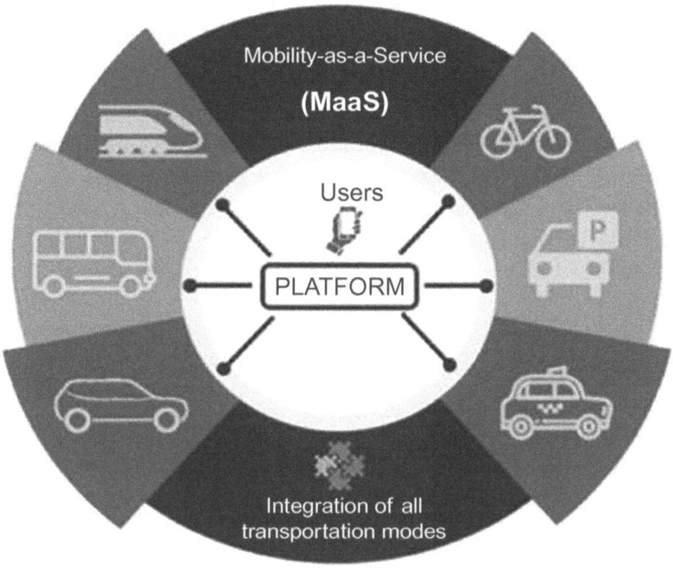

FIG. 1 Schematic diagram of Mobility-as-a-service. *(Source: Guangming web: https://m.gmw.cn/baijia/2019-11/07/33301241.html.)*

2.2 Object of data mining

The types of data can be structured, semi-structured, or even heterogeneous. The methods of discovering knowledge can be mathematical, non-mathematical, or inductive. Finally, the discovered knowledge can be used for information management, query optimization, decision support, and data maintenance.

The object of data mining can be any type of data. It can be a relation database, which includes data sources with structured data. It can also be a data warehouse, text, multi-media data, spatial and temporal data, web data, which mainly refers to data sources with semi-structured or heterogeneous data.

The methods of knowledge discovery can be digital, non-digital, or inductive. Finally, the discovered knowledge can be used for information management, query optimization, decision support, and data maintenance.

2.3 Classical steps of data mining

Before starting data mining, it is very necessary to confirm what steps to follow, what to do in each step, and what objectives to achieve. Here we need to understand that a good plan can ensure the orderly execution and goal realization of data mining. Many data mining consulting companies tend to provide some process models of data mining to guide their clients to conduct data mining stepby step.

The steps of data mining process mainly include the definition of the problem, the establishment of data mining database, data analysis, data preparation, model establishment, model evaluation, and implementation. Let's see the details of each step as follows:

(1) *Define problems*. Before knowledge discovery, the first and most important requirement is to understand the problems of data and business. A clear definition of the goal for data mining is very necessary, which determines what to do next.
(2) *Establish data mining database*. The establishment of data mining database includes the following steps: data collection, data description and selection, data quality evaluation and data cleaning, merging and integration, building metadata, loading and maintaining a data mining database.
(3) *Analyze data*. The purpose of data analysis is to find out the data fields that have the greatest impact on the output prediction and to decide whether it is necessary to export data fields. If the data set includes hundreds of data fields, then browsing and analyzing these data is time-consuming and boring. Therefore, a tool software with a good interface and powerful functions is required to complete these things.
(4) *Prepare data*. This is the last step of data preparation before data modeling, and can be divided into four parts: variable selection, record selection, new variable creation, and variables transformation.
(5) *Data modeling*. Data modeling is a core step in data mining. It utilizes quantitative models to analyze and extract the trend, characteristics, and potential patterns of data, so as to provide useful information for users. In short, to obtain the expected outputs, the data modeling problem is to process the inputs by an appropriate model according to the demand, which uses the processed data as input. The key lies in finding the mapping relationship between inputs and outputs. The modeling problems can be further divided into regression problems, classification problems, and clustering problems.
(6) *Model evaluation*. When the model is set up, it is necessary to evaluate the obtained results and explain the values of the model. The accuracy obtained from the test set only makes sense for the data used for building the model. In practice, we have to further understand the types of errors and the related costs. Experience has shown that an effective model is not necessarily the correct one. The direct reason is the various assumptions used in the model. Therefore, it is very important to test the model directly in the realistic world.
(7) *Implementation*. When the model is established and verified, there are two main paths to use the model. The first is to provide references for project analysts, and the other applies this model to different datasets (Fig. 2).

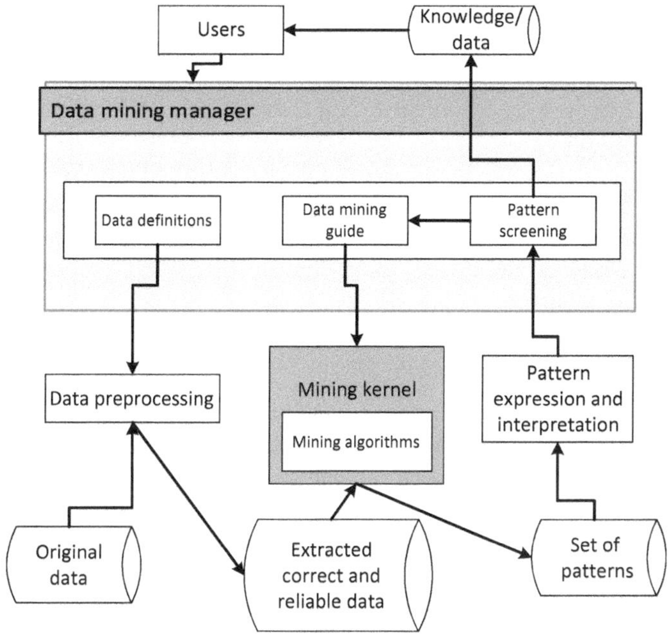

FIG. 2 System model of data mining.

2.4 Types of transportation data

2.4.1 Static data

This type of data does not change in the short term, and usually provides the basis for other data analysis and relevant research. It mainly includes:

Road network data: road network data mainly include road network GIS data, CAD data, traffic zone data, urban functional area data, and so forth. Based on various road network data, urban planning information such as the spatial distribution of road networks and urban functional zones can be extracted.

Public transport network data: public transport network data include public transport lines, station data, and so on, which is the basis of network performance evaluation and the operation state evaluation of public transport systems.

2.4.2 Fixed detector data

The fixed detector data only utilize the detectors fixed at certain locations to collect data, which are obtained based on consecutive sampling at a single location. The widely-used data type includes:

Traditional detector data: traditional detector data usually are obtained by inductive loop, microwave, radar, and so on. This type of data has higher temporal coverage, but lower spatial coverage due to a large amount of investment and maintenance difficulties. Such data can be applied to the establishment and validation of traffic flow models, the calibration of simulation models, the research on incomplete data technologies, traffic flow prediction models, and so on.

New type of detector data: this type of data is collected from the new type of detector on road networks. For example, the speed and license plate data of vehicles can be obtained by image processing technology analyzing video data. In addition, based on radio frequency identification (RFID) technology, road traffic data are obtained.

2.4.3 Mobile detector data

Such data is usually collected by specific devices, and can continuously sample to a single object. With the popularity of related devices, its temporal-spatial coverage rates keep increasing.

GPS trajectory data: GPS trajectory data take samples to the vehicle positions at a certain time interval based on the onboard positioning devices (buses, taxis, online car-hailing, and other vehicles) to obtain the longitude and latitude of the vehicles. The data is widely used for the estimation of road network traffic flow, identification and prediction of congestion, operation evaluation, and short-term traffic prediction based on public transport GPS data.

Cellphone signaling data: compared to other data, due to the popularity of cellphones, one of the advantages cellphone signaling data enjoy an advantage in high sampling rate, which can accurately reflect the complete characteristics of the full data sample. One important application scenario of cellphone signaling data is to replace the traditional means for traffic surveys when analyzing the travel characteristics of urban residents.

Electric vehicle data: electric vehicle data mainly include vehicle and charging pile data. Vehicle data can record vehicle track position information, battery status information, and so on, while charging pile data include the position of charging piles and charging information (charging vehicle ID, charging time, and so on.). Electric vehicle data can be used to study the charging behavior of EV drivers, the location layout and number optimization of charging piles, and the optimization of important parameters for EVs.

2.4.4 Operation data

Operation data refer to the data generated in the operation process of public transport systems and taxi systems, which are relevant to the user information of mobile phones. This type of data may play a very important role in the optimization of operation and management for the system.

Public transport card data: Such data include metro card data (or AFC data) and bus card data. The Metro card data is mainly used to analyze the OD flow distribution of stations, so as to conduct the study of metro timetable compilation. The bus card data refer to the data of getting on the bus, which can be used for the further exploration of the identification of alighting stations, the analysis of the metro line operation, the metro network planning, and so forth.

Customized public transport data: Customized public transport refers to the flexible urban public transport service provided by public transport companies according to the dynamic needs of passengers, which provides point-to-point service in a way of demand response and online booking. The customized bus data include the trajectory data of the bus, the number of passengers at each bus station, and so on. Customized bus data can be used for bus network planning, operation evaluation, and other relevant research.

Taxi operation data: Such data include the time and location of a passenger getting on and off the taxi, distance, cost, and other information, which can be used to analyze the behavioral characteristics of passengers choosing a taxi and the spatial distribution of boarding and alighting.

Public bicycle data: Public bicycle data include IC card data, public bicycle rental point data, and so forth, which can be used for the study of the operation, rental characteristics, location selection, scheduling for public bicycles.

3 Methodologies of data mining technologies used in MaaS system

3.1 Support vector machine

Support vector machine (SVM) is a generalized linear classifier that classifies data in the way of supervised learning [2]. The purpose of SVM is to find the best classification line or classification surface, and its decision boundary is the maximum margin hyperplane of solutions for learning data samples [3].

SVM uses the hinge loss function to calculate empirical risk and adds a regularization term to the solution system to optimize structural risk. It is a sparse and robust classifier. SVM can be classified by kernel method, which is one of the common kernel learning methods.

SVM was proposed in 1964 and developed rapidly after the 1990s. A series of improved and extended algorithms have been derived. It has been applied in pattern recognition problems such as portrait recognition and text classification.

3.1.1 linear SVM in linearly separable case

For binary classification problems, training set $T = \{(x_1, y_1), (x_2, y_2), \ldots (x_N, y_n)\}$, its classification $y = y_i \in \{0, 1\}$, linear SVM can obtain a separation hyperplane:

$$w.x + b = 0$$

Corresponding classification decision function is shown below:

$$f(x) = \text{sign}(w.x + b)$$

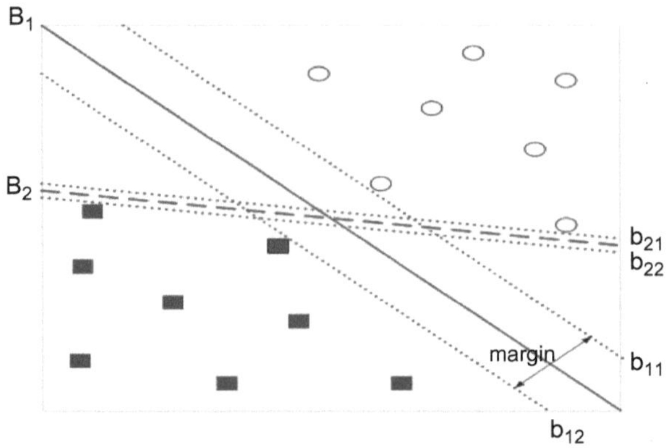

FIG. 3 The separating hyperplanes.

Separation hyperplanes are shown in Fig. 3, and which hyperplane is better for classification?

Intuitively, the classification effect of hyperplane B_1 is better. In this chapter, the two different types of sample points which are nearest to the separation hyperplane are called support vectors, these can form two long bands paralleling to the separation hyperplane, and the distance between them is called margin [4]. Obviously, if the margin is larger, the classification accuracy is higher (the distance from the hyperplane indicates the classification accuracy, and the further the distance is, the higher the classification accuracy is). It is easy to obtain the following results by calculation:

$$\text{margin} = \frac{2}{||w||}$$

As can be seen from Fig. 3, the sample points outside have no contributions to the determination of the separation hyperplane, in other words, SVM is determined by important training samples (support vectors). The SVM classification problem can be described as maximizing $\frac{2}{||w||}$ (equivalent to minimizing $\frac{||w||}{2}$) when all the classifications are correct. Constrained optimization problem of linear classification can be shown below:

$$\min_{w,b} \frac{1}{2}|w|^2 \tag{1}$$

$$\text{s.t. } y_i\left(w^T.x_i + b\right) - 1 \geq 0 \tag{2}$$

Lagrange multipliers are introduced for each inequality constraint $\alpha_i \geq 0$, $i = 1, 2, \ldots, N$. Constructing Lagrange function, and then according to Lagrange duality, the original constrained optimization problem can be equivalent to the minimax dual problem:

$$\max_a \min_{w,b} L(w, b, \alpha) = \frac{1}{2}|w|^2 + \sum_{i=1}^{m} \alpha_i \left[1 - y_i\left(w^T.x_i + b\right)\right]$$

$L(w,b,\alpha)$ derivates w, b and makes them equal to zero:

$$\frac{\partial L}{\partial w} = w - \sum_{i=1}^{N} \alpha_i y_i x_i = 0 \rightarrow w = \sum_{i=1}^{N} \alpha_i y_i x_i$$

$$\frac{\partial L}{\partial b} = \sum_{i=1}^{N} \alpha_i y_i = 0 \rightarrow \sum_{i=1}^{N} \alpha_i y_i = 0$$

Substituting the above equations into Lagrange function, the dual problem can be transformed into:

$$\max_a \sum_{i=1}^{m} a_i - \frac{1}{2} \sum_{i=1}^{m} \alpha_i \alpha_j y_i y_j x^T{}_i x_j$$

$$\text{s.t.} \sum_{i=1}^{N} \alpha_i y_i = 0 \quad \alpha_i \geq 0, i = 1, 2, \ldots, m$$

The problem is a quadratic programming problem, which can be solved by a quadratic programming program. However, when the number of samples is large, the number of parameters to be optimized is very large, so the sequential minimal optimization (SMO) algorithm can be used. The detailed process of the algorithm will not be described here. The readers can be referred to Zhiyuan Liu's book [5].

3.1.2 linear SVM in linearly inseparable case

Linearly inseparability means that some sample points do not satisfy the constraint condition (2). To solve this problem, a slack variable $\xi_i \geq 0$ is introduced for each sample, so the constraint condition transforms:

$$y_i\left(w^T.x_i + b\right) \geq 1 - \xi_i$$

The objective function becomes:

$$\min_{w,b,\xi} \frac{1}{2}|w|^2 + C\sum_{i=1}^{N} \xi_i$$

where C is the penalty function, and the objective function has two meanings:

- the margin should be as large as possible;
- there are few misclassification samples.

C is a parameter to adjust two objectives. By constructing the Lagrange function and solving partial derivation, the equivalent dual problem can be obtained:

$$\max_a \sum_{i=1}^{m} a_i - \frac{1}{2}\sum_{i=1}^{m}\sum_{j=1}^{m} \alpha_i\alpha_j y_i y_j x^T{}_i x_j \tag{3}$$

$$s.t. \sum_{i=1}^{m} \alpha_i y_i = 0 \quad 0 \le \alpha_i \le C, i = 1,2,\dots,m \tag{4}$$

Compared with the linear separable dual problem, only a constraint condition α_i is changed, and the method of solving the problem is similar.

3.1.3 Nonlinear SVM

As for nonlinear SVM problems, the method to solve linear SVM is no longer applicable, and nonlinear SVM is needed to solve such problems [6]. The way to solve the nonlinear classification problems is based on space transformation ψ. (In general, a low dimensional space is mapped to a high dimensional space $x \rightarrow \psi(x)$.) In the example shown in Fig. 4, the separation surface of the ellipse in the left figure can be transformed into a straight line in the right figure by spatial transformation.

In the equivalent dual problem of SVM, the objective function has the inner product of sample points $x_i^T . x_j$, which can be transformed into $\psi(x_i)^T(\psi(x_j)$ after spatial transformation.

In this case, since the increase of the number of the dimension leads to an increase in computation costs, the kernel function is used to transform the mapped inner product of high-dimensional space into a function of low-dimensional space.

$$k(x, z) = \psi(x).\psi(z)$$

Substitute it into the objective function of the generalized SVM learning algorithm. The optimization problem of nonlinear SVM is obtained

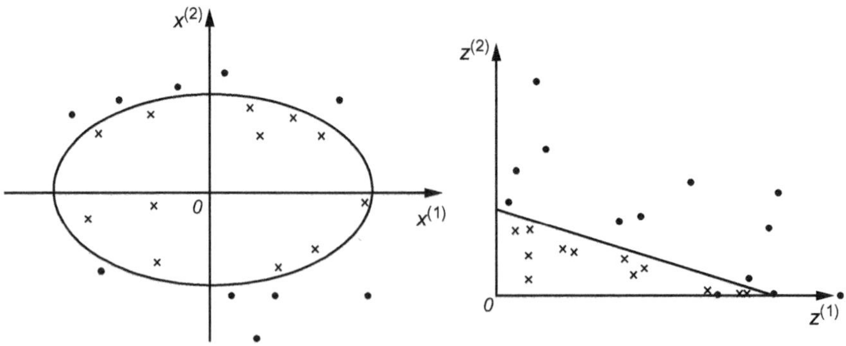

FIG. 4 Example of spatial transformation.

$$\max_a \sum_{i=1}^{m} a_i - \frac{1}{2}\sum_{i=1}^{m}\sum_{j=1}^{m} \alpha_i\alpha_j y_i y_j \psi(x_i)^T (\psi(x_j)) \tag{5}$$

$$\text{s.t.} \sum_{i=1}^{m} \alpha_i y_i = 0 \quad 0 \le \alpha_i \le C, i = 1,2,\ldots,m \tag{6}$$

The detailed solve algorithms and interpretation can be referred to Zhiyuan Liu's book.

3.2 Linear regression

Linear regression model is the simplest and most basic supervised learning model in machine learning [7,8]. Although it is not complicated, many complex models are built based on such models.

Linear model presents the relationship between characteristics and target values in a way of "linear" combination and is one of the most commonly used and important machine learning models in the transportation field [9]. Supposing there is a set of eigenvectors $x = \{x_1; x_2; \ldots; x_d\}$, composed of d characteristics, where x_d is the value of the dth characteristic. If we know the linear relationship between the f eigenvectors x and the target value y, then we can construct a linear function based on d characteristics to predict y:

$$y = w_1 x_1 + w_2 x_2 + \ldots + w_d x_d + b = w^T x + b$$

This is the standard form of linear model, $w = \{w_1; w_2; \ldots; w_d\}$ is the parameters to measure the relationship between x and y, and b is the constant term. Once the parameters w and b are determined, the linear model can be determined, so the learning process of model is the process of optimization for these two groups of parameters.

Although linear model is simple, it contains many important ideas in machine learning. Many powerful nonlinear models are based on linear models, such as hierarchical structure or high-dimensional mapping. The weight W provides good interpretability for the linear model.

3.2.1 Least square method

Given a dataset $D = \{(x_1, y_1), (x_2, y_2), \ldots, (x_m, y_m)\}$, $x_i = (x_{i1}; x_{i2}; \ldots; x_{id})$, $y_i \in \mathbb{R}$. We are trying to set up linear regression model $f(x_i) = w^T x_i + b$ to describe the relationship between x_i and y_i, and conduct predictions based on new datasets. How to determine the parameters w and b?

Considering the case of univariate linear regression, namely, $d = 1$, only one input variable exists. The dataset including m samples can be denoted as D. theoretically, we can find an arbitrary number of different parameter combinations (w, b) to simulate the distribution of the sample dataset, but the purpose of linear regression is to find a group of optimal parameters (w^*, b^*) to ensure that $f(x_i)$ is close to y_i as much as possible, in other words, to accurately predict the realistic objective values. Therefore, the loss function can be used (shown in Eq. 7). The process of determining the parameters of linear regression can be described as a search for a group of optimal parameters so as to minimize the value of the loss function.

$$\mathcal{L}(w,b)=\sum_{i=1}^{m}(f(x_i)-y_i)^2 \tag{7}$$

The method that determines the parameters of the linear regression model by loss function is a classical least square method:

$$(w^*,b^*)=\arg\min_{(w,b)}\mathcal{L}(\mathrm{w},\mathrm{b})=\arg\min_{(w,b)}\sum_{i=1}^{m}(f(x_i)-y_i)^2 \tag{8}$$

Least square is a mathematical optimization method, which can be used to find a straight line in the linear regression problem. It is the least square sum of error between the prediction data and the sample data. The above process is also called the least-squares parameter estimation of linear regression. The model (x) is a convex programming model without constraints, thus we set the partial derivations of $\mathcal{L}(w,b)$ to w and b as zero to calculate the optimal (w^*,b^*).

$$\frac{\partial\mathcal{L}(w,b)}{\partial w}=0$$

$$\frac{\partial\mathcal{L}(w,b)}{\partial b}=0$$

It is easy to obtain the algebraic expression of (w^*,b^*), which means that we obtain the optimal regression line by LSM based on sample data.

$$w^*=\frac{\sum_{i=1}^{m}y_i(x_i-\overline{x})}{\sum_{i=1}^{m}x_i^2-\frac{1}{m}\left(\sum_{i=1}^{m}x_i\right)^2},\quad \overline{x}=\frac{1}{m}\sum_{i=1}^{m}x_i,$$

$$b^*=\frac{1}{m}\sum_{i=1}^{m}(y_i-w^*x_i)$$

3.2.2 Maximum likelihood estimation

Different from the LSM method, the core idea of maximum likelihood estimation (MLE) assumes that the overall distribution of the model is known, and use the known sample results to deduce the model parameter values which are most likely (maximal probability) to cause such results, that is, the value of probability density function is the largest at the sample value. In other words,

MLE provides a method to evaluate the model parameters based on the observation data. The model is determined but the parameters are unknown.

Back to the objective of univariate linear regression, the linear model $f(x_i)=wx_i+b$ is obtained by learning, to make prediction $f(x_i)$ close to the target values y_i, while the relationship between these two values can be depicted by introducing random error ε:

$$y_i = f(x_i) + \varepsilon_i = wx_i + b + \varepsilon_i$$

In practice, the random error term represents the comprehensive reflection when the independent influence of many factors cannot be observed. According to the central limit theorem, error term ε_i, $i=1, 2, \ldots, m$. can obey Gaussian distribution (normal distribution) with mean value of 0 and variance of δ^2, namely, $\varepsilon_i \sim N(0, \delta^2)$. when $\varepsilon_i = \varepsilon$ the probability when $\varepsilon_i = \tilde{\varepsilon}$.

$$p(\varepsilon_i = \tilde{\varepsilon}) = \frac{1}{\sqrt{2\pi}\,\delta} e^{-\frac{(\tilde{\varepsilon})^2}{2\delta^2}}$$

Given sample x_i and target value y_i, the conditional probability of y_i under x_i and parameter (w, b) is as follows:

$$p(y_i|\, x_i;\, (w, b)) = \frac{1}{\sqrt{2\pi}\,\delta} e^{-\frac{[y_i-(wx_i+b)]^2}{2\delta^2}}$$

For the dataset with m samples, using MLE method for parameter estimation is to obtain the optimal parameter array the maximum likelihood function $\mathcal{L}(w, b)$ for the optimal parameter array (w^*, b^*). The Likelihood function is shown below:

$$\mathcal{L}(w, b) = \prod_{i=1}^{m} p(y_i|\, x_i;\, (w, b)) = \prod_{i=1}^{m} \frac{1}{\sqrt{2\pi}\,\delta} e^{-\frac{[y_i-(wx_i+b)]^2}{2\delta^2}}$$

The log-likelihood function $\mathcal{LL}(w, b)$ can be obtained by further calculating logarithm:

$$\mathcal{LL}(w, b) = \log \mathcal{L}(w, b) = \log \prod_{i=1}^{m} p(y_i|\, x_i;\, (w, b))$$

Since the log function is monotonically increasing, and compared with $\mathcal{L}(w, b)$, the log-likelihood function does not change its monotonicity but is easier to derive. The optimal parameter array can be determined by maximizing the value of the log-likelihood function, that is, solving the partial derivation of the above equation and setting the derivation value as zero.

$$\mathcal{LL}(w, b) = m \log \frac{1}{\sqrt{2\pi}\,\delta} - \frac{1}{2\delta^2} \sum_{i=1}^{m} [y_i - (wx_i + b)]^2$$

MLE is not only suitable for univariate linear regression, but also for multivariate linear regression and even generalized linear regression.

3.3 Decision tree

Decision tree model is widely used for classification and regression in the field of transportation. Due to its special tree structure, it is named decision tree. It utilizes a tree-like graph or decision for decision-making, and determines a strategy that is most likely to achieve goals.

Decision tree model does not rely on knowledge in other fields and parameter assumptions, it can clearly show the importance of characteristics and then generate classification rules that are easily understood, which has advantages of simple principle and strong interpretability. In addition, due to its excellent performance, decision tree can also be used for both classification and regression analysis, so it can be frequently used as the basic model for combination algorithms.

A milestone of decision tree model is the famous ID3 algorithm proposed by Quinlan [10], which lays the foundation for the decision tree algorithm. However, at the beginning stage, the algorithm did not consider many issues such as ignorance of continuous attributes, attribute missing, and noise, so Quinlan proposed another C4.5 algorithm in 1993 and introduced the pruning methods for optimal decision tree. Breiman et al. proposed CART algorithm in 1984, which uses binary tree as a model structure, continuously divides decision attributes through recursive operation and optimizes tree model by using validation data.

3.3.1 The structure of decision tree

A decision tree generally consists of nodes and directed edges. In general, a tree contains a root node, multiple internal nodes, and multiple leaf nodes. The root node and the internal node represent the test conditions of a feature or attribute, which can be used to separate records with different characteristics. The leaf node corresponds to the decision result (Fig. 5).

3.3.2 Attribute partition selection

In order to improve the learning efficiency, each step of the decision tree requires selecting the optimal feature. The selection criteria are the selection of the features with the best classification ability for the remaining features, that is, the samples contained in each subset (each value of the feature) can be divided into the same class as much as possible under the feature space division.

Information entropy

Information entropy is a commonly used index to measure the purity of sample set. If there are a pair of transport modes here, some of these you want to use and some of these you prefer not to take, and if there are only those transport modes

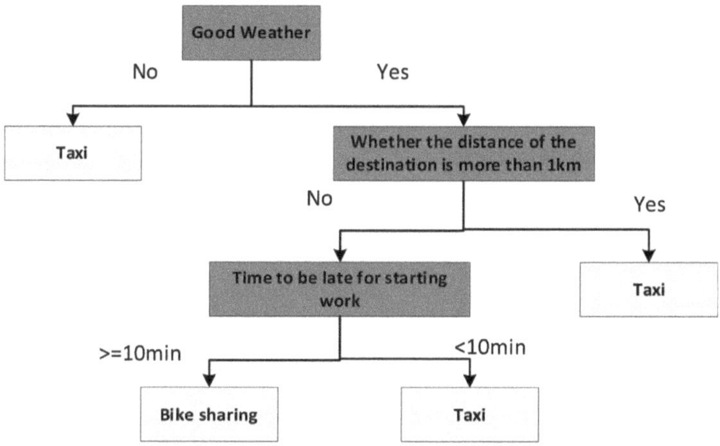

FIG. 5 Decision tree for classifications: choice between taxi and bike-sharing.

you want to or don't want to use, then we may say that the purity of the dataset is very high, since these data all belong to the same category.

If the proportion of kth class samples in sample D is $p^k(k=1,2,\ldots,|y|)$, then the information entropy of dataset D is:

$$\text{Ent}(D) = -\sum_{k=1}^{|y|} p^k \log_2 p^k$$

Based on the above equation, we can see that the smaller the value of information entropy, the higher the purity of sample set D.

Information gain

What is the relationship between information entropy and attribute choice? Assuming we have dataset D with discrete attributes. Now we try to divide these discrete attributes successively, and finally calculate their information gain, and we determine to use specific attributes according to the information gain of these attributes.

The definition of information gain is presented as follows:

$$\text{Gain}(D,a) = \text{Ent}(D) = -\sum_{v=1}^{V} \frac{|D^v|}{|D|}\text{Ent}(D^v)$$

Here, a represents a discrete attribute, Gain (D,a) is the information gain to attribute a, $|D|$ represents the total number of samples in the dataset, $v=(1,2,3\ldots,V)$ denotes the V values of a, $|D^v|$ represents a subset of samples when a takes v; generally speaking, the greater the information gain, the greater the "purity improvement" obtained by using attribute a for division. Therefore, the information gain is calculated by traversing all the attributes, and the attribute with the largest information gain will be regarded as the partition attribute.

The attribute partition of ID3 decision tree algorithm uses the information gain.

Rate of information gain

In fact, according to the equation of information gain, when v takes more values, that is, when a has more values, the Gain (D,a) becomes larger. In other words, when using information gain to select partition attributes, it tends to select those attributes with more values for partition. Apparently, such a partition does not have better generalization ability, so we have to use the gain rate to choose the partition attribute. The gain rate is defined as follows:

$$\text{Gain}_{\text{ratio}} = \frac{\text{Gain}(D,a)}{IV(a)}, \quad \text{where } IV(a) = -\sum_{v=1}^{V} \frac{|D^v|}{|D|} \log_2 \frac{|D^v|}{|D|}$$

$IV(a)$ is called the "intrinsic value" of attribute a. The more the number of a values, the larger it is, which balances the preference that information gain tends to choose attributes with more values. However, the information gain rate has a preference for the attributes with smaller numbers, so there is a heuristic rule: firstly, to find the attributes with higher information gain than the average level from the candidate partition attributes, and then to select the attribute with the highest gain rate.

C4.5 decision tree algorithm adopts the rate of information gain index to select the partition attributes.

Gini index

In addition to entropy, the purity of sample set can also be measured by Gini value:

$$\text{Gini}(D) = \sum_{k=1}^{|y|} \sum_{k' \neq k} p_k p_{k'} = 1 - w \sum_{k=1}^{|y|} p_k^2$$

Gini value reflects inconsistent probability for labels when two samples are randomly selected from the dataset. Therefore, the smaller the Gini value is, the higher the purity of the dataset has.

The Gini index of attribute a is defined as:

$$\text{Gini\_index}(D,a) = -\sum_{v=1}^{V} \frac{|D^v|}{|D|} \text{Gini}(D^v)$$

Therefore, when selecting the partition attribute, the attributes that make the Gini index minimum after the partition should be selected as the partition attribute.

Classification and regression tree (CART) uses Gini index as the attribute division index [11].

At last, does it use Gini index or entropy to select attributes? In fact, in most cases, the decision trees generated by the two attribute partition methods are basically identical, but the calculation for Gini index is more effective. Therefore, the default attribute partition method in Sklearn, which is a powerful machine learning repository based on Python, is Gini index, and the default decision tree model is CART. However, the Gini index tends to isolate classes with a larger number and assign them to one leaf node; while the entropy method tends to build a balanced tree, that is, the classes with a larger number may be dispersed into different leaf nodes.

3.4 Clustering analysis

When dealing with big data problems in the field of transportation, we cannot directly obtain the corresponding tags all the time, or the acquisition of tags needs to pay a large amount of cost. In this case, we usually conduct unsupervised learning. Different from supervised learning, the goal of unsupervised learning is to reveal the inherent nature and laws of data via learning unlabeled training samples, so as to provide a foundation for further data analysis. Clustering is the most widely used algorithm in such learning methods [12].

Clustering analysis can be used not only as a separate step to find the internal distribution structure of data but also as the precursor process of other learning tasks such as classification. Clustering analysis usually attempts to automatically divide the samples in the dataset into several disjoint subsets, each of which can become a "cluster." After such division, each cluster may contain potential concepts. It is worth noting that these concepts are unknown to the clustering algorithms, and the corresponding concepts behind them still require to be analyzed by users according to the actual situations, so as to carry out self-understanding and analysis.

Clustering analysis is one of the areas in machine learning with the fastest emerging algorithms [13]. One important reason is that there are no objective standards for clustering, but only the modeling ideas of clustering exist. Therefore, similarity measurement and dataset partition have no unified and objective labels for modeling quality analysis. Given a dataset, we may find that the previous algorithms cannot be covered by some standards all the time from a certain perspective, so as to design new algorithms. The classic K-means algorithm is achieved from many different fields. For example, Steinhaus [14], Ball and Hall [15], and Macqueen [16]. As one of the earliest widely used clustering algorithms, many other improved algorithms are derived based on K-means. Hierarchical clustering was first proposed by Johnson [17]. Hinneburg and Keim [18] proposed DBSCAN clustering algorithm based on density. Following this, Jain introduced the commonly used clustering algorithms and the challenges and key problems in the algorithm design. The methods and theoretical framework of clustering analysis are relatively mature, and it has been widely used in many fields, such as traffic big data analysis. For example, we can use clustering analysis to recognize the urban traffic congestion according to the speed, density, and other indicators, or study the driver's accident propensity according to the driver's characteristics, or identify the black spot of the accidents based on the road accident information.

The definition of cluster analysis is as follows:

Clustering analysis is the task of grouping a set of data objects based on the information describing objects and their relationships existing in data. Its purpose is to make sure the objects in one group are similar (related) to each other, while the objects in different groups are different (unrelated). If the greater similarity within groups, and the greater differences between groups, which indicates the clustering effect is significant.

The effect of clustering analysis depends on two factors: (1) similarity measurement and (2) clustering algorithms. Therefore, this chapter introduces clustering analysis from these two aspects

3.4.1 Similarity measurement

In the process of clustering algorithm modeling and performance evaluation, the calculation of the similarity is very important. The methods of similarity measurement can be listed as follows:

Distance measurement. Assuming each observation sample has d attributes, and dataset D can be represented by $m \times d$ dimensions matrix:

$$\begin{bmatrix} x_{11} & \cdots & x_{1d} \\ \vdots & \ddots & \vdots \\ x_{m1} & \cdots & x_{md} \end{bmatrix}$$

Distance mainly refers to Minkowski distance. The algorithm is shown in Eq. (9):

$$\text{dist}(x_t, x_s) = \sqrt[q]{\sum_{j=1}^{d} \left(x_{tj} - x_{sj}\right)^q} \tag{9}$$

When $q=1$, that is Manhattan distance. When $q=2$, that is Euclidean distance, which is the most widely used.

Similarity coefficient. It mainly refers to angle cosine and correlation coefficients. Correlation coefficient is widely used, and its main advantage is that it is not impacted by the original linear transformation, and can be easily converted into distance. However, when the dimension of the dataset is very high, its computation speed is much slower than distance measurement.

Kernel function (x,y). It is defined in a binary function. In essence, it also can reflect the distance between x and y, and the kernel function is to project the data from lower dimensional space to higher-dimensional space.

Dynamic time warping (DTW). DTW can be regarded as a very special distance algorithm, and only calculates the distance between two vectors with different lengths. It also can be used for matching two groups of data in two different time periods. DTW is mainly used for the cases involving time series data, here we do not make a specific analysis.

Numerical variable

But the disadvantage of the above distance calculation is that the importance of each variable $x_1, x_2, \ldots$ Is the same, but in fact, some variables are more important, and some are not so important. At this time, weight is introduced in the calculation of distance (e.g., when measuring the difference between a person's academic performance and another person's academic performance, the score outside the number of words is obviously more important than that of art and music), so

3.4.2 *Clustering algorithms*

K-means

The process of K-means algorithm can be shown as follows:

a. Select K initial centroids, which can be randomly selected, and each centroid can be regarded as a class.
b. Assign each observation to its nearest centroid to form a new class with the centroid.
c. Recalculate the centroid of each class. The so-called centroid is the average vector of all observations in a class (it is called vector due to the fact that each observation contains many variables, so we regard each observation as a multidimensional vector, and the number of dimensions is determined by the number of variables).
d. Repeat steps b and c.
e. Until the centroids do not change or reach the maximum number of iterations.

Objective function The core idea of K-means algorithm is to minimize the sum of squares (SSE) of the errors from obtained clusters $\{E_l | l = 1, 2, \ldots, k\}$ based on the given dataset $D = \{x_1, x_2 \ldots, x_m\}$. The calculation equation of SSE is as follows:

$$\mathrm{SSE} = \sum_{l=1}^{k} \sum_{x \epsilon E_l} \mathrm{dist}(\mu_l, x)^2$$

where k is the number of clusters, E_l is the first cluster, x is the sample, and μ_l is the cluster center of cluster E_l.

From the calculation process of K-means clustering algorithm, we can see that K-means algorithm is very sensitive to outliers in the dataset. If there are significant outliers in the sample data, the distribution of the data could be distorted greatly. K-Medoids algorithm is proposed to eliminate such sensitivity. The main difference between k-means and K-medoids algorithm is the selection method of center points. We do not provide detailed introductions here.

Hierarchical clustering

Hierarchical clustering is another commonly used clustering algorithm. The core idea of this algorithm is to regard every single observation as a class, then calculate the distance between different classes, select two most similar classes, and merge them into one. Following this, the new classes continue to compute the distance and merge into the nearest two classes. In such a way, only one class is left. A tree diagram is used to record the process, which contains the necessary information we need.

Algorithm The process of the algorithm is presented below:

a. Calculate the distance between different classes and record it with proximity matrix.
b. Merge two nearest classes into a new class.
c. Update the proximity matrix based on the new class.
d. Repeat b and c.
e. Stop until only one class left.

Density-based spatial clustering of applications with noise (DBSCAN)

DBSCAN is a commonly used density clustering method. Its core idea is to find out the scattered dense areas in the data. In short, this algorithm is to draw a circle. There are two parameters to define, one is the maximum radius of the circle, and the other is the minimum number of points contained in the circle.

Algorithm The process of the algorithm is presented below:

a. Randomly select core points from the dataset.
b. Take a core point as the center of the circle, draw a circle with radius V, and select the core points whose number of the points within the circle meets the density threshold, thus these points are core objects, and the points in the circle form a cluster, in which the density of core points can reach other surrounding solid origin points directly.
c. Merge these overlapping clusters.

Grid-based clustering

The principle of this algorithm is to divide the data space into grid cells, map the data objects to the grid cells, and calculate the density of each cell. Based on the given threshold to judge whether each grid cell is highly dense, the "class" is composed of adjacent dense cells.

Algorithm We may see the process of the algorithm as follows:

- **i.** Divide the data space into grid cells.
- **ii.** Based on the given threshold, determine whether the grid cells are dense or not
- **iii.** Merge adjacent dense grid cells into one class

4 Case study of data mining for MaaS: Bike sharing in Beijing during Covid-19 pandemic

In the previous section, we mainly introduced several commonly used data mining technologies in the field of transportation. In this section, we are going to present a case study of bike-sharing in Beijing during the Covid-19 pandemic, so as to demonstrate the role of data mining in urban traveler behaviors.

The data used in this study are obtained from the Traffic Information Center of Beijing Municipal Commission of Transport (BMCT) and mainly fall into the area within 6th ring road in Beijing. Therefore, the study area can be shown in Fig. 6.

To observe the travel behaviors of bike-sharing users, we plot the average trip time during the outbreak of the pandemic in Fig. 7. It can be observed that the average trip time of bike-sharing was approximately 1330–1360s before the pandemic. However, since human-to-human transmission of COVID-19 and the lockdown of Wuhan city were announced, the average trip time began to increase gradually, and then reached the highest on 6th Feb, approximately 1600s. Thereafter, the average trip time decreased to a relatively stable level, which was still higher than that before the pandemic.

The spatial distributions of origins and destinations from 14th Jan to 10th Mar for all bike-sharing trips are presented in Fig. 8 by Geographical Information System (GIS) software. As can be observed, most trips take place in Beijing's central areas, such as Dongcheng, Xicheng, Chaoyang, and Haidian districts.

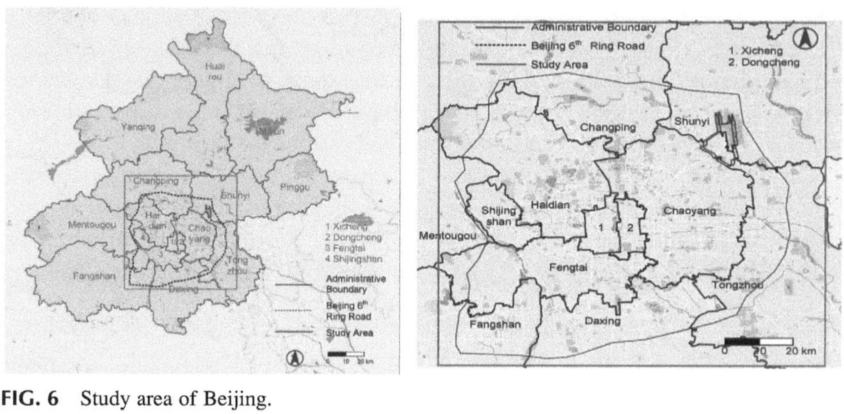

FIG. 6 Study area of Beijing.

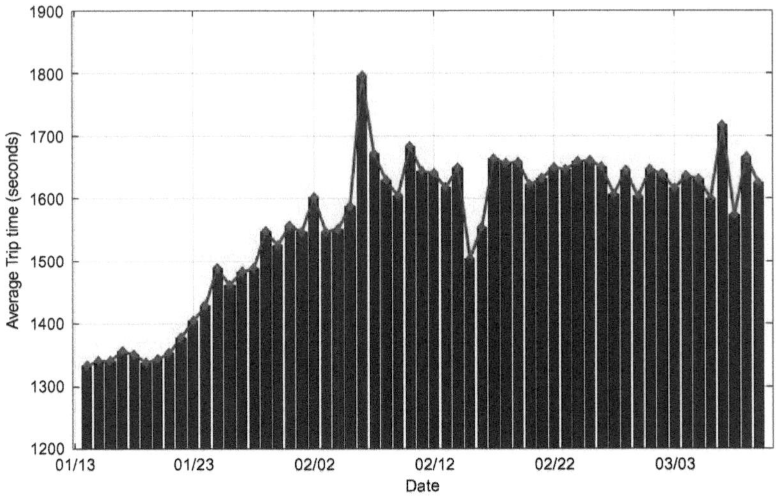

FIG. 7 Average trip time of bike-sharing in Beijing during the outbreak of the pandemic.

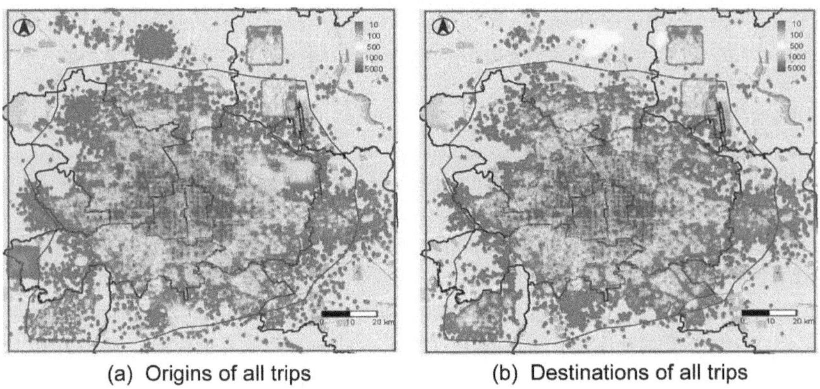

FIG. 8 The distribution of origins and destinations of all bike-sharing trips.

In order to visually observe the variations of spatial distribution of origins and destinations for bike-sharing trips during the COVID-19 pandemic, we select four different typical days to exhibit the trip distribution of bike-sharing at different stages of the COVID-19 outbreak, as shown in Fig. 9.

On 17th Jan, the Spring Festival travel season had begun, but there were still many trip orders of bike-sharing in Beijing. As can be seen from Fig. 9, most of the trip origins and destinations are distributed in central areas of Beijing, such as Dongcheng, Xicheng, the parts of Haidian and Chaoyang districts. It is expected that people's behaviors remain normal state before the announcement of human-to-human transmission and Wuhan lockdown. Thereafter, the COVID-19 panic widely spread in China, the government extended the holiday to 2nd Feb accordingly, so the users of bike-sharing in Beijing decreased greatly

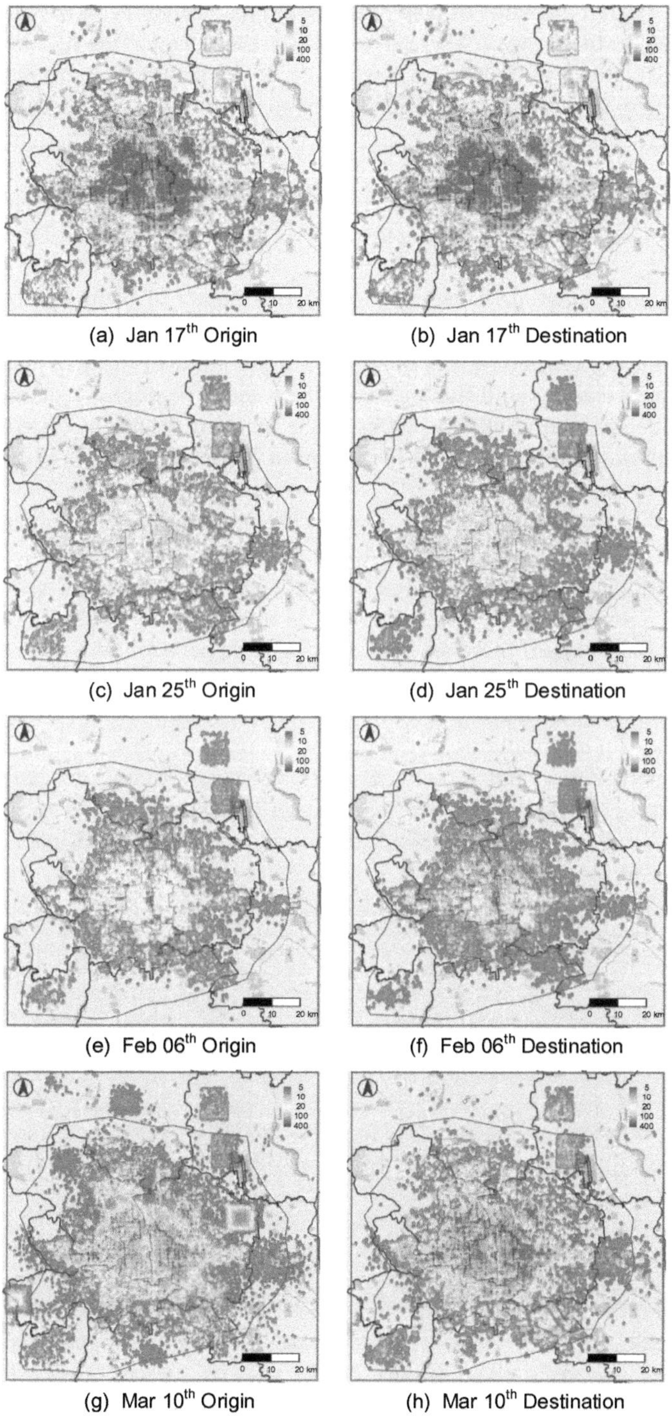

FIG. 9 Spatial distribution of origins and destinations for bike-sharing under COVID-19 pandemic.****

after 20th Jan. At the stage of pandemic spreading, we select the distribution of origins and destinations on 25th Jan to demonstrate the decline of bike-sharing orders in Beijing, and the trips at the city center were significantly reduced, as observed in graphs (c) and (d) of Fig. 9. Due to the strong measures taken by the Chinese government, such as restrictions on public transport, quarantining infected cases, and promoting wearing masks, this COVID-19 pandemic was effectively and quickly controlled. However, the number of bike-sharing trips did not recover but decreased to a lower level even after the end of the holiday on 2nd Feb.

At the early stage of work resumption, the impact of pandemic panic still existed, as can be observed from graphs (e) and (f) of Fig. 9, the trip number for bike-sharing kept going down. Afterward, the number of bike-sharing trips slowly climbed along with fluctuations, which showed people were reluctant to increase their traveling due to shocks of the pandemic, and the distribution of origins and destinations on 10^{th}Mar shown in graphs (g) and (h) of Fig. 9 had confirmed this point.

Afterward, we utilize complex network theory to explore the impacts of the pandemic on the user behaviors of dockless bike-sharing in Beijing. To build the network, firstly we divide the study area into small lattices with the size of 1 km × 1 km; following this, these lattices are defined as the nodes of the network, the links exist if there are bike-sharing trips between two nodes, and the number of trips within the lattices is the weight of the links. The complex networks based on bike-sharing trips are built, as shown in Fig. 10. We attempt to present the trip distributions in the proposed complex network by using links with different colors, but it does not work very well. Therefore, here we use node heatmap to exhibit the trip distribution of bike-sharing in Beijing from the perspective of complex network theory, which is presented in Fig. 11.

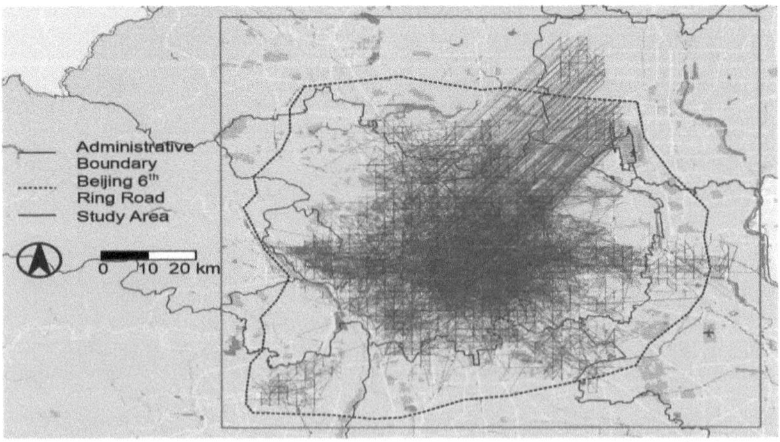

FIG. 10 Topology of the complex network based on bike-sharing trips in Beijing.

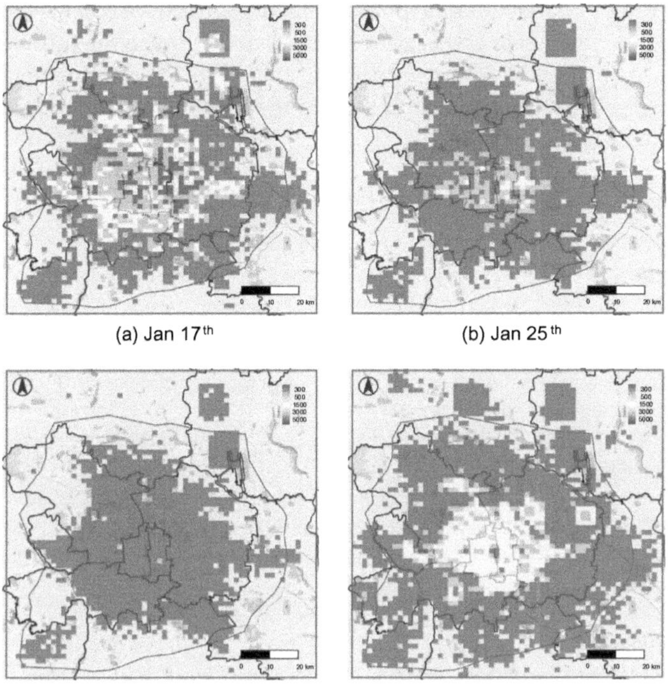

(a) Jan 17th (b) Jan 25th

(c) Feb 6th (d) Mar 10th

FIG. 11 Trip distribution of bike-sharing in the complex networks.

In Fig. 11, we can see that the trip distributions in the proposed complex networks are significantly different at different stages as the COVID-19 pandemic evolves in China.

Here we just present the case study of data mining for bike-sharing briefly, as, for the detailed description and analyses, we refer readers to Wen-Long Shang et al.'s paper "Impacts of user behaviors and environmental benefits of bike-sharing: A big data analysis" [19].

5 Summary of chapter

This chapter introduces the development, definition, and purpose of MaaS briefly, and puts forward the significance of data mining technologies for MaaS. In addition, this chapter briefly reviews the definition of data mining, its processing objects, classical steps and processes, and types of traffic big data. Following this, according to the characteristics of traffic big data, this chapter mainly focuses on the introduction of the support vector machine, linear regression, decision tree, and clustering analysis, which are commonly used in the field of transportation. Finally, a case study of data mining technology used for bike-sharing in Beijing during the Covid-19 pandemic is presented to demonstrate the role of data mining technologies in travel behaviors.

In the era of big data, data mining technologies have become an important means to obtain valuable and useful information and make scientific decisions making in many fields. It is worth noting that with the development of data mining technologies, various methods and theories have been very abundant. The content introduced in this chapter cannot fully cover all aspects of data mining due to the limitation of space. Interested readers can further learn and explore based on the clues provided in this chapter.

References

[1] Tan P-N, Steinbach M, Karpatne A, Kumar V. In: Introduction to data mining. 2nd ed; 2005. https://www-users.cs.umn.edu/~kumar001/dmbook/index.php.

[2] Schölkopf B, Smola A. Learning with kernels: support vector machines, regularization, optimization, and beyond. Cambridge: MIT Press; 2001.

[3] Liu J, Yan F, Yang X. The integration of Internet of things and logistics management and control. China Fortune Press; 2017.

[4] Cortes C, Vapnik V. Support-vector networks. Mach Learn 1995;20(3):273–97.

[5] Liu Z, Zhang W. Traffic big data: theory and methods. Zhejiang University Press; 2021.

[6] Quinlan JR. Learning efficient classification procedures and their application to chess end games. Mach Learn 1983;1.

[7] Vapnik NV. Statistical learning theory. vols. 10–11. New York, NY: Wiley; 1998. p. 10401–1492.

[8] Zhou Z. Machine learning. Beijing: Tsinghua University Press; 2016. p. 121–39.

[9] Géron A. Hands-on machine learning with Scikit-learn and TensorFlow: concepts, tools, and techniques to build intelligent systems. 2nd ed. O'Reilly Media; 2019.

[10] Quinlan JR. Simplifying decision trees. International Journal of Man-Machine Studies 1987;27(3).

[11] Breiman L. Classification and regression tree. New York: Routledge; 2017.

[12] Shang W-L, Chen Y, Bi H, Zhang H, Ma C, Ochieng YW. Statistical characteristics and community analysis of urban road networks. Complexity 2000;2020.

[13] Li H. Statistical learning methods. 7. Beijing: Tsinghua University Press; 2012. p. 95–135.

[14] Steinhaus H. Sur La division des corp materials en parties. Bull Acad Polon Sci 1956;1 (804):801.

[15] Ball GH, Hall DJ. ISODATA, a novel method of data analysis and pattern classification. Menlo Park, CA: Stanford Research Institute; 1965.

[16] Macqueen J. Some methods for classification and analysis of multivariate observation. In: Proceedings of the fifth Berkeley symposium on mathematical statistics and probability; 1967. p. 281–97.

[17] Johnson HC. Hierarchical clustering scheme. Psychometrika 1967;32(3):241–54.

[18] Hinneburg A, Keim DA. An efficient approach to clustering in large multimedia database with noise. KDD 1998;58–65.

[19] Shang WL, Chen J, Bi H, Sui Y, Chen Y. Impacts of user behaviors and environmental benefits of bike sharing: a big data analysis. Appl Energy 2021;285:116429.

Chapter 8

MaaS and IoT: Concepts, methodologies, and applications

Hongbin Xie[a], **Xuan Song**[a,b], **and Haoran Zhang**[b]
[a]Department of Computer Science and Engineering, South University of Science and Technology, Nanshan, China, [b]Center for Spatial Information Science, The University of Tokyo, Kashiwa, Chiba, Japan

1 Introduction

With the rapid development of the urban economy and the rapid increase of urban population, urban road traffic is facing increasing pressure, which often results in road congestion due to the lag in the construction of traffic infrastructure and improper traffic scheduling management. Increased emissions will also aggravate environmental pollution [1–3]. The increasingly serious traffic problems not only affect people's travel experience but also seriously hinder the development of the city's economy.

Regarding how to alleviate urban traffic problems, in addition to increasing investment and construction of transportation infrastructure to meet people's increasing travel needs, rational dispatch and management through the integration of existing traffic resources is a more direct and effective solution at present. In this regard, Mobility as a Service (MaaS) is a new concept proposed in recent years. The core concept is based on the travel needs of users, integrating transportation resources through science and technology, providing user-centric transportation services, and achieves the sustainable development of the whole urban traffic ecology. In recent years, many articles have given the concept and definition of MaaS. Hietanen [4] first proposed the concept of MaaS, hoping to provide users with customized and convenient travel services for a unified entrance by integrating different transportation methods. Jittrapirom et al. [5] pointed out that MaaS is a brand-new transportation solution that integrates different transportation modes and mobile services. Holmberg et al. [6] emphasized that MaaS needs to pay attention to users' travel needs and preferences, and allows users to independently make travel plans and choose different payment methods. Atkins et al. [7] pointed out that MaaS can provide a combination of different transportation modes and present it to users in a

Big Data and Mobility as a Service. https://doi.org/10.1016/B978-0-323-90169-7.00009-9

complete way, giving users more flexible, personalized, and on-demand transportation services.

MaaS hopes to improve the utilization of traffic resources and improve the quality of traffic services through the integration and sharing of traffic resources. Traffic data is an important part of traffic resources. There are massive traffic sensors that generate a lot of traffic data in cities every day. Effective use and management of these massive traffic data is a major problem in current urban traffic management [8–11]. By using the Internet of Things (IoT) technology, a large number of traffic sensors are connected to the Internet to realize the rapid upload, preprocessing, and sharing of traffic data. These traffic data can be used to analyze the travel needs of users and carry out the scheduling and management of traffic resources, so as to provide users with more comfortable and convenient intelligent transportation services [12, 13].

In this chapter, we will review and introduce the general concepts of MaaS and IoT in Section 2, as well as the challenges that IoT will face when applied to MaaS. Section 3 will introduce the key technologies and methods involved in the application of IoT in MaaS, with a detailed description of each technology. In Section 4, we will introduce a case study of IoT in MaaS. Finally, we will summarize the whole article and look forward to the future application of IoT in MaaS.

2 Overview of the concept

2.1 Overview of the general concept

The main goal of MaaS system based on IoT technology is to realize the interconnection of mass transportation equipment through IoT technology, the MaaS system can realize the interconnection and intercommunication of massive traffic equipment, improve the quality of traffic data collection and analysis, and effectively integrate traffic resources to provide users with better traffic services.

In the whole transportation ecosystem, transportation places, vehicles, and transportation equipment are the important components, and they are also the basic carrier of transportation services, which produce a massive amount of traffic data every day. It plays a vital role for MaaS to provide intelligent transportation services by obtaining and analyzing these traffic data to understand the travel scenarios and users' travel habits. MaaS system can use IoT technology to connect and manage traffic equipment while using IoT communication protocol to reduce the cost of traffic data transmission, and using IoT-based cloud computing, edge computing, big data, machine learning, block chain, and other IoT-based technologies for traffic data processing and analysis, and finally provide general users, governments, transportation departments, and research institutions with integrated intelligent transportation services such as travel ticketing, travel planning, travel epidemic prevention, traffic scheduling, data analysis, equipment management, and manual consultation on the same MaaS platform.

By combining IoT technology, the MaaS system can truly connect the entire transportation ecosystem, breaking down the cooperation barriers of traditional transportation modes and ending the market situation where each transportation system operates independently and is difficult to communicate with each other. In addition, it is more important to promote the transformation of the whole transportation service from platform construction oriented to user demand-oriented, so as to create a convenient transportation service construction system. In the long run, the MaaS system based on IoT technology not only solves the pain points of people's travel difficulties, but also realizes green travel through the integration, scheduling, and optimization of transportation resources, so it has a long-term sustainable development prospect.

2.2 Challenges of IoT application in MaaS

IoT is one of the core technologies for building a MaaS system. The combination of the two mainly faces the following challenges:

Equipment connection and management: In the MaaS system, all transportation resources need to be integrated, so there will be a large number of transportation equipment that needs to be interconnected.

Heterogeneity: it is necessary to consider the differences between different traffic fields, different traffic equipment, different network communication protocols, and different traffic data, and how to integrate and analyze the correlation is also an aspect that needs to be concerned [14].

Large-scale data: The MaaS system generates massive amounts of traffic data at all times, and IoT technology needs to have such a large-scale traffic data processing capability [15].

Security: In MaaS, the access of transportation equipment to the Internet and the interoperability of different transportation platforms mean that they may be attacked and invaded by viruses on the Internet. The IoT needs a complete security protection mechanism and strong response capability [16].

Distributed: The MaaS system has a wide range of communication, a large amount of calculation, and a complex network structure. It is necessary to establish a distributed system with the help of IoT technology to achieve resource sharing, rapid communication, system scalability, and high fault tolerance.

3 Key technologies and methodologies

3.1 Intelligent transportation equipment

Traffic signal lights, charging piles, gates, and other intelligent traffic devices are the most basic components of the MaaS system and the basic equipment of IoT technology. In addition to maintaining traffic conditions and providing

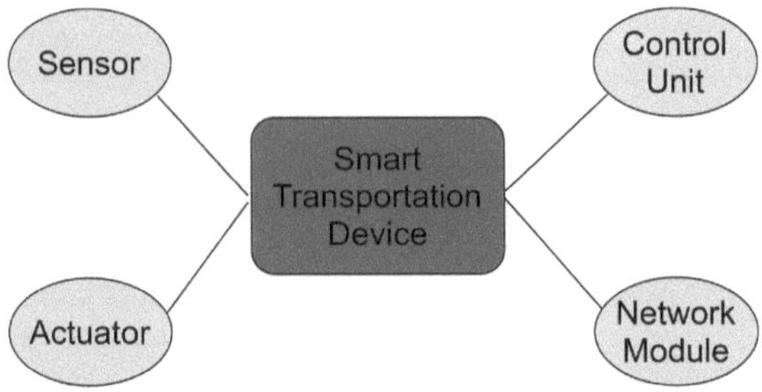

FIG. 1 The composition structure of intelligent transportation equipment.

traffic services, they are also responsible for sensing and collecting traffic data. Fig. 1 describes the composition structure of intelligent transportation equipment. Like other IoT devices, intelligent transportation equipment usually consists of four parts: sensor, control system, actuator, and network module, with the characteristics of low power consumption, small storage, and weak computing power [17–19].

Sensor: It can detect the information related to the traffic state and transform it into electrical signals that can be read or recorded according to certain rules, which is the primary link to realize intelligent traffic control.

Control unit: Responsible for receiving and processing signals generated by sensors and issuing control instructions for execution.

Actuator: Responsible for receiving the execution instructions of the control system and generating corresponding execution actions, generally divided into three types of actuators: pneumatic, hydraulic and electric.

Network module: It is responsible for transmitting the information generated by the sensor to the remote control center, and transmitting the instructions and data issued by the remote control center at the same time. The transmission mode is divided into wired and wireless.

3.2 Communication protocols for the Internet of Things

There are numerous intelligent transportation devices in the MaaS system, which need to realize information interaction with remote servers through the IoT communication protocol framework. IoT communication protocol model can be divided into the physical layer, data link layer, network layer, transmission layer, and application layer [20–23]. The communication protocol of each layer is shown in Fig. 2.

3.3 Microservices based on the Internet of Things

Compared with the traditional transportation system, MaaS system aggregates the services of different transportation systems and platforms, so the functions of the whole system will become very large and complex. In the design of

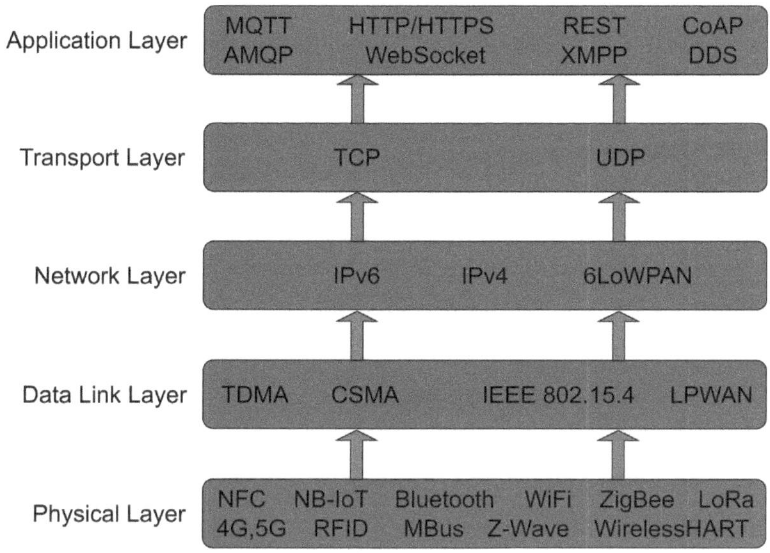

FIG. 2 Communication protocol of IoT.

system software architecture, the traditional transportation system adopts monolithic pattern. With the increase of system functions, the functions of monolithic systems will be closely coupled, unable to effectively and quickly troubleshoot problems, and it is difficult to achieve rapid iterative development. Moreover, it is unable to effectively cope with the sudden increase of call flow by increasing the on-demand scalability of temporary servers. Therefore, the maintenance cost of the entire monolithic system is very high, which seriously hinders business development. It can be seen that the traditional monolithic pattern is difficult to support and satisfy the complex MaaS system.

The emergence of microservice can effectively solve the defects and problems of a monolithic system. Microservices was first proposed by Martin Fowler and James Lewis in 2014 [24]. Microservices architectural style is an approach to developing a monolithic application using a set of small services, each running in its own process and communicating using a lightweight mechanism, usually HTTP API. These services are built based on business capabilities and can be deployed independently through automated deployment mechanisms. These services are implemented using different programming languages and different data storage technologies and maintain a minimum of centralized management [25–28]. Microservices have the advantages of service decoupling, independent deployment, rapid iteration, strong scalability, and high reliability. They can well match and satisfy the MaaS system architecture design in the selection of system architecture.

In the design of IoT technology-based microservice architecture of MaaS system, it can be divided into five layers: infrastructure layer, unified service layer, service governance layer, service distribution layer, and application layer. The infrastructure layer mainly contains resources such as computing, storage, network, and middleware. The responsibility of the unified service layer is to plan and coordinate the development of services, which can provide travel booking, travel planning, traffic signal control, payment, invoice, and other transportation services centrally. The service governance layer is mainly responsible for the registration, configuration, discovery, and management of services, and provides efficient service invocation functions. The service distribution layer realizes the load balance of the system call by managing the authority authentication and flow control of the service call to ensure the stable operation of the system. The application layer realizes that clients such as transportation vehicles, PC browsers, APPs, and traffic devices can call the services on the MaaS system through the network.

3.4 Cloud computing based on the Internet of Things

The rapid development of cloud computing technology in the past decade has greatly promoted the development of the Internet [29, 30]. The combination of cloud computing and the IoT can effectively support the construction of the MaaS system. Cloud computing based on the IoT can more effectively manage and analyze the massive traffic data of different structures, and it can also effectively improve and increase the transmission bandwidth of the entire transportation network, as well as providing powerful computing power and better security protection measures. Cloud computing mainly has three different service modes: Infrastructure as a Service (IaaS), Platform as a Service (PaaS), and Software as a Service (SaaS). IaaS mainly provides users with computing resources, such as network, storage, server, and virtual machine, by paying on-demand [31, 32]. PaaS mainly provides users with a basic environment for building applications and provides end-to-end distributed software development, deployment, operating environment, and complex application hosting as services to users [33]. SaaS can provide users with application services and allow their users to connect to applications and access applications through the Internet [34–36].

The adoption of cloud computing technology based on the IoT in the MaaS system is a major trend in the future. The advantages of cloud computing such as dynamic scalability, flexibility, on-demand use, high reliability, and security can be utilized to provide better transportation services. In the design of MaaS system of cloud computing based on the IoT, it can be divided into perception layer, IaaS layer, PaaS layer, SaaS layer, and client layer. In the perception layer, the IoT technology is used to connect a large number of intelligent transportation equipment and upload the collected traffic data to the cloud. The IaaS layer can provide basic computing resources such as networks and servers. In the PaaS layer, it is mainly divided into two parts: application support platform and external support platform. Application support platform can provide middleware, operating system, message management components, GIS service components, and other resources. External support platforms can provide data resources of different transportation platforms by integrating systems and platforms in the whole transportation field, such as bus platforms, taxi platforms, railway platforms, civil aviation platforms, shipping platforms, and bike-sharing platforms. The SaaS layer aggregates multiple transportation services based on different vehicles, such as trip planning, ticketing service, one-click payment, traffic signal control, operational analysis, etc. The client layer includes various terminals such as APP, PC browser, command center, and electronic screen. In addition, the whole cloud computing system will also support cloud management services, including security mechanism, resource management, billing management, and monitoring services.

In terms of deployment mode, cloud computing based on the IoT can also be divided into private cloud, public cloud, and hybrid cloud [37, 38]. Private cloud is a cloud service specially built for the internal use of a customer or organization, with high security and independence. The public cloud refers to the cloud services provided by third-party providers for ordinary users, who can generally access and use them through the Internet. The hybrid cloud combines the private cloud with the public cloud. Users can store their internal important data in the private cloud, and at the same time take advantage of the abundant computing resources in the public cloud.

3.5 Edge computing

Although the cloud computing technology based on the IoT provides good scalability and reliability for the MaaS system, due to the wide area and quantity of transportation equipment, the centralized cloud computing service center will have the problems of large computing pressure, insufficient bandwidth, and low delay. With the development of autonomous driving technology, terminal traffic equipment needs more powerful computing power and faster response speed, so as to provide better traffic services for users. How to solve the calculation demand of traffic terminal equipment is also the main problem that the MaaS system needs to face. By using edge computing technology, a platform integrating network, computing, storage, and application core capabilities is built on the side of traffic equipment [39–41], which can produce faster response speed and reduce the load of cloud MaaS system.

Edge computing service systems based on IoT technology can effectively improve the terminal computing capacity of MaaS, which mainly includes the perception layer, edge service center, and remote cloud service center. In the perception layer, the intelligent transportation device can upload data to the nearest edge service center, thus reducing the distance of data transmission. At the same time, the edge service center can issue instructions to the intelligent transportation equipment at a faster speed and improve the response speed. The edge service center includes functional modules such as small data center, authorization, and authentication, IoT middleware, small server, data processing analysis, etc., which can undertake the computing tasks of the edge devices of the MaaS system, and also greatly reduce the pressure of the remote cloud service center. The edge service center regularly uploads and downloads data to the remote cloud service center, and the remote cloud service center provides advanced services such as resource management, big data center, and machine learning to the edge service center. Edge computing technology can effectively solve the calculation problem of the MaaS system in the last mile, and while improving the response speed, it also reduces the cost of the system.

3.6 Security technologies for the Internet of Things

IoT technology realizes the interconnection and intercommunication of a large number of intelligent transportation equipment in the MaaS system, but at the same time, it also exposes these transportation equipment to the Internet, making it face more network attacks and threats.

As more and more traffic equipment will involve the user's travel data, once attacked, the user's private data will be leaked. What's more serious is that the entire MaaS system will stop service due to attacks, causing traffic chaos throughout the city.

Table 1 describes the common security problems and solutions of IoT [42–45]. In the dimension of the data transmission chain, IoT security issues can be divided into three categories: physical layer issue, network layer issue, and application layer issue. The physical layer security issues refer to that the attacker launches attacks against the communication process on the physical layer and the data link layer, mainly including Jamming adversaries, spoofing attacks, and physical interface attacks. Network layer security issues refer to attacks launched by attackers against the network layer and transport layer of the IoT, mainly including RPL routing attacks, fragment replication attacks, and buffer retention attacks. Application layer security issues refer to the attack launched by the attacker against the application program running on the IoT, mainly including COAP security problem and application layer interface attack.

Security protection is the top priority of the MaaS system, which determines whether the entire system can operate normally. In addition to avoiding the aforementioned IoT security issues and vulnerabilities, it is also necessary to establish a security prevention and control system during the establishment of the MaaS system, and establish a security management mechanism and emergency plan during the daily maintenance and operation of the system to ensure the normal operation of MaaS.

TABLE 1 Common security problems and solutions of IoT.

Issue type	Security issue	Issue description	Solution
Physical layer issue	Jamming adversaries	Attackers emit radio frequency signals to interfere with the normal receiving data of wireless device	Compare with the threshold of attack detection by detecting signal strength
	Spoofing attacks	Attackers disguise themselves as other devices by forging MAC values, and consume network resources to the full so that legitimate nodes are denied access	Detection by signal strength and channel evaluation
	Physical interface attacks	Attackers control devices and launch attacks by hacking into insecure physical interfaces	Hardware avoids the use of the USB interface when unnecessary, and avoids the access of the debugging tool to the hardware device
Network layer issue	RPL routing attacks	Attackers exhaust resources and eavesdrop on information by attacking the IPv6 routing protocol of the low-power and Lossy networks	Strengthen identity verification and monitor abnormal behavior of nodes
	Fragment replication attacks	Attackers interfere with normal legitimate data assembly and exhaust network resources by copying a large number of IPv6 packet fragments	In order to prevent replay attacks, data fragments can be verified by using hash chains
	Buffer retention attacks	Receiving nodes often need to reserve buffers to receive and process data. Attackers attack buffers by sending incomplete data packets, resulting in a buffer overflow and unable to process other normal data packets, and ultimately resulting in the consequences of denial of service	Prevent attacks by establishing a mechanism to verify the complete transmission of fragment information in the buffer
Application layer issue	COAP security problem	Attackers cause network congestion and denial of service by targeting vulnerabilities in the COAP protocol for the Internet of Things	TLS/DTLS and HTTP/CoAP mapping
	Application layer interface attack	Attackers steal private data by attacking the IoT application layer service interface	Perform security scanning and vulnerability detection on the interface, and use the https protocol at the same time

4 Application and case study

4.1 Background introduction

At the end of 2019, a major pandemic of novel coronavirus (COVID-19) broke out around the world, seriously affecting the lives and health safety of people around the world. It also has a great impact on the social economy [46–48]. The COVID-19 is very contagious, mainly through close contact with droplets spread. Therefore, during the pandemic period, people are most concerned about how to avoid the key areas of the pandemic and COVID-19 patients, as well as the risk of infection when traveling. At the same time, the relevant government departments are also thinking about how to quickly and accurately find out the close contacts of the confirmed patients during their travels. Thus it can be seen that the outbreak of COVID-19 has seriously affected people's travel safety [49–51], and it is extremely urgent to find a set of solutions that can effectively solve people's travel safety problems during the pandemic.

Here we proposed a comprehensive solution for travel safety based on IoT and MaaS technologies. By using Bluetooth matching technology of mobile devices to perceive the contact between users and others, users can grasp the risk of infection of COVID-19 in real-time in the travel process. This comprehensive solution use AI and big data technology to model and analyze the spread of infectious diseases, to provide a reference for relevant departments to formulate different public epidemic prevention policies. Through this comprehensive solution, users and relevant government departments can use the service functions of epidemic travel safety on a MaaS system, which can effectively meet the needs of users and government departments for travel safety and epidemic prevention during the epidemic.

4.2 System framework

The goal of this comprehensive solution for epidemic prevention and travel safety is to design a MaaS system based on IoT technology for travel safety requirements during the epidemic. The framework of this comprehensive solution is divided into five layers in total. The first layer describes the travel scene during the epidemic. When a user has close contact with others during travel, their mobile device can be connected to other people's mobile devices through Bluetooth scanning. At the same time, the system will record the contact's ID, contact time, and contact distance in the background. In addition, Bluetooth sensing devices can also be installed at high-frequency contact points in public transportation or airports and other public travel places. In addition, Bluetooth sensing devices can also be installed at high-frequency contact points of public transportation vehicles or airports, and other public travel places. When users appear in these public transportation vehicles or public travel places, their mobile devices will connect with these Bluetooth sensing devices and generate secret contact records, which can be used to assess the infection risk of customers' current travel. The second layer is the data layer, which collects sensor data such as Bluetooth signal strength, acceleration data, and GPS data generated by the user's mobile device during the trip to judge the user's mobile state and close connection state during the trip. The third layer is the computing layer, which realizes the storage, calculation, and analysis of data through big data, cloud computing, machine learning, and security technologies. The fourth layer is the service layer. Through the calculation and analysis of data, travel safety services such as health passcode, surrounding real-time risk, infection risk of the past 14 days, travel data analysis and policy simulation will be provided on the same system platform. The fifth layer is the user layer. Ordinary users can use the health code of the MaaS system to pass quickly when entering public transport places during the epidemic, as well as to inquire about the risk of infection when traveling. In this layer, CDC, hospitals, and relevant government departments can quickly and accurately check close contacts through the system, which solves the traditional epidemic investigation methods' pain points of low accuracy, long time, and high cost. Transportation departments can use the system to conduct data analysis and policy simulation for traffic control and human flow distribution during epidemics.

4.3 Core function

The MaaS system based on travel safety needs during the epidemic includes an application program running on the mobile device side (Referred to as "APP"), and a simulation and deduction platform for infectious disease transmission based on human flow big data and AI technology (Referred to as "Deduction platform").

Users use the health code on the homepage of APP to take public transportation and know their risk of infection in the past 14 days. Green means safety, and red means risk of infection. By clicking on the real-time surrounding risk function of the APP, users can check the surrounding infection risk in real-time during the travel, and at the same time check the surrounding air quality, so as to avoid places with poor air quality and reduce the risk of infection. In addition, users can also query the contact history and infection risk of a certain day in the past. Through the announcement function, the system can push relevant travel and epidemic prevention notices to all users. In terms of privacy protection, the APP uses a variety of advanced technologies to protect users' private data. Users do not need to provide any personal information throughout the entire process, achieving true anonymity.

Based on the close connected data and mobile status data collected by APP, through the integration, processing, and analysis of multimodal human travel big data with AI algorithm, the deduction platform has realized the fine-grained modeling and iterative training of the spread of the epidemic virus and the infected population, and finally can output the visual results of related predictions and simulations. At the same time, it realizes the quantitative evaluation of implementation results of different epidemic prevention measures individually or in combination, providing a reference for relevant government departments to formulate different public epidemic prevention policies.

5 Conclusion and future directions

The combined application of MaaS and IoT is an interdisciplinary research field with great research value and development prospects. This paper describes the current development status of the combination of MaaS and IoT, and focuses on the key technologies and methods involved in the application of IoT in MaaS. At the same time, it introduces a MaaS system based on IoT technology and the safety requirements of travel prevention and control during the epidemic. Finally, this paper concludes by listing several future research directions and open questions.

Autonomous driving: With the development of technology, autonomous driving is the development trend of human travel in the future [52], and it is also the key research direction of the combination of IoT and MaaS in the future, but how to use smarter sensors to perceive the environmental information around the vehicle and the state of the vehicle itself, how to improve the calculation speed for large and complex data during driving, and how to improve the accuracy of driving decision-making, these are the core challenges faced by autonomous driving.

5G technology: More and more scenarios in the MaaS system require faster data transmission speed and response speed, such as automatic driving. The emergence of 5G technology greatly improves the speed and bandwidth of data transmission, and truly meets the low delay requirements of the MaaS system. 5G technology can well promote the application of IoT technology in the MaaS system and make the MaaS system more intelligent, which is also worth paying attention in future research on the combination of IoT and MaaS [53].

Block chain: In the application process of IoT in MaaS, many sensors will involve sensitive data such as the location information and movement status of users. To ensure the privacy security of users is also one of the core functions of MaaS. By introducing block chain technology and taking advantage of its features such as decentralization, anonymity, consensus mechanism, and encryption function. It can effectively assist the MaaS system to achieve user privacy protection, which is also an important research direction of MaaS in the future.

AR/VR: The increasing demand for travel experience is a major challenge for the MaaS system. The introduction of AR and VR technology is expected to greatly improve the user experience with MaaS. AR technology can provide real-time three-dimensional space navigation experience by combining the real environment and GPS navigation function, and can also provide entertainment and leisure content based on augmented reality during travel, as well as assist people to adapt to the transition from manned driving to unmanned driving. VR can simulate vehicle driving and travel scenarios through virtual reality technology, so as to improve users' driving safety technology and travel experience.

Standard formulation: During the application of IoT in MaaS, on the one hand, due to large number of manufacturers of intelligent transportation equipment, the specifications of the same device will be quite different. On the other hand, there are different types and standards of IoT communication protocols, leading to increased communication selection costs. These different standards will undoubtedly increase the difficulty of connection and integration in the MaaS system, and will greatly hinder the application and development of IoT on MaaS. Therefore, how to develop a set of IoT standards suitable for the MaaS field is a major challenge in the future.

References

[1] Wen Y, Zhang S, Zhang J, Bao S, Wu X, Yang D, et al. Mapping dynamic road emissions for a megacity by using open-access traffic congestion index data. Appl Energy 2020;260:114357.

[2] Wu Y, Zhang S, Hao J, Liu H, Wu X, Hu J, et al. On-road vehicle emissions and their control in China: a review and outlook. Sci Total Environ 2017;574:332–49.

[3] Qian Y, et al. Review of the state-of-the-art of particulate matter emissions from modern gasoline fueled engines. Appl Energy 2019;238:1269–98.

[4] Hietanen S. Mobility as a Service. New Transp Model 2014;12(2):2–4.

[5] Jittrapirom P, Caiati V, Feneri AM, Ebrahimigharehbaghi S, González MJA, Narayan J. Mobility as a service: a critical review of defifinitions, assessments of schemes, and key challenges. Urban Plan 2017;2(2):13.

[6] Holmberg P-E, et al. Mobility as a Service—MaaS: describing the framework. Tuesday; February 16, 2016.

[7] Anon. Atkins. In: Journeys of the future. Introducing Mobility as a Service; 2015. Retrieved from: http://www.atkinsglobal.com//media/Files/A/Atkins-Corporate/uk-and-europe/uk-thought-leadership/reports/Journeysofthefuture_300315.pdf.

[8] Chen Y-Y, Lv Y, Li Z, Wang F-Y. Long short-term memory model for traffic congestion prediction with online open data. In: Proc IEEE 19th int conf Intell Transp Syst (ITSC), November; 2016. p. 132–7.

[9] Ma X, Tao Z, Wang Y, Yu H, Wang Y. Long short-term memory neural network for traffic speed prediction using remote microwave sensor data. In: Transp Res C Emerg Technol, 54; 2015. p. 187–97.

[10] Han L, Zheng K, Zhao L, Wang X, Wen H. Content-aware traffic data completion in ITS based on generative adversarial nets. IEEE Trans Veh Technol 2020;69(10):11950–62.

[11] Lv Y, Duan Y, Kang W, Li Z, Wang F-Y. Traffic flow prediction with big data: a deep learning approach. IEEE Trans Intell Transp Syst 2015;16(2):865–73.

[12] Majumdar S, Subhani MM, Roullier B, et al. Congestion prediction for smart sustainable cities using IoT and machine learning approaches. Sustain Cities Soc 2021;64:102500.

[13] Syed AS, Sierra-Sosa D, Kumar A, et al. IoT in smart cities: a survey of technologies, practices and challenges. Smart Cities 2021;4(2):429–75.

[14] Cheng S, Lu F, Peng P, et al. Short-term traffic forecasting: an adaptive ST-KNN model that considers spatial heterogeneity. Comput Environ Urban Syst 2018;71:186–98.

[15] Liu Y, Yang C, Sun Q. Thresholds based image extraction schemes in big data environment in intelligent traffic management. IEEE Trans Intell Transp Syst 2020.

[16] Khan MA, Salah K. IoT security: Review, blockchain solutions, and open challenges. Futur Gener Comput Syst 2018;82:395–411.
[17] Tang J, Sun D, Liu S, et al. Enabling deep learning on IoT devices. Computer 2017; 50(10):92–6.
[18] Martinez B, Monton M, Vilajosana I, et al. The power of models: Modeling power consumption for IoT devices. IEEE Sensors J 2015;15(10):5777–89.
[19] Huh S, Cho S, Kim S. Managing IoT devices using blockchain platform. In: 2017 19th international conference on advanced communication technology (ICACT). IEEE; 2017. p. 464–7.
[20] Firouzi F, Chakrabarty K, Nassif S, editors. Intelligent Internet of Things: from device to fog and cloud. Springer Nature; 2020.
[21] Al-Sarawi S, Anbar M, Alieyan K, et al. Internet of Things (IoT) communication protocols. In: 2017 8th International conference on information technology (ICIT). IEEE; 2017. p. 685–90.
[22] Song T, Li R, Mei B, et al. A privacy preserving communication protocol for IoT applications in smart homes. IEEE Internet Things J 2017;4(6):1844–52.
[23] Andy S, Rahardjo B, Hanindhito B. Attack scenarios and security analysis of MQTT communication protocol in IoT system. In: 2017 4th international conference on Electrical Engineering, Computer Science and Informatics (EECSI). IEEE; 2017. p. 1–6.
[24] Lewis J, Fowler M. Microservices—a definition of this new architectural term, http://martinfowler.com/articles/microservices.html; 2014. [Accessed 29 June 2016].
[25] Thönes J. Microservices. IEEE Softw 2015;32(1):116.
[26] Dragoni N, Giallorenzo S, Lafuente AL, et al. Microservices: yesterday, today, and tomorrow. Present Ulterior Softw Eng 2017;195–216.
[27] Zimmermann O. Microservices tenets. Comput Sci Res Dev 2017;32(3):301–10.
[28] Jamshidi P, Pahl C, Mendonça NC, et al. Microservices: the journey so far and challenges ahead. IEEE Softw 2018;35(3):24–35.
[29] Hayes B. Cloud computing; 2008.
[30] Varghese B, Buyya R. Next generation cloud computing: new trends and research directions. Futur Gener Comput Syst 2018;79:849–61.
[31] Bhardwaj S, Jain L, Jain S. Cloud computing: a study of infrastructure as a service (IAAS). Int J Eng Inform Technol 2010;2(1):60–3.
[32] Manvi SS, Shyam GK. Resource management for Infrastructure as a Service (IaaS) in cloud computing: a survey. J Netw Comput Appl 2014;41:424–40.
[33] Beimborn D, Miletzki T, Wenzel S. Platform as a service (PaaS). Bus Inf Syst Eng 2011;3 (6):381–4.
[34] Cusumano M. Cloud computing and SaaS as new computing platforms. Commun ACM 2010;53(4):27–9.
[35] Palos-Sanchez PR, Arenas-Marquez FJ, Aguayo-Camacho M. Cloud computing (SaaS) adoption as a strategic technology: results of an empirical study. Mob Inf Syst 2017.
[36] Rani D, Ranjan RK. A comparative study of SaaS, PaaS and IaaS in cloud computing. Int J Adv Res Comput Sci Softw Eng 2014;4(6).
[37] Diaby T, Rad BB. Cloud computing: a review of the concepts and deployment models. Int J Inform Technol Comput Sci 2017;9(6):50–8.
[38] Jadeja Y, Modi K. Cloud computing-concepts, architecture and challenges. In: 2012 international conference on Computing, Electronics and Electrical Technologies (ICCEET). IEEE; 2012. p. 877–80.
[39] Shi W, Cao J, Zhang Q, et al. Edge computing: Vision and challenges. IEEE Internet Things J 2016;3(5):637–46.

[40] Abbas N, Zhang Y, Taherkordi A, et al. Mobile edge computing: a survey. IEEE Internet Things J 2017;5(1):450–65.
[41] Varghese B, Wang N, Barbhuiya S, et al. Challenges and opportunities in edge computing. In: 2016 IEEE international conference on Smart Cloud (SmartCloud). IEEE; 2016. p. 20–6.
[42] Khan MA, Salah K. IoT security: review, blockchain solutions, and open challenges. Futur Gener Comput Syst 2018;82:395–411.
[43] Dvir A, Holczer T, Buttyan L. Vera—version number and rank authentication in RPL. In: 2011 IEEE eighth international conference on mobile ad-hoc and sensor systems; 2011. p. 709–14. https://doi.org/10.1109/MASS.2011.76.
[44] Brachmann M, Keoh SL, Morchon OG, Kumar SS. End-to-end transport security in the ip-based Internet of Things. In: 2012 21st international conference on Computer Communications and Networks (ICCCN); 2012. p. 1–5. https://doi.org/10.1109/ICCCN.2012.6289292.
[45] Xu W, Wood T, Trappe W, Zhang Y. Channel surfing and spatial retreats: Defenses against wireless denial of service. In: Proceedings of the 3rd ACM workshop on wireless security, WiSe'04, ACM, New York, NY, USA; 2004. p. 80–9. https://doi.org/10.1145/1023646.1023661. http://doi.acm.org/10.1145/1023646.1023661.
[46] Velavan TP, Meyer CG. The COVID-19 epidemic. Tropical Med Int Health 2020;25(3):278.
[47] Oliver N, et al. Mobile phone data for informing public health actions acrossthe COVID-19 pandemic life cycle. Sci Adv 2020;6. eabc0764.
[48] Pfefferbaum B, North CS. Mental health and the Covid-19 pandemic. N Engl J Med 2020;383 (6):510–2.
[49] Chen K, Wang M, Huang C, et al. Air pollution reduction and mortality benefit during the COVID-19 outbreak in China. Lancet Planet Health 2020;4(6):e210–2.
[50] Chinazzi M, Davis JT, Ajelli M, et al. The effect of travel restrictions on the spread of the 2019 novel coronavirus (COVID-19) outbreak. Science 2020;368(6489):395–400.
[51] Aloi A, Alonso B, Benavente J, et al. Effects of the COVID-19 lockdown on urban mobility: empirical evidence from the city of Santander (Spain). Sustainability 2020;12(9):3870.
[52] Nagel HH, Enkelmann W, Struck G. FhG-Co-Driver: from map-guided automatic driving by machine vision to a cooperative driver support. Math Comput Model 1995;22(4-7):185–212.
[53] Javaid N, Sher A, Nasir H, et al. Intelligence in IoT-based 5G networks: opportunities and challenges. IEEE Commun Mag 2018;56(10):94–100.
[54] Wang H, You F, Chu X, et al. Research on customer marketing acceptance for future automatic driving—a case study in China City. IEEE Access 2019;7:20938–49.

Chapter 9

MaaS system visualization

Chuang Yang, Renhe Jiang, and Ryosuke Shibasaki
Center for Spatial Information Science, The University of Tokyo, Kashiwa, Chiba, Japan

1 Overview of the general concept

With the development of Information and Communications Technology (ICT) and the rise of the sharing economy, the human travel experience has undergone a radical change. Smart cards (i.e., IC cards) have been widely equipped in public transportation systems (PTS). A wide range of navigation, car-hailing, and ticketing software has emerged and gradually integrated into people's daily lives. The consumption of transportation tools (e.g., bicycles, private cars, etc.) has also started to shift away from purchase ownership to rent sharing, which has dramatically increased the efficiency of transportation tool usage. However, different transportation systems (e.g., city bus system and taxi-hailing system) are often independent. People often need to switch back and forth between various apps or IC cards during a complete trip.

Mobility as a Service (MaaS) is an emerging concept in the transportation domain, which treats the mobility demand as a service, hoping to liberate individuals from ownership-based transportation to an access-based paid service [1]. It aims to integrate existing transportation modes seamlessly (e.g., public transit, online car-hailing, bicycle sharing, car-sharing, railways, and airlines) into one unified platform and provide the services of one-click trip planning, booking, and payment. It has been attracting much attention from all over the world since proposed.

Data visualization is a method of data representation in a visual form. Given the raw data, visualization maps it into a graphical representation, making it easier to perceive, understand, and utilize (such as the statistical charts we commonly use). MaaS, as the future one-stop mobility solution, needs the cooperation of multiple actors, including the demanders of mobility, providers of transport services, platform managers, and city administrators [1]. Thus, it needs to integrate massive amounts of data from different data providers. For example, routes, stations, schedules, and vehicle GPS data from bus operators, travel demand data from passengers. Besides, with the MaaS system running, new data is generated (such as GPS trajectory data of moving objects, daily

Big Data and Mobility as a Service. https://doi.org/10.1016/B978-0-323-90169-7.00010-5

TABLE 1 Heterogeneous multi-source data of transportation system.

Static data	• *Routes and planned schedules*: PTS (bus, metro, light rail, tram, railway, airline, ship) • *Stops*: bike sharing, PTS • *Map data*: road network, point of interests (POI)
Dynamic data	• *GPS trajectories data*: passenger, taxi, car-hailing, bike sharing, PTS. • *Transaction data*: smart card data, taxi trip data, online ride-hailing data, bike sharing data • *Real schedules*: PTS • *Demand data*: passenger (OD + preference data) *Monitoring data*: lane detector data, lane camera data

delay data of metros, and ridership data). Table 1 summarizes these heterogeneous multisource data. These massive amounts of data contain a considerable amount of information, knowledge, and insights, giving rise to the demand to perceive, understand, and utilize them, which visualization technology has native potential for such scenarios.

However, in this era, the development of MaaS is in the early stage, and the local MaaS systems are in the pilot phase. There is no well-established visualization solution for current MaaS systems. Nevertheless, visualization technologies are pervasive and well-researched in the traditional transportation domain, which built a solid foundation for future MaaS systems. For instance, given the origin-destination (OD) data, Google Maps[a] visually helps users plan their trips by displaying the alternative routes on the map and enabling users to revise the trips by themselves, such as adding waypoints and setting time preferences based on the OD data of London's sharing bike, Wood et al. [2] devised a series of visualization views to perceive and understand the dynamics of sharing bicycle scheme, which is helpful to guide the station rebalancing and service maintenance; driven by the taxi trajectory data, Wang et al. [3] proposes a visual reasoning approach to help the city manager evaluate the transportation of urban roads. All the above works could be treated as references and integrated into the future MaaS system for a better user travel experience. Thus, in this paper, we summarize the existing visualization technologies of the current transportation system into a multiview frame, covering the perspective of the three MaaS's leading actors, i.e., demanders of mobility, suppliers of transport services, and city managers. For each actor, we analyze the requirements and explicate how data visualization aids it.

The remaining part of this paper is structured as follows. In Section 2, we elaborate on the key visualization technologies for each kind of stakeholder.

a. https://www.google.com/maps/

In Section 3, three real-world application cases from different actor perspectives are introduced. Besides, we recommend some open-source visualization tools and libraries for references. Finally, in Section 4, we conclude the current research works and propose some future challenges and directions.

2 The key visualization technologies in MaaS for different stakeholders

2.1 The perspective of demanders of mobility

From the perspective of demanders of mobility, the visualization requirements mainly focus on trip planning (i.e., designing and developing practical and user-friendly trip planning tools and views). Map providers such as Google, Yahoo!,[b] Baidu,[c] and Tencent[d] have already put much effort into trip planning tools. These tools build a bridge between complex user requirements (Table 2) and feasible trip plans, allowing the user to express their requirement through a visual way (i.e., well-designed UI). Then, the tools translate the requirements into understandable language for the computer (e.g., SQL). Once the query results are obtained, these tools visualize the candidate routes for users to decide. The visualization of the candidate routes consists of two parts. One part contains

TABLE 2 User demand data including OD data and preference data.

OD data	The origin and destination location of the trip. Usually entered manually or marked on the map directly
Preference data	Expressing the concrete requirements of users. Usually set directly in the user interface (UI), including: • *Travel mode preference*: driving, public transport, walking, and cycling. For each selected mode, further settings are supplied. For example, under the public transport mode, Google Maps provides the options like vehicle types (bus, train, subway, tram, light rail, etc.), fewer transfers, less walking, and wheelchair accessibility for selection • *Time preference*: intended arrival time, intended departure time, and fastest route • *Price preference*: cheapest route • *Waypoints preference*: expected path points. Users can edit waypoints on the map directly or input manually

b. https://map.yahoo.co.jp/
c. https://map.baidu.com/
d. https://map.qq.com/

only semantic information, which draws the timeline of each route in combination with critical features (travel duration, expected cost, transfers, etc.), giving the user a quick and intuitive perception of the route's characteristics. The user initially filters routes through this view and then makes decisions in conjunction with the map view. The map view presents route-specific spatial displacement information, allowing users to sense the spatial movement and perform secondary planning (e.g., interactive adding waypoints on the map). To further improve the user experience, some works have optimized the visualization method of candidate routes. For example, Wunderlich et al. [4] designed a candidate routes visualization method that considers delay uncertainty and its impact, on which we will elaborate in Section 3.

In addition to the above trip planning tools for general purposes, there are also efforts concentrating on specific purposes, such as for tourists. Vansteenwegen et al. [5] devised a tourists-oriented route planning application for touring attractions, allowing the route planning for five cities in Belgium. Not restricted to the OD and start and end time setting, the application allows the user to take the rest (i.e., the expected rest periods, rest duration) and the level of interest for different attractions (four interest levels available) into consideration. After submitting the personalized demand data, the server returns a constraint satisfied route with a detailed schedule. However, the work does not support transportation planning between attractions. Perhaps in the future, the MaaS system (e.g., Whim[e]) could integrate such a function for a more comfortable tourist service.

Besides trip planning, guiding users to start and complete trips is also required for the MaaS system. Supposing a user who has purchased a monthly unlimited mobility package of Whim wants to ride a bicycle or drive a car to someplace (or merely wants to hang around by driving a car). In that case, the MaaS system needs to guide him/her to find the available vehicles nearby, which is a fundamental visualization technique that all car-sharing apps possess. This function needs to display the vehicles nearby the user in real-time (Fig. 1A). If the vehicles are dense, it also needs to use hierarchical clustering visualization (Fig. 1B) or heat map (Fig. 1C) methods to avoid spatial clutter.

2.2 The perspective of supplier of transportation service

As the operation of the service, a large amount of data was generated and collected (such as the smart card data, GPS trajectory data, delay data, etc.). For transportation service suppliers, it is beneficial to perceive and understand these data for improving and enhancing their service and business. We will discuss how data visualization functions toward this target from two aspects: monitoring (Section 2.2.1), analysis, and optimization (Section 2.2.2).

---

e. https://whimapp.com/

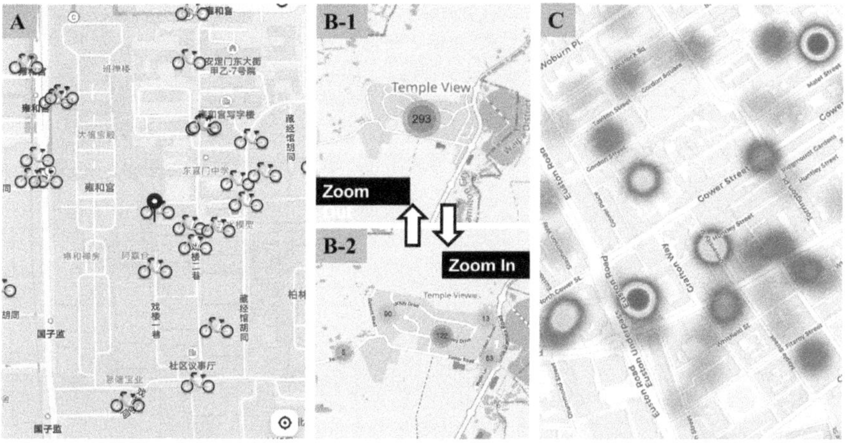

FIG. 1 (A) The available bicycles nearby the user in real-time; (B) Marker Cluster (https://github.com/Leaflet/Leaflet.markercluster) View for hierarchical clustering displaying; and (C) Heatmap (https://github.com/Leaflet/Leaflet.heat) of GPS points.

2.2.1 Monitoring

Object movement monitoring

With the prevalence of GPS technology, the technology to visualize the real-time location of moving objects has been widely studied and applied in practice. Bast et al. [6] presents an advanced frontend rendering and efficient backend response technology for smoothly visualizing the massive public transit objects' movement on the map in real-time. If the user clicks on an object, the detailed information of that object will be displayed, and the movement of the selected object will be tracked. Fig. 2A exhibits an example of such monitoring in New York.[f] In Fig. 2, we also provide the real-time monitoring cases of the railway in Munich[g] and the London Underground[h] for reference. In addition, Pu et al. [7] designed a Cell-Glyph to monitor the taxi's real-time movement, in which both the real-time and historical information is encoded. The cell's size and fill color are associated with the vehicle's instant speed, and the inner arc of the cell encodes the taxi status (loaded/vacant). It is very helpful for the taxi company to survey the operating conditions of the vehicles in real-time.

Operation status monitoring

It is crucial to monitor the operation status in the operation process, by which the problems can be dealt with timely once occurred. Here we summarize two common types of monitoring which employing visualization.

f. https://tracker.geops.ch/
g. http://s-bahn-muenchen.hafas.de
h. https://traintimes.org.uk/map/tube/

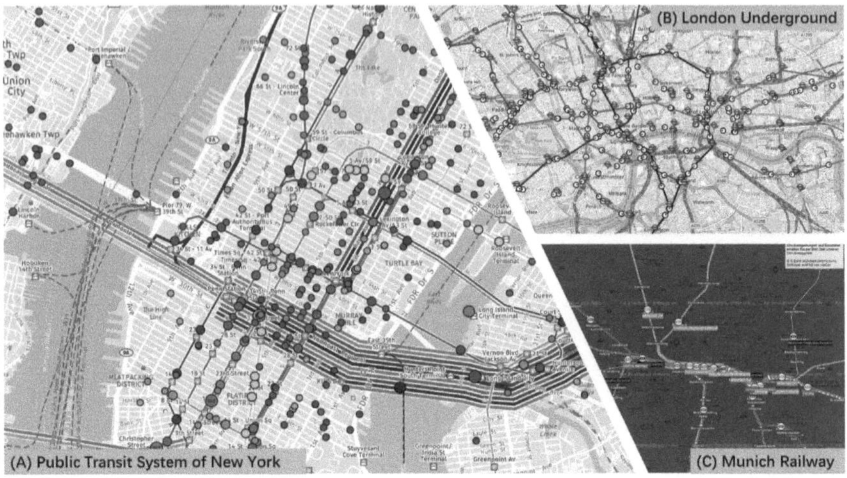

FIG. 2 The real-time location monitoring pages of (A) New York PTS, (B) London Underground, and (C) Munich Railway, respectively.

Delay monitoring. Delay is one of the most prevalent problems in public transportation systems. Most of the commonly used delay visualization methods are derived from a variant of the Marey Graph (a public transportation schedule visualization method first used by Marey Etienne-Jules in 1885). Fig. 3 shows a simple instance of the Marey Graph. The horizontal axis represents the time, and the vertical axis expresses the stations (the distance between stations corresponds to the actual geographic distance). Each continuous line denotes one complete service, with the slope of a single line depicts the speed between stops (the steeper, the faster). Through this chart, users can visually perceive the departure frequency and the speed between stations. Ochiai et al. [8] modified the Marey Graph by coloring the single lines between stations according to the delay time. Importing the historical running data, the user can locate the stations and lines that face severe delay problems. Graffagnino [9] presented a use case of Marey Graph at Swiss Federal Railways, where they

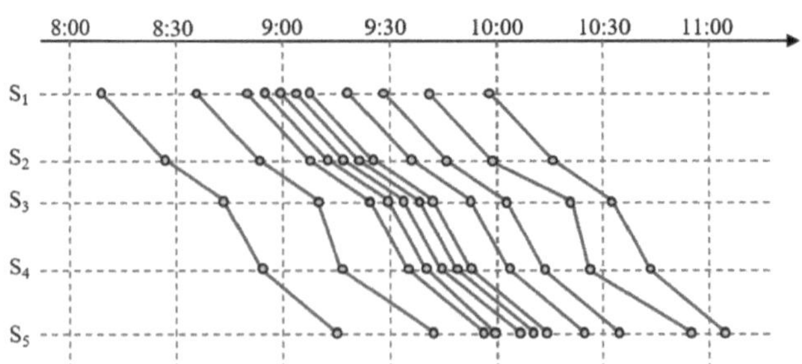

FIG. 3 A case of Marey Graph.

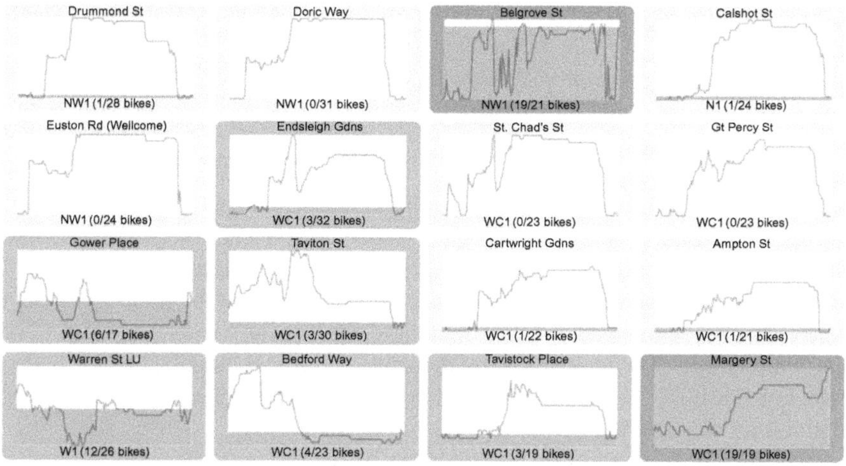

FIG. 4 The monitoring view of bike station.

designed a variation of it to monitor compliance with schedules of specific routes. Besides, they also proposed a novel box plot view to top-down monitor the delay at the station, line, and vehicle level. Ushida et al. [10] presented a series of variant Marey Graphs re-designed by Tokyo Metro Co. Ltd. for monitoring delay-related characteristics, with which at a glance to know how severe the delays are, where they appear and disappear, and how widely they are propagated.

Station and line status monitoring. Inspired by Marey Graph, Palomo et al. [11] devised a novel visualization method, with a similar form of Marey Graph, but utilizing kernel density estimation (KDE) to plot heat maps of different attribute values (such as station reliability, waiting time, and inter-station speed). It can be used to measure the difference between planned services (scheduled) and real-service, assisting the maintenance staff of PTS in identifying problems and anomalies at the station and route level. Wood et al. [2] proposed a novel visualization view that can be used to monitor the real-time status of the bike stations in London (Fig. 4). They use a sink-like glyph to represent a station, and the current number of the station's docked bikes is summarized as the height of the blue bar in the "sink." In this way, the users can overview the whole scheme's status at any point at any time. Besides, the glyph also shows the curve of available bikes number for the last 24 h (the black curve in Fig. 4) for understanding the past trend. It is worth mentioning that they also found the apparent tidal phenomenon with stations in this view.

2.2.2 *Analysis and optimization*

The final purpose of monitoring is to identify problems and solve them. The visualization-assisted analysis and optimization of services have been widely studied for such purposes by the community and will be useful for future MaaS systems.

Analysis and optimization system: Based on the collected bus data (bus stop data, bus route data, and bus fare card records), Weng et al. [12] devises a bus network visual analytics system. It enables the user to interactively discover the deficient bus routes and find the optimal replacements from the complex bus network. Ning et al. [13] presents Cheetah VIS, a visual analytical system built upon Spark[i] for efficient massive urban bus data analysis. It integrates the route analysis function, enabling the time-vary passenger flow analysis at the top level, and hourly forward/backward bus number analysis at the route level. Ferreira et al. [14] built a visualization system to explore the trip data of New York city interactively. Using this system, they can explore the imbalance in taxi service. For example, they find the lack of taxi service in Harlem and apply further exploration to show the potential causes.

Analysis view: In addition, the OD flow visualization technologies are also compelling for transportation system analysis. Using a devised OD flow map, Wood et al. [2] discover hub stops from a large number of historical trajectories of bikes. In the section III-D of Ref. [15], the existing OD flow visualization technologies are thoroughly summarized. Moreover, data is the foundation for visual analytics, yet data is not perfect and has many problems (e.g., data missing, location shifting, and false alarm). Hamad and Quiroga [16] uses geovisualization to analyze data integrity and the spatial distribution of data quality, allowing users to have a deeper and intuitive understanding of the quality of the collected sensor data. Station-level volume analysis is also helpful in detecting anomalies and improving the service quality. Mike Barry and Brian Card designed an interactive visual analysis view[j] toward the historical stational volume of the Boston subway. Users can understand station traffic patterns and detect anomalies intuitively and efficiently. Besides, they also designed a view to analyze the relationship between passenger flow and delay.

2.3 The perspective of city manager

With the operation of MaaS, the dynamic generation of large amounts of data is also helpful for city management, thus improving the human travel experience. For example, trajectory data has been widely used for urban planning [17], emergency management [18], and traffic jam monitoring [19]. However, due to these data's highly complex spatial-temporal characteristics and heterogeneous nature, it is not easy for city managers to perceive, understand, and utilize these data directly. This section will exhibit some visualization works from three-scale (road intersection, road, and city scale) that can assist city managers in better maintain the city's travel services.

---

i. https://spark.apache.org/
j. http://mbtaviz.github.io/

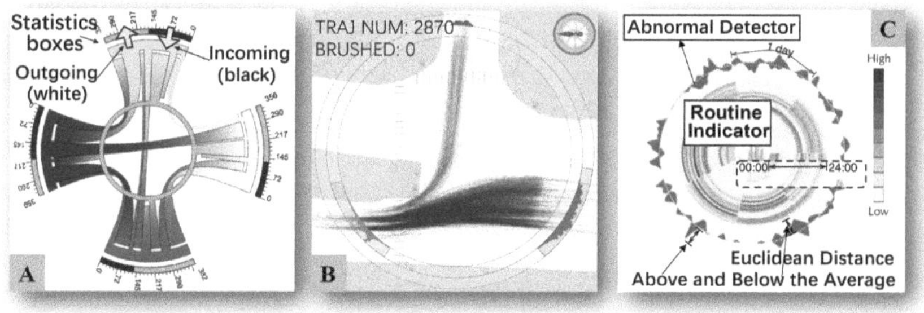

FIG. 5 (A) Interchange ciroco diagram. (B) Traffic view in TripVista. (C) Road fingerprint view.

Road intersection scale: Zeng et al. [20] proposed a novel visualization view (the interchange ciroco diagram, Fig. 5A) to reveal the redistribution of moving objects derived from massive trajectory data. It can help the city managers to perceive and understand the spatial interchange patterns at the road intersection level in a compact way. Based on the traffic data collected through laser scanners and other auxiliary devices at the road intersection level, Guo et al. [21] proposed an interactive visual analytics system (TripVista) to explore and analyze microscopic traffic patterns and abnormal behaviors (e.g., different ways of cornering dangerous and illegal movement behaviors).

Road scale: Wang et al. [3] proposed a visual reasoning approach for the road-level transport assessments (e.g., distribution of different driving behaviors, lane-level distribution of vehicle speed and flow). Pu et al. [7] designed a road fingerprint view (Fig. 5C) to abstract the temporally related attribute distribution and variations of road. Users can efficiently find patterns and abnormalities in the bound temporal attribute.

City scale: City-scale analysis is a broad topic. Here we list three main areas.

- *Traffic jam analysis*. Wang et al. [22] presents an interactive visualization system for analyzing traffic jams and their propagation in different roads based on taxi GPS trajectories. Yang et al. [23] proposes a star topology view for event correlation analysis, which can be used to trace traffic congestion at the city scale.
- *Traffic accident analysis*. Anwar et al. [24] proposes a visualization method (Traffic Origin) to aid the traffic management controllers in understanding the relationship between traffic accidents (road accidents and vehicle breakdown) and traffic conditions. VanDaniker [25] designs a tool (the transportation incident management explorer) to visualize the real-time and archived traffic accident data, assisting the accident manager capture the complete picture of the event within the shortest time.
- *Urban planning analysis*. Based on the vast amount of real-world travel data, Zeng et al. [26] proposes a set of visual views to evaluate the travel efficiency of PTS (accessibility, alternative routes, etc.) in different

locations, which is helpful for urban planning. Huang et al. [27] proposed a new visual analytics approach toward large-scale taxi trajectory data, revealing the importance of the various parts of the traffic network (such as time-varying hubs and backbones) for urban planning optimization.

3 Real-world application and case study

In this section, we first present three real-world cases for different kinds of stakeholders, respectively. Then some practical open-source tools and libraries are introduced according to our experience. Hoping to provide helpful insight into developing the visualization technologies for future MaaS systems.

3.1 Case for demanders of mobility

Trip delays are a common dilemma that people encounter when traveling. It disrupts the planned trips and results in a series of cascading influences, e.g., bringing additional waiting time, even missing the scheduled transfer, and essential appointments. If users could refer to the historical delays information when planning the trip, it would be possible to avoid the above cascading influences. However, existing trip planners did not contain such information or merely shown them literally (e.g., The flight/train has a historical late arrival rate of 10%, the average late time is 30 min). Thus, the user may be (1) directly unaware of the possible delay or (2) spend too much extra time on planning with the limited delay information.

Here, we introduce a novel visualization view devised for such pain points designed by Wunderlich et al. [4]. The proposed view conveys the historical/expected delay uncertainty, its impact, and potential alternative simultaneously in an intuitive manner, as shown in Fig. 6. The well-designed user studies demonstrated the superiority of their proposal. Given the origin

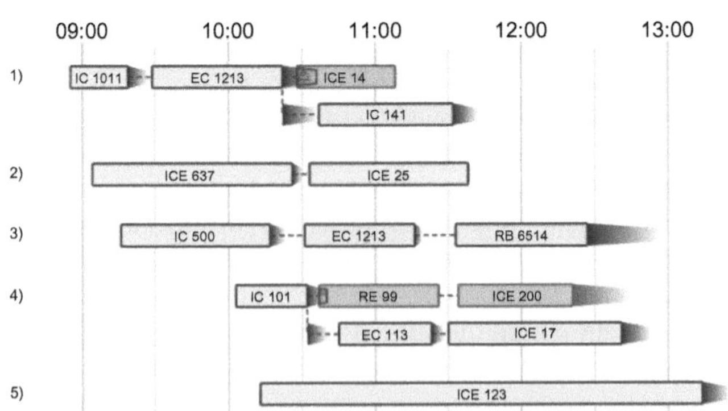

FIG. 6 Candidate trip plans with cumulative delay information integration.

and destination stations, it displays the possible trip plans row by row (totally five displayed in this Fig. 6), including the basic trip plan information and delay information.

Basic trip plan information: A trip plan may include one or several trip units (i.e., trains), represented by the dark blue bordered rectangles and connected by the dashed line. The text in the rectangle shows the train number. The horizontal position of the left and right border depicts the departure and arrival time of that train, respectively. Sometimes, several alternative transfers exist, and it is clarified by the branching dashed line.

Delay information: The historical delay uncertainty is encoded in the trailing gradient trapezoid, representing the cumulative delay probabilities in this case (both cumulative and non-cumulative supported in the study)—dark blue to white means the probability from high to low. If the trailing gradient trapezoid overlaps with the connection trains, the connection train may be missed due to the delay. The color of connecting trains will be set to grey under the circumstances, and a warning mark is drawn for duplicated encoding to emphasize the impact of such delay. Through this design, the authors neatly conveyed the message of delay and its potential impact. According to the case shown in Fig. 6, if the user had a crucial appointment at 12:30 PM, trip plan 2 should be the best choice. However, if there is no delay information, the user probably selects trip plan 3 or 4.

In summary, this case demonstrates a novel visualization design for better train trip planning, considering the delay uncertainty and its impact. The author performs a rigorous visual view generation process in their paper: *Requirements Gathering → Resulting Design Goals → Propose Candidates Design → Experiment Design and Evaluation*, which is very referential for the UI design of future MaaS systems.

3.2 Case for supplier of transportation service

In the field of public transportation, there are often deviations between planned and actual services. There are many reasons for these deviations: episodic, such as traffic accidents, or structural, such as a new residential development in a specific area. It is beneficial and essential for transport providers to perceive such differences, identify problems in the actual service, and develop appropriate solutions (e.g., schedule updates, adding additional vehicles.). With the development of ICT technology, the operating conditions of vehicles can be tracked in detail. This difference can be revealed by comparing historical data with planned data. However, historical data is often vast and complex, containing multiple routes, stops, and dates, and cannot be used directly. Service providers are not professional computer scientists, for whom simple and usable visualization tools are the better choices. The traditional Marey Graph cannot handle this problem because the considerable amount of data can cause severe

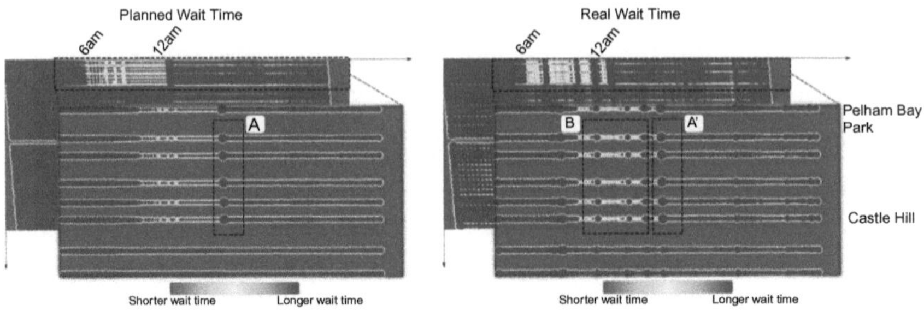

FIG. 7 Planned and actual wait time visualization of New York Subway line 6 on weekdays.

clutter. Therefore, based on the Marey Graph, Palomo et al. [11] proposes a method to visualize actual and planned operation information.

Fig. 7 displays an example of this visualization method. It compares the planned and actual wait time at station level of New York Subway line 6 on weekdays. Like the Marie graph, the vertical axis represents the time, and the horizontal axis represents the station. However, no line is drawn between the two consecutive stations but only points because this example is for station-level feature visualization (actually, the author provides a trip-level visualization of the view where there are links between stations. We just throw in a few ideas here). Besides, the main difference with Marey Graph is that this view utilizes a heat map to plot points/lines, preventing the clutter problem of classical Marey Graph when drawing massive historical actual schedules (e.g., 1-month data). The heat degree of the points in the heatmap is calculated using kernel density estimation (KDE), supporting various feature values (e.g., wait time, delay). Thus, we can observe the wait time difference between stations.

Let us go back to this case. The color from blue to red mapping the heat degree from low to high, thus revealing the wait time length. Compared to the planned graph, there are three intuitively different periods of waiting time: 8 a.m., 10 a.m., and 11 a.m., in which the waiting time is longer than planned. Besides, for the other periods of region B, the waiting time is less than planned. With these insights, the user could dig out the wait time problem at some stations and periods, then do some further investigation.

In summary, the main features of their work are as follows:

- Support planned and actual service comparison of multiple operational parameters at trip-level (Deviation and Speed) and station levels (Delay, Wait time, and Reliability).
- Is a variant of Marey Graph combining heatmap and KDE, supporting visualization of large-scale public transportation historical data.
- Interactive with good interactive rates can filter data by time range and station; allows custom heatmap color scheme; and can set heatmap detail scale (coarser to finer).

3.3 Case for city manager

In this task, we demonstrate how does a city manager explores the traffic flow between two main districts of Shenzhen using Vaite system [23]. Vaite is a visualization-assisted interactive big urban trajectory exploration system could help users find interesting insights from large-scale urban trajectories interactively.

The manager first draws two polygons roughly to specify Futian District and Longhua District, as the polygons shown in Fig. 8A. Vaite incurs a sub-trajectory extraction operator (extracting all the sub-trajectories between A and B), and the results are visualized in Fig. 8A. As show in it, the heavier the color in that road, the more sub-trajectories pass it. Thus, the manager could know which are the main paths between two districts and their corresponding popularity them. One more step further, the manager could analyze the traffic flow of a chosen road (i.e., the left side road in Fig. 8B), the visualization result is shown in Fig. 8C. The blue and brown lines indicate the traffic flow of the chosen road segment over one week on both two directions, and the red line indicates the net flow.

The results could help the manager to understand the characteristics of the resident travel between two districts. For example, the peak of road flow appears at 22:00 every day, by contrast, there are less trip appear during working

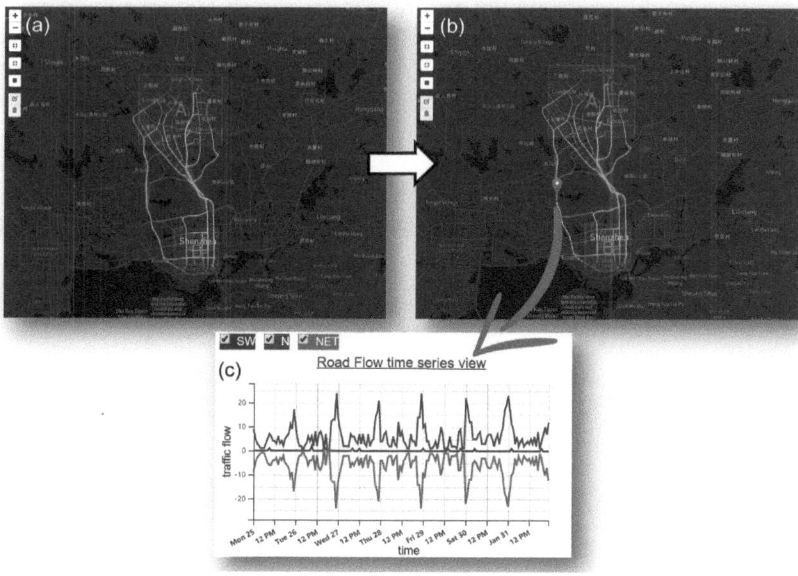

FIG. 8 Traffic flow exploration. (A) Popular paths between Futian District and Longhua District. (B) Select a road segment for traffic flow exploration. (C) Traffic flow of the selected road segment in (B).

hours (i.e., 9:00–17:00), and nearly all of the trips on this road are from Futian to Longhua. While exploring the traffic flow on the other popular paths, the results are similar. That is because Futian is a central business district of Shenzhen, most people work there but less live in. People prefer to choose public transportation to the office in the morning (brown line) and take a taxi home after normal working hours (blue line). This condition may have a relation with the 996-incentive system, the taxi fee will be paid by the company if people work overtime.

3.4 Open-source visualization tools and libraries

At the end of this section, we introduce some practical open-source tools and libraries for future MaaS systems.

Mobmap[k]: Mobmap is a tool that specializes in visualizing and analyzing trajectory data (both living and historical data are supported), developed by Shibasaki Lab of the University of Tokyo and has been used internationally and in a variety of applications (e.g., real-time movement monitoring, road network reconstruction, velocity map generation). For satisfying different requirement scenarios, Mobmap (1) allows users to bind the attribute value of moving objects with the display status (color, size), (2) supports customized color schemes, (3) enables setting the length of the trailing tail of moving objects, and (4) support the customized visualization of historical trajectories. For example, by streamingly importing the historical trajectories of 664 electric taxis in Shenzhen, China, Mobmap reproduces their real-time movement at that time. Fig. 9A shows the snapshot of them at 9 AM, 10/22/2014. The color (from white to blue) of the moving objects represents the velocity (from high to low), and the trailing tail's length is set to 180s, drawing the locations of the past 3 min. By which users can not only observe the real-time location but also perceive the spatial distribution of velocity. For more details, please refer to our website. *Potential Usages for MaaS:* The supplier of transportation service can use this

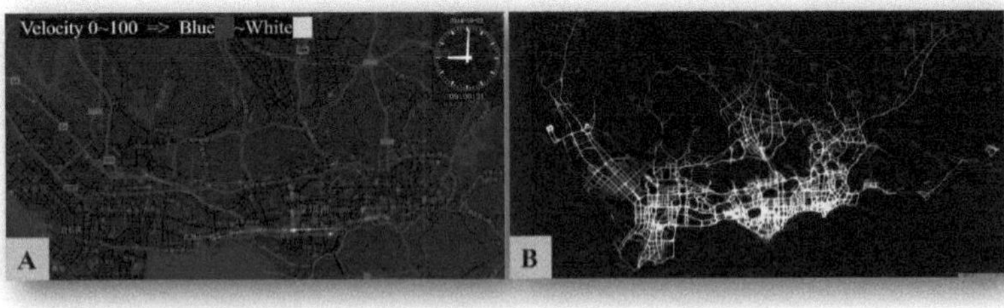

FIG. 9 (A) The real-time movement monitoring case using Mobmap. (B) Road network reconstruction case using Mobmap.

k. http://webapp.mobmap.net/

tool to real-time monitoring their vehicles. The city manager can use this tool to perceive and analyze the rhythm of the city. Besides, road network reconstruction could be helpful for urban planning (Fig. 9B).

Flowmap.blue[l]: Flowmap.blue is a tool for visualizing the geographic flow data that contains the origin and destination (OD) attributes. It has been used for representing the commuting patterns, migrations, transportation trips, etc. Users can upload their data satisfying the required format to visualize online. Fig. 10 shows an example of visualizing the metro OD flow network of the Tokyo Metropolitan Area, which we will combine for the introduction:

- *Basic information*: The circles represent nodes in the network, which could be a subway station or multiple, depending on the scaling. Zoom-in or zoom-out operation (Fig. 10A and B) that leads to the scaling change will trigger the node splitting or merging. Thus, users can observe the flow trend with different spatial levels. The arrow direction encodes the flow direction. The shade and the width of the arrow duplicate-encode the magnitude, which prominently reduces the clutter.
- *Interactions*: Users can click the node to filter all the flow of it (Fig. 10C). Moreover, filtering by drawing a polygon is also supported. When the mouse hovers on the node, its detailed information will be displayed for reference.

Potential usages for MaaS: The supplier of transportation service can use this tool to perceive, understand, and analyze their service network, finding the drawbacks. The city manager can use this tool to analyze the city-wide transportation network for urban planning.

D3.js[m]: D3 is a widely used JavaScript visualization library that supports a wide variety of visual charts. Users can devise secondary designs on-demand based on existing charts. Here, we introduce two visualization charts for

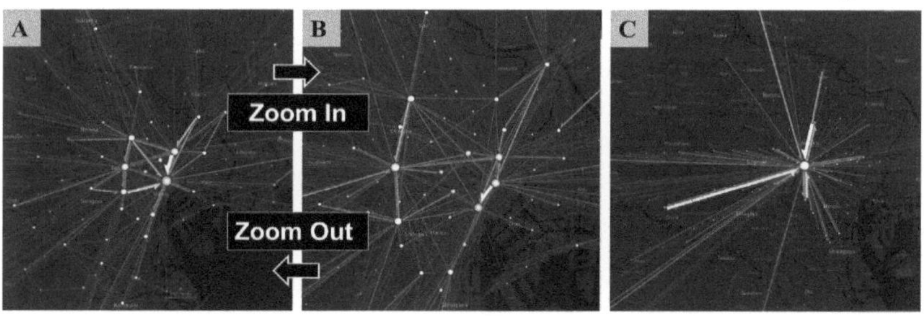

FIG. 10 (A) Visualize the OD flow network of the Tokyo Metropolitan Area's metro by Flowmap. (B) The result of further zoom-in on (A). (C) The OD flow visualization of Shijuku Station.

l. https://flowmap.blue/
m. https://d3js.org/

plotting the temporal distribution of data, which is helpful for perceiving the location/road/region-based time-varying attributes (e.g., hourly mobility demand over 1 week). For more detail, please refer to their website.

- *Calendar chart*: Generally, the horizontal and vertical axes represent two different time units (e.g., hours on the horizontal axis and weeks on the vertical axis). Given a color scheme, the data trend over a period of time could be observed with the cumulative effect on axes. For example, Wang et al. [22] proposes a road speed view to display information about the temporal distribution of speed on the road. Fig. 11A shows an example of such a chart. It is a calendar chart of traffic jam events, i.e., the taxis had a relatively slow speed (i.e., <5 km/h) and movement distance larger than 100 m at Dongmen Middle Rd, Shenzhen. It includes 7 × 24 grids or pixels, and the number of aggregated events in each period (i.e., hourly in Fig. 11A) are shown by varying the degree of color. A data analyst could identify the outlier or discover trends among these traffic jam events through this view. For example (1) 1:00–5:00 on Tuesday is abnormal as the other days in this week had traffic jam events at that time. It provides insights for data analysts for further exploration. (2) As shown in Fig. 11A, 20:00–2:00 (next day) is a peak period in Dongmen Middle Rd as Dongmen is one of the most popular night markets in Shenzhen, it has many snack bars and pubs. Thus, the traffic jam at night is heavier than daytime.

Circular heat chart: A circular heat chart encodes the periodic time distribution of data in a series of concentric rings, with each sector representing a time unit. Given a color scheme, the data trend could be observed with the cumulative effect on the circular axis over a period of time. For example, Liu et al. [28] creates a variant called heatmap style fingerprint, measuring the taxi-related attributes at a location (e.g., getting on/off taxis number). Fig. 11B shows an example of such a chart. The time coordinate is displayed on the circular axis (day of the week) and the outermost circumference (hour of the day).

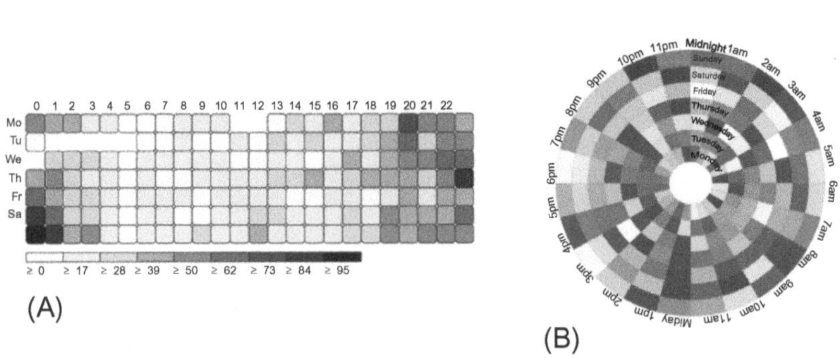

FIG. 11 (A) The calendar chart of traffic jam events. (B) Circular heat chart of random data.

The sector's position encodes the time, and the color of the sector encodes the value at that time.

Deck.gl[n]: Deck.gl is a library for visualizing large-scale data with high performance, especially for geographic data. We conclude three main characteristics about it here: (1) Supporting a wide variety of geographic data, e.g., GPS point data, trajectory data, geo-polygon data. (2) Multiple views to choose from, e.g., heatmap view, spatial clustering view, trip view. (3) Interactive, customized, and easy to start. Inherently supporting a range of visual parameters setting, which allows users to quickly customize visual views; and inherently supporting interactive visualization; users who are familiar with frontend programming could get up and running quickly.

4 Conclusion and future directions

The current development of MaaS is still in its infancy. Here, we propose some possible future directions and challenges for reference.

Data volume problem: As the cost of data collection and storage decreases, more and more data responding to MaaS operation is collected. Visualizing large-scale data in real-time is a considerable challenge. The most intuitive solution is sampling. However, this can lose much information, such as ignoring distant areas or areas with poor signal quality because of these areas' low data volume. A promising direction may be to develop better visualization-oriented sampling methods for different geographic data types.

Multiple-mode transportation data visualization: Currently, most of the visualization techniques are focused on a specific type of traffic, e.g., public transportation systems, taxis, sharing bikes, due to the data collected mode. There is no unified data collection and management platform, and each stakeholder is independent from each other. However, the future MaaS will string together the whole range of travel modes, including bike sharing, public transportation, car-sharing, online taxi-hailing, etc. Emerging visualization techniques to analyze the complete trip process will be a very new and promising direction.

Privacy issues: MaaS collects a large amount of personal trajectory data, how to ensure privacy issues in data visualization, for example, visualizing someone's travel model, home location, work location should be forbidden or severely restricted. What standards should be set, and what technologies should be developed to prevent this issue. One possible solution is to introduce permission management for the use of different categories of visualization views.

In summary, we first analyze the relationship between MaaS system and data visualization in this Chapter. Then introduce some key visualization technologies of current transportation domain into the multi-view frame of MaaS,

n. https://deck.gl/

including Demanders of Mobility, Supplier of Transportation Service, and City Manager. After that, three real-world application cases from different actor perspectives are presented. To help readers get started quickly, we also introduce some practical visualization tools and libraries that could be useful for MaaS visualization. Finally, we propose some future challenges and directions.

References

[1] Jittrapirom P, Caiati V, Feneri AM, Ebrahimigharehbaghi S, Alonso-González MJ, Narayan J. Mobility as a service: a critical review of definitions, assessments of schemes, and key challenges. Urban Plan Jun. 2017;2(2):13–25. https://doi.org/10.17645/up.v2i2.931.

[2] Wood J, Slingsby A, Dykes J. Visualizing the dynamics of London's bicycle hire scheme. Cartographica 2011;46(4):239–51.

[3] Wang F, et al. A visual reasoning approach for data-driven transport assessment on urban roads. In: VAST'14; Feb. 2015. p. 103–12. https://doi.org/10.1109/VAST.2014.7042486.

[4] Wunderlich M, Ballweg K, Fuchs G, von Landesberger T. Visualization of delay uncertainty and its impact on train trip planning: a design study. Comput Graph Forum Jun. 2017;36(3):317–28. https://doi.org/10.1111/cgf.13190.

[5] Vansteenwegen P, Souffriau W, Vanden Berghe G, Van Oudheusden D. The city trip planner: an expert system for tourists. Expert Syst Appl Jun. 2011;38(6):6540–6. https://doi.org/10.1016/j.eswa.2010.11.085.

[6] Bast H, Brosi P, Storandt S. TRAVIC: a visualization client for public transit data. In: GIS: proceedings of the ACM international symposium on advances in geographic information systems; Nov. 2014. p. 561–4. https://doi.org/10.1145/2666310.2666369. 4–7 November 2014.

[7] Pu J, Liu S, Ding Y, Qu H, Ni L. T-watcher: a new visual analytic system for effective traffic surveillance. In: Proceedings—IEEE international conference on mobile data management, vol 1; 2013. p. 127–36. https://doi.org/10.1109/MDM.2013.23.

[8] Ochiai Y, Nishimura J, Tomii N. Punctuality analysis by the microscopic simulation and visualization of web-based train information system data. In: WIT transactions on the built environment, 135; Jun. 2014. p. 537–49. https://doi.org/10.2495/CR140441.

[9] Graffagnino T. Ensuring timetable stability with train traffic data. In: WIT Transactions on the Built Environment, vol 127; 2012. p. 427–38. https://doi.org/10.2495/CR120361.

[10] Ushida K, Makino S, Tomii N. Increasing robustness of dense timetables by visualization of train traffic record data and Monte Carlo simulation. In: World congress railway research; 2011.

[11] Palomo C, Guo Z, Silva CT, Freire J. Visually exploring transportation schedules. TVCG'16 Jan. 2016;22(1):170–9. https://doi.org/10.1109/TVCG.2015.2467592.

[12] Weng D, et al. Towards better bus networks: a visual analytics approach. IEEE Trans Vis Comput Graph Feb. 2021;27(2):817–27. https://doi.org/10.1109/TVCG.2020.3030458.

[13] Ning W, et al. CheetahVIS: a visual analytical system for large urban bus data. In: Proceedings of the VLDB endowment, vol 13 (12); Aug. 2020. p. 2805–8. https://doi.org/10.14778/3415478.3415480.

[14] Ferreira N, Poco J, Vo HT, Freire J, Silva CT. Visual exploration of big spatio-temporal urban data: a study of New York city taxi trips. IEEE Trans Vis Comput Graph 2013;19(12):2149–58. https://doi.org/10.1109/TVCG.2013.226.

[15] Andrienko G, Andrienko N, Chen W, Maciejewski R, Zhao Y. Visual analytics of mobility and transportation: state of the art and further research directions. TIST'17 Aug. 2017;18(8):2232–49. https://doi.org/10.1109/TITS.2017.2683539.

[16] Hamad K, Quiroga C. Geovisualization of archived ITS data-case studies. IEEE Trans Intell Transp Syst Jan. 2016;17(1):104–12. https://doi.org/10.1109/TITS.2015.2460995.

[17] Zheng Y, Capra L, Wolfson O, Yang H. Urban computing: concepts, methodologies, and applications. ACM Trans Intell Syst Technol Sep. 2014;5(3). https://doi.org/10.1145/2629592.

[18] Song X, Zhang Q, Sekimoto Y, Shibasaki R. Prediction of human emergency behavior and their mobility following large-scale disaster. In: Proceedings of the ACM SIGKDD international conference on knowledge discovery and data mining; Aug. 2014. p. 5–14. https://doi.org/10.1145/2623330.2623628.

[19] Zheng K, Zheng Y, Yuan NJ, Shang S, Zhou X. Online discovery of gathering patterns over trajectories. IEEE Trans Knowl Data Eng Aug. 2014;26(8):1974–88. https://doi.org/10.1109/TKDE.2013.160.

[20] Zeng W, Fu C-W, Arisona SM, Qu H. Visualizing interchange patterns in massive movement data. Comput Graph Forum Jun. 2013;32(3 Pt 3):271–80. https://doi.org/10.1111/cgf.12114.

[21] Guo H, Wang Z, Yu B, Zhao H, Yuan X. TripVista: triple perspective visual trajectory analytics and its application on microscopic traffic data at a road intersection. In: IEEE pacific visualization symposium 2011, PacificVis 2011—Proceedings; 2011. p. 163–70. https://doi.org/10.1109/PACIFICVIS.2011.5742386.

[22] Wang Z, Lu M, Yuan X, Zhang J, Van De Wetering H. Visual traffic jam analysis based on trajectory data. IEEE Trans Vis Comput Graph 2013;19(12):2159–68. https://doi.org/10.1109/TVCG.2013.228.

[23] Yang C, Zhang Y, Tang B, Zhu M. Vaite: a visualization-assisted interactive big urban trajectory data exploration system. In: ICDE'19, 2019-April; Apr. 2019. p. 2036–9. https://doi.org/10.1109/ICDE.2019.00233.

[24] Anwar A, Nagel T, Ratti C. Traffic origins: a simple visualization technique to support traffic incident analysis. In: IEEE pacific visualization symposium; Mar. 2014. p. 316–9. https://doi.org/10.1109/PacificVis.2014.35.

[25] VanDaniker M. Visualizing real-time and archived traffic incident data. In: 2009 *IEEE international conference on information reuse and integration, IRI* 2009; 2009. p. 206–11. https://doi.org/10.1109/IRI.2009.5211552.

[26] Zeng W, Fu CW, Arisona SM, Erath A, Qu H. Visualizing mobility of public transportation system. TVCG'14 Dec. 2014;20(12):1833–42. https://doi.org/10.1109/TVCG.2014.2346893.

[27] Huang X, Zhao Y, Ma C, Yang J, Ye X, Zhang C. TrajGraph: a graph-based visual analytics approach to studying urban network centralities using taxi trajectory data. IEEE Trans Vis Comput Graph Jan. 2016;22(1):160–9. https://doi.org/10.1109/TVCG.2015.2467771.

[28] Liu S, Pu J, Luo Q, Qu H, Ni LM, Krishnan R. VAIT: a visual analytics system for metropolitan transportation. TITS'13 Dec. 2013;14(4):1586–96. https://doi.org/10.1109/TITS.2013.2263225.

Chapter 10

MaaS for sustainable urban development

Xiaoya Song[a], Rong Guo[a], and Haoran Zhang[b]
[a]Key Laboratory of Cold Region Urban and Rural Human Settlement Environment Science and Technology, Ministry of Industry and Information Technology, School of Architecture, Harbin Institute of Technology, Harbin, China, [b]Center for Spatial Information Science, The University of Tokyo, Kashiwa, Chiba, Japan

1 Introduction

The latest United Nations report points out that currently 55% of the world's population lives in urban areas, and it is estimated that 68% of the world's population could be urbanized by 2050 [1]. The urban scale has been expanding with the acceleration of the social economy and urbanization process [2, 3]. The continuous increase in the urbanization rate has triggered a series of environmental and development problems, such as the inadequacy of urban public services [4], increased time costs, urban environmental pollution [5], and urban road congestion. Moreover, as cities carry more and more people, the motorization rate, car ownership, and traffic accidents in cities worldwide continue to increase. At the same time, the deterioration of the transportation environment is also hindering sustainable urban development [6, 7]. Sustainable urban development has expanded to multiple dimensions such as society, economy, equity, and culture. Moreover, the realization of sustainable development is also closely related to national policies, technology, planning, and crowd behavior.

As the foundation of urban infrastructure, transportation is the basic framework of urban spatial form. Therefore, the urban transportation system has always been focused on sustainable urban development. Hence, the urban transportation system endeavors to meet residents' travel requirements [8] and ensure the smoothness of transportation under the premise of energy saving. The urban transportation system needs to focus on the requirements of different transportation facilities on urban roads [9], such as walking, bicycles, public transportation, private cars, and rail transportation. These transportation facilities need to be comprehensively analyzed and strive to promote the ability to provide matching support conditions for sustainable transportation.

Big Data and Mobility as a Service. https://doi.org/10.1016/B978-0-323-90169-7.00011-7

In order to reduce the ecological footprint of cities and control the growth of vehicles and related fuel consumption, increasing cities are beginning to explore sustainable transportation systems [10]. Transportation under the sustainable urban development means a relatively sounder transportation system, short travel time, and less transportation carbon emissions. It also implies a broader integration of public interests. In the context of the rapid development of the Internet of things, cloud computing, big data, artificial intelligence (AI), and other high-tech technologies, countries and regions have followed the informatization intending to break the traditional urban development model [11]. Sustainable urban development and intelligent technology research have gradually crossed over and extensively. Mobility as a service (MaaS) has first begun to be proposed at the European Union ITS Conference in Helsinki, Finland, in 2014. After nearly 7 years of development, MaaS has gradually become systematic. It is prone to gather and integrate dispatching information and data from multiple transportation operators, such as buses, subways, rentals, and shared bicycles, to form a large-scale dispatching system [12, 13]. Furthermore, it obtains and manages traffic-related services through an interactive electronic interface to meet the travel requirements [14].

MaaS aims to understand individual travel needs deeply. Based on the unified service system and platform of integrating various transportation modes, big data are utilized to make decision making by real-time control, allocating optimal resources [15], optimizing dispatching systems, and optimizing Seam connection. Thus, MaaS further increases the public transportation sharing rate, promotes urban green travel, alleviates urban traffic congestion [16], and improves the quality of public travel services.

2 MaaS interacted with urban traffic and space

MaaS covers several concepts that have been widely discussed in the transportation field in the past decades, such as transportation service integration, interconnection, service optimization, intelligent travel, and sustainability. As an innovative travel service solution, MaaS could assist transportation operators in obtaining user needs promptly and meeting user needs that traditional transportation operators cannot meet [17]. The ultimate vision is to achieve cooperation and interconnection in the transportation market and to provide users with a seamless travel service and a convenient, economical, and sustainable alternative to vehicle travel [16]. The characteristic differences between the MaaS service model and the traditional transportation service model are reflected in many aspects [18], including account management, travel planning and reservation, ticket payment methods, travel information services, travel service evaluation, etc.

Given that the requirements of sustainable urban development, it is needed to go beyond the transportation system itself, broaden the analysis dimension, and pay attention to people-oriented thinking. The sustainable development of urban space and transportation from a more diversified perspective is also considered. As a complex giant system [19], a city can be divided into three spaces: physical space, human space, and cyber space (Fig. 1).

First, the physical space refers to the physical environment of the city and the physical system corresponding to urban traffic. The optimal allocation of traffic resources and the accessibility of urban traffic are challenges faced by physical space. In addition, in the physical space, urban issues show differences on different scales. As future urban transportation pays more attention to human feelings, the research scale of future physical space needs to focus on refined units such as streets and intelligent communities. Second, the human space refers to the urban social space [20] with government, market, and individuals as the main body. The issue of rights and responsibilities in sustainable urban development needs to focus on the human space [21]. From an individual point of view, as a service, human behavior and perception are essential to transportation, and providing high-quality services is one of transportation and urban sustainable development goals [22]. Finally, cyber space refers to the digital-driven information space constructed by urban transportation. With the rapid development of communication network technology, more and more urban traffic information is stored in a digital and informative manner, and the industry chain of traffic informatization is formed. Information collection, data analysis, and decision support are all critical aspects of urban traffic. The intervention of MaaS has reconfigured the elements of urban transportation and promoted the interconnection and coordinated development of urban transportation networks and external transportation more effectively. For example, mobile phone signaling, bus IC card, and GPS data could realize the travel behavior perception of individuals and groups [23]. Unlike traditional public transportation, the MaaS concept pays more attention to the travel efficiency, traffic safety, comfort and overall experience of traffic users [13], and more reasonable costs. Therefore, it is bound to bring changes to the urban spatial structure with changes in traffic patterns.

2.1 Urban traffic structure

The core concept of MaaS is to provide services instead of occupying vehicles and occupy fewer social resources by transforming ownership rights into usage rights [24]. MaaS allocates the high cost of buying a car to each trip, significantly reducing travel costs, improving the efficiency of car use, and reducing ineffective trips and idle vehicles. From the current successful cases in North America and Europe, it can be seen that the promotion of MaaS has indeed reduced the number of private cars, reflecting the social effect of environmental protection and saving. MaaS encourages users to use more types of public transportation. Furthermore, with the increase in the number of public transportation users, public transportation stations in the traditional sense may be operated on notable routes along with public transportation stations with relatively

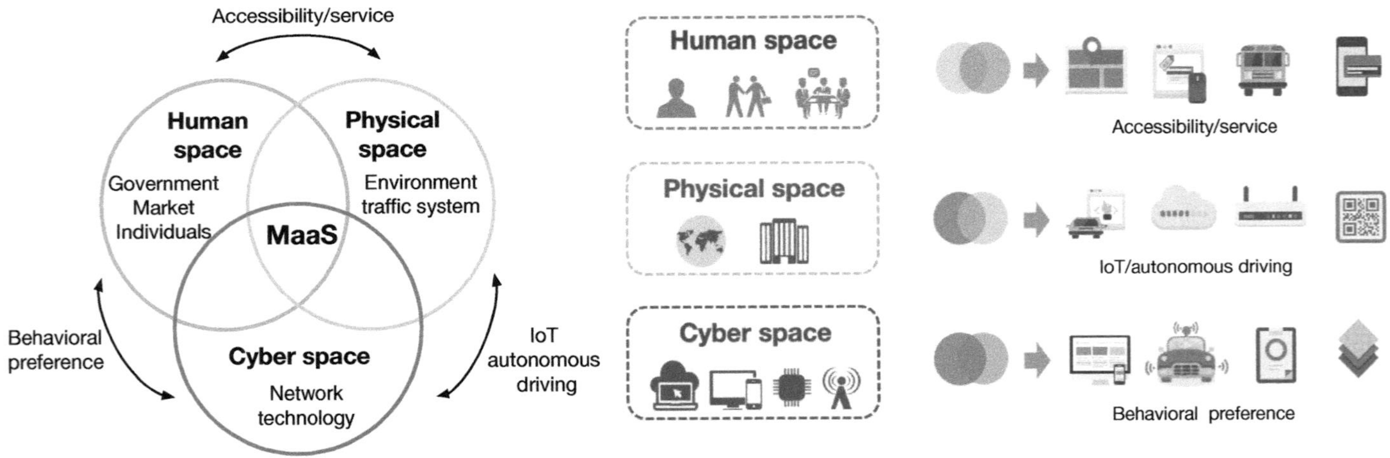

FIG. 1 Urban spaces of MaaS.

concentrated users after big data analysis and calculation. This measure avoids overloading or empty driving of traditional buses. The improvement of transportation efficiency and the decrease of empty vehicle rate could help reduce traffic congestion and shorten travel time and help further improve traffic efficiency and save social resources.

2.2 Urban spatial structure

With the coverage and promotion of the MaaS, traffic accessibility, convenience, and comfort have been improved, and efficient travel saves time and cost. At the same time, the intelligent combination of multiple transportation tools meets the comprehensive travel needs of multiple nodes and multiple goals. The promotion of MaaS makes it easier for people to reach multiple urban central nodes and increases the density and the dispersion of urban vitality centers. Furthermore, the overall urban land layout under the MaaS system could be more balanced, and the boundaries between urban core areas and remote areas could gradually weaken [25]. Relieving traffic could also shift to guided traffic, thereby increasing the use and participation of all citizens. The transportation nodes with shared travel service areas as the core could drive the development of surrounding urban land and changes in land use nature. The development model centered on rail transit hubs could gradually shift to a mobility-oriented development model.

MaaS emphasizes the integrated operation of multiple transportation carriers and the integration of transportation information and service resources so that passengers could use transportation resources more efficiently, comfortably, and conveniently. Even the private vehicle near the job-living space could significantly improve travel efficiency and comfort as a part of the MaaS system. The coordinated utilization of vehicles at close distances can significantly relieve the pressure of commuting traffic, balance the tidal traffic in the work and residence space, and effectively solve the separation of living space and workspace [15]. As the popularization and promotion of MaaS increase, the degree of vehicle sharing, vehicle idleness, and parking lots could be controlled accordingly. Therefore, parking lots at the center of urban vitality could be reduced or partly transferred to urban areas. The MaaS could increase the scale and proportion of urban public space to a certain extent. Some original parking lots and transportation service facilities could be transformed into urban public spaces. The parking spaces of some public buildings could also be transformed into urban public spaces, street parks, green spaces, and other public spaces.

3 Strategies for MaaS in urban sustainable development at multiple scales

Land use mainly concentrates on the process of managing the natural environment and transforming it into a built environment. It is closely related to disciplines such as urban planning, transportation, and geography [26]. In urban

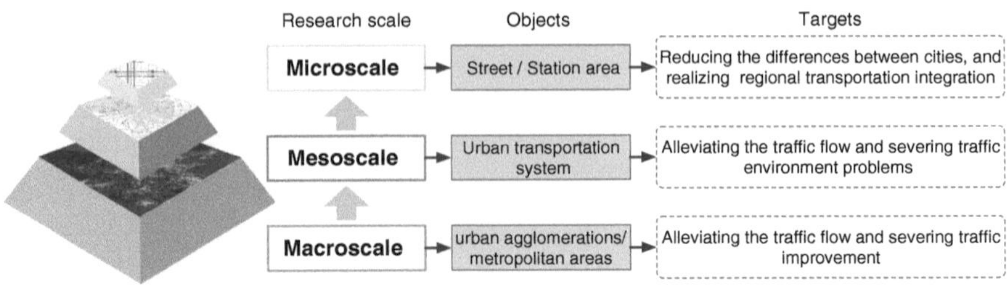

FIG. 2 MaaS in urban sustainable development at multiple scales.

planning, MaaS research also focuses on the integrated development of urban land use and transportation and its interactive effects (Fig. 2). Macroscopic research focuses on the impact of land use on urban traffic accessibility and crowd mobility and the impact of rail transit-based public transport development [25] and utilization patterns. The research on relatively small scale focuses on the analysis of community land mix, residential density, resident satisfaction, and travel behavior from the level of residential communities [27]. In recent years, analysis tools and methods for MaaS-related research such as urban dynamics, urban economics, and microsimulation analysis have been added.

3.1 Macroscale: Synergy between urban agglomerations and metropolitan areas

From the perspective of global trends, urban agglomerations have become the core areas that support the development of major economies in the country [28]. The future urban transportation system is facing the challenge of regional integration. The future urban development is no longer limited to the development of individual cities but has a trend of global urbanization, forming urban agglomerations radiating around the metropolis. As a vital lifeline of urban operation, transportation is also the primary carrier connecting the economy and service functions between cities and is a crucial infrastructure supporting the emergence and development of urban agglomerations. Therefore, on a macro scale, MaaS mainly concentrates on promoting urban agglomerations, reducing the differences between cities, and realizing and developing regional transportation integration [25]. For example, based on mobile phone signaling data, analysis of passenger flow distribution, urban spheres of influence, spatial corridors, and radiation competitiveness are used to support the collaborative relationship between rail and city. However, the current application of MaaS at the macroscale still has problems such as inconsistent data in time and space and uneven accuracy. Furthermore, subject to cost and calculation reasons, it is not easy to achieve continuous tracking. Therefore, MaaS should realize the observation and analysis of the characteristics of the crowd, vehicle movement, and migration for the macrospatial level, such as urban agglomerations and metropolitan areas.

3.2 Mesoscale: Optimization of internal resources in cities

The city is regarded as a spatial agglomeration. The "center-periphery" theory of new economic geography believes that even if two symmetrical areas have the same initial conditions, as long as there are increasing returns to scale and transportation costs, the economy could be agglomerated in self-evolution [29]. Moreover, these two areas could, respectively, develop into core and surrounding areas. Due to the positive factors such as population, resources, and information, the centripetal force for development could be formed by the scale economies effect. However, when the urban scale exceeds a specific limit, it could turn to its opposite: diseconomies of scale or crowding effects could appear.

In consequence, the loss of urban operation efficiency, pollution, traffic congestion, crime, and other negative factors are highlighted, forming the centrifugal force of urban development. When the urban scale expands to a certain extent, to reduce the congestion effect and overcome the diseconomies of scale, part of the economic activities and population could be scattered to the periphery and concentrated again, forming a polycentric pattern. The "multicenter" urban structure has a more vital carrying capacity than the "single center" and can achieve spatial balance and sustainability [30]. Therefore, the spatial structure of a "distributed and compact" multicenter city could be relatively in accordance with the space requirements of a sustainable city. In order to alleviate the increasing traffic flow and severe traffic environment problems in single-center cities and to guide the development of the city to a multicenter spatial structure, MaaS, as a carrier of sustainable transportation development, is gradually promoted at the mesoscale.

3.3 Microscale: The refinement of urban streets

As a critical urban public space, streets are in urgent need of refined and quantitative traffic analysis. The collection of multisource data such as street view data, mobile phone signaling data, POI, and multichannel public participation are applied to the microlevel. MaaS uses multisource data to analyze facilities, street characteristics [31], and citizen's activities, identify blind areas of accessibility, and provide decision support for street-scale traffic connections and spatial quality improvement. For example, AI technology is used to structure streetscape data and evaluate streetscape quality; citizens' proposals for street-crossing facilities, pedestrian spaces, bicycle lanes, bicycle parking, and other issues are collected to strengthen public participation in the refined urban governance. In addition, to better highlight the requirements of sustainable development, the design of green traffic space at the street scale should be emphasized. Based on avoiding the impact of green belts and related green plants on traffic safety, the green area needs to be increased to meet the requirements of sustainable development.

4 Case study

Lately, given that transportation has become a leading component of urban GHG, combining built environment with big data of mobility plays a pivotal role for urban sustainable performance assessment, which is crucial to urban sustainability. Existing studies reveal that the unbalance of urban resource allocation in the high-density areas is inevitable and mainly reflected when developing transportation infrastructure and the corresponding urban functional facilities. However, few studies aimed at measuring the urban sustainable performance in the perspective of the interaction between urban development and human mobility on microscale. This case study aims to measure the sustainable urban performance in the interaction between urban development and transportation and provides more specific support for policies. The unbalance between urban development and development of transportation facilities is proposed to be investigated through the travel attraction, and their internal relations is further explored on microscale.

In this case study, an MaaS model was proposed to measure and discuss the problem focusing on entire urban CAZs mentioned earlier. In addition, since applying GPS data to urban event analysis, travel modes detection, traffic emissions, etc. could contribute to the understanding of urban activities and human mobility [32], GPS data are proposed to be adopted as reference to evaluate the vehicle emissions in this case. As shown in Fig. 3, the method consists of four steps: (1) study area and function identification; (2) potential traffic attraction assessment; (3) transport emissions calculation; and (4) principal component clustering analysis.

In this study, Tokyo special wards are chosen as our study area, endeavoring to identify which areas have the potential for further sustainable redevelopment, and to support the promotion on low-carbon travel mode. In order to analyze the attraction of different facility types in each region, in this research, road network is collected from Digital Road Map road network and PoI data from Tele-Point data. The number of CAZs in this study is extracted from the whole zones with the famous DBSCAN clustering method. In Tokyo special wards, there are 413 CAZs. For each CAZ, the facilities are extracted by a 500 m shed and group the facilities into five clusters. For the vehicle emissions, the result in our previous research is used in this study [11]. "Konzatsu-Tokei (R)" data [10] refer to people flows data collected by individual location data sent from mobile phone under users' consent, through applications provided by NTT DOCOMO, INC [10]. Those data are processed collectively and statistically in order to conceal the private information. Original location data are GPS data (latitude, longitude) sent in about every a minimum period of 5 min and do not include the information to specify individual.

First, the study area shed was identified by the road network, and the PoI data were classified. Second, a metric named PTA was proposed to measure the urban sustainable performance based on the improved gravitational model.

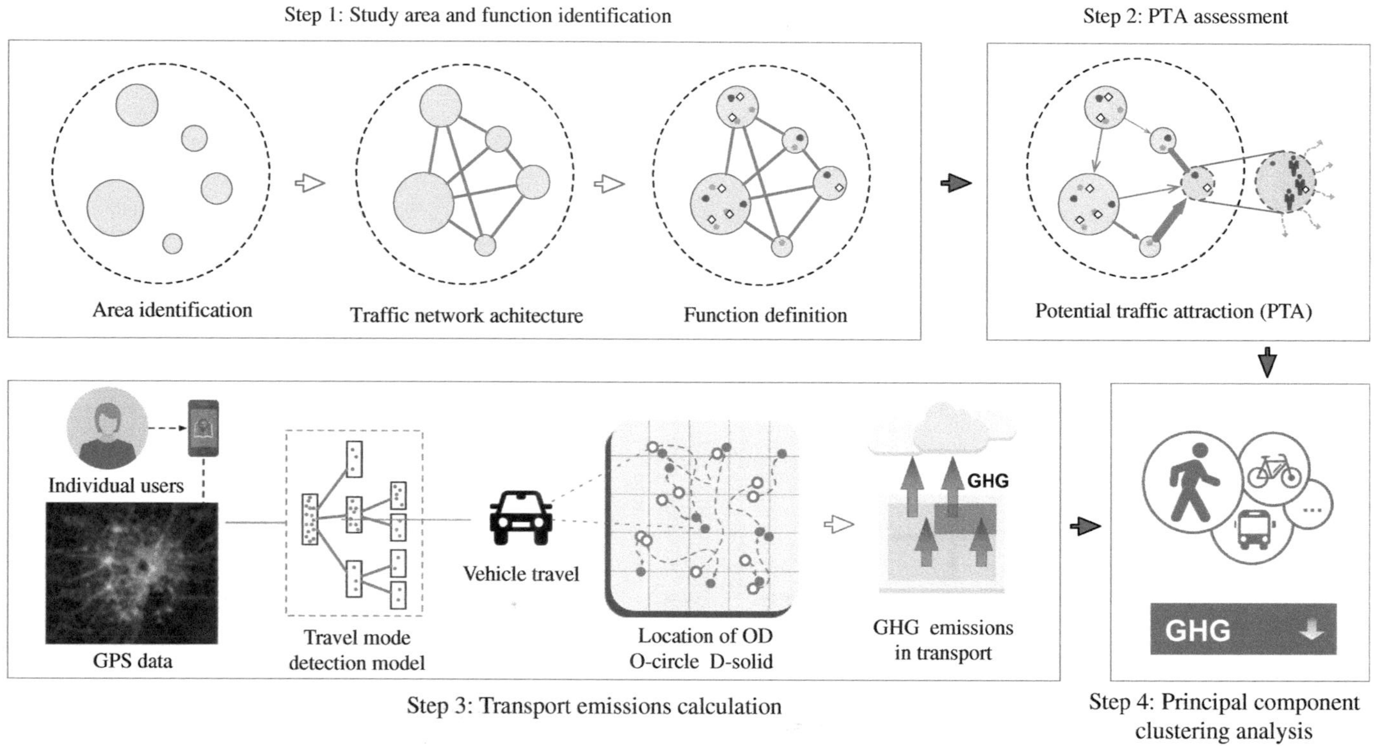

FIG. 3 Framework of potential traffic attraction model and MaaS utilization.

Then, the trajectories of vehicle traffic were detected and extracted using emission model. After that the principal component of the impact on vehicle emission was confirmed by the correlation analysis. Finally, a cluster analysis on the potential traffic attraction is proposed to identify the status of the integrated development. Although the selection criteria of the CAZs could vary according to the development state of each city, the procedure mentioned earlier could be a general technique and be applicable to various cities.

To conduct the further comparative analysis at the ward level, we averaged the five indicators of all CAZs in each ward as shown in Fig. 4. The high potential travel attraction in Tokyo generally is caused by lacks in functions of "restaurant, entertainment, and hotel." City centers (Chiyoda, Shinjuku, and Chuo) characterize more comprehensive urban function than remote areas. In addition, the degree of regional vehicle emissions increases radially from the center areas to outward. Then, the correlation between potential traffic attraction and vehicle emissions was confirmed with 0.563 of R (Fig. 5).

Based on the normalized potential travel attraction, the CAZs are classified into four clusters illustrated in Fig. 6: integrated CAZs, enhanced CAZs, potential CAZs, and under-supplied CAZs. To reduce integral vehicle emissions, relevant strategies are suggested accordingly in different sustainable development status. The principal directions on the sustainable development of the city were proposed according to the classification of CAZs. (1) Integrated CAZs: the facilities need to be improved in a refined and sustainability-oriented manner to maintain the balance between transportation and functions; (2) enhanced CAZs: the facilities in F_{CFR}, and F_{AS} need to be enhanced in the future; (3) potential CAZs: the synergistic redevelopment of F_{REH}, F_{CFR}, and F_{AS} is crucial for future development; (4) under-supplied CAZs: the primary focus

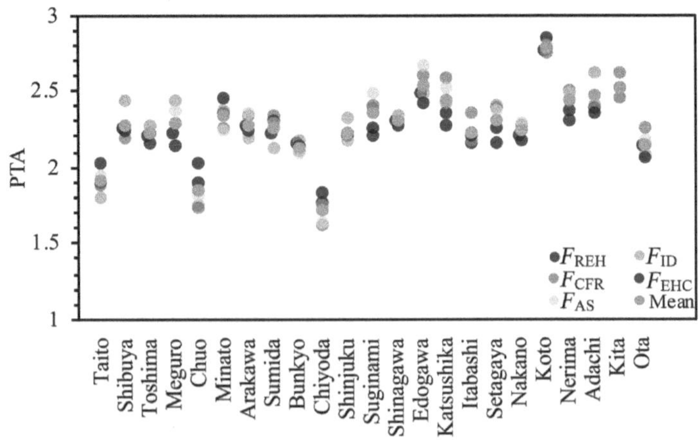

FIG. 4 Potential travel attraction of Tokyo special wards.

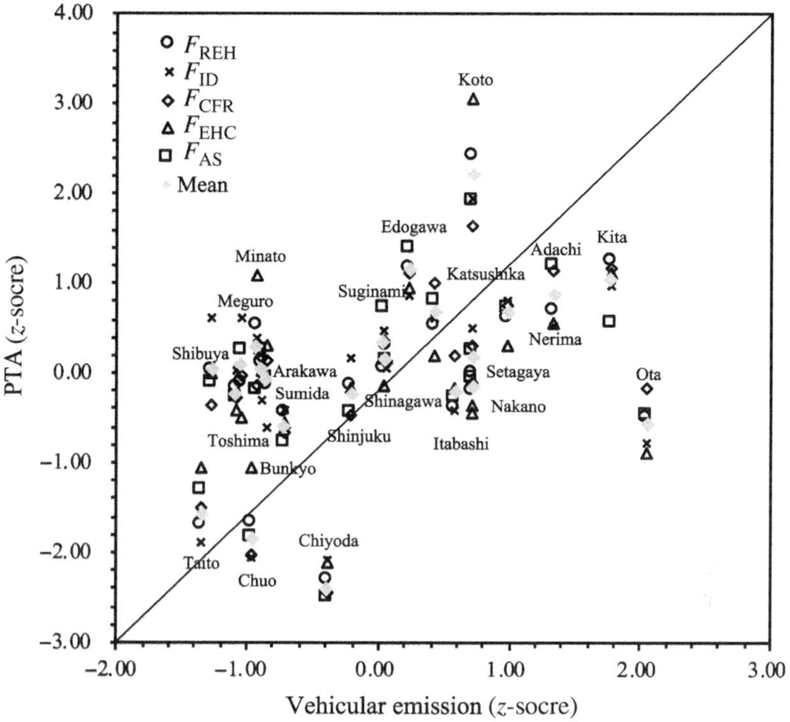

FIG. 5 PTA-vehicle emissions continuum of Tokyo special wards.

FIG. 6 Distribution of the four CAZs clusters in Tokyo.

should be on the urban basic functions as "restaurant, entertainment, and hotel" instead of other aspects.

The interaction between different CAZs development stages and the vehicle emissions reveals that from under-supplied CAZs to integrated CAZs, the proportion of low-carbon transportation increases gradually, whereas long-distance and high-emission vehicle travel decreases (Fig. 7). Although multiple essential dimensions of urban development unbalance and substantial data sets for transport emissions are introduced in this work, the proposed method still has limitations. This study does not distinguish the travel trajectories between weekdays and weekends as well as the different travel modes. In future research, a more comprehensive spatiotemporal analysis is planned to explore the urban development unbalance and transport emissions. Such a study would provide more specific guidelines for urban planning and MaaS.

5 Conclusion

MaaS mainly concentrates on integrating transportation services, sharing transportation, and the low carbonization of transportation methods. In the field of urban planning, the challenges faced by the application of MaaS to sustainable urban development include the application of monitoring, analysis, evaluation, and optimization in sustainable urban spatial structure, transportation infrastructure, ecosystems, and urban service connections, as well as ongoing modeling, simulation, decision support, and forecasting work. MaaS aims at sustainable urban development on the macro-, meso-, and microscales and carries out urban technological innovation. It optimizes the design and real-time monitoring of urban transportation systems and communication systems by developing simulation models and optimization algorithms. It provides more efficient and higher-quality public services and transportation services. Requirements are the starting point for selecting tools and taking appropriate actions. Therefore, MaaS for sustainable urban development is to develop intelligent systems and satisfy the reality of people and society. The application of MaaS is ultimately for the sustainable development of cities, in which the sustainability of transportation and the sustainability of society are two crucial components. Thus, the MaaS concept provides new ideas for improving the transportation service system, and it also raises more questions and considerations for the transformation of the city's public transportation system. Sustainable urban development could continue to explore human activities with the possibilities brought by new technologies such as MaaS.

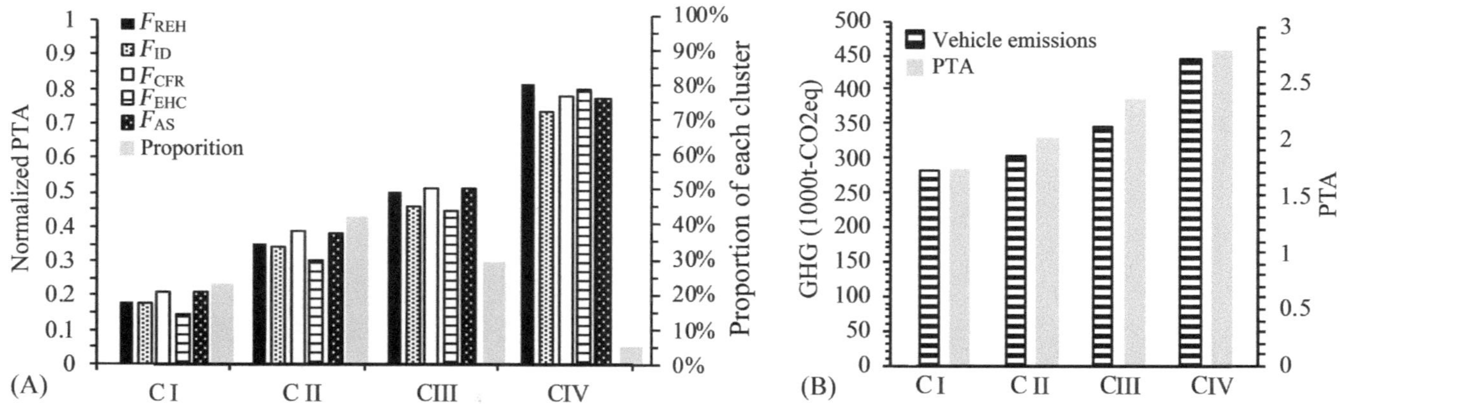

FIG. 7 Potential travel attraction and GHG emissions in four TOD clusters. (A) PTA and proportion of four clusters. (B) Vehicle emissions and PTA of four clusters.

References

[1] UN. Department of Economic and Social Affairs; 2018. https://www.un.org/development/desa/en/news/population/2018-revision-of-world-urbanization-prospects.html.

[2] Klopp JM, Petretta DL. The urban sustainable development goal: indicators, complexity and the politics of measuring cities. Cities 2017;63:92–7.

[3] Valencia SC, Simon D, Croese S, Nordqvist J, Oloko M, Sharma T, et al. Adapting the sustainable development goals and the new urban agenda to the city level: initial reflections from a comparative research project. Int J Urban Sustain Dev 2019;11(1):4–123.

[4] Song X, Guo R, Xia T, Guo Z, Long Y, Zhang H, et al. Mining urban sustainable performance: millions of GPS data reveal high-emission travel attraction in Tokyo. J Clean Prod 2020;242:118396.

[5] Guo R, Song X, Li P, Wu G, Guo Z. Large-scale and refined green space identification-based sustainable urban renewal mode assessment. Math Probl Eng 2020;2020.

[6] Ma CX, Peng FL. Monetary evaluation method of comprehensive benefits of complex underground roads for motor vehicles orienting urban sustainable development. Sustain Cities Soc 2021;65:102569.

[7] Balogun AL, Marks D, Sharma R, Shekhar H, Balmes C, Maheng D, et al. Assessing the potentials of digitalization as a tool for climate change adaptation and sustainable development in urban centres. Sustain Cities Soc 2020;53:101888.

[8] Wangai AW, Rohacs D, Boros A. Supporting the sustainable development of railway transport in developing countries. Sustainability 2020;12(9):3572.

[9] Wołek M, Wolański M, Bartłomiejczyk M, Wyszomirski O, Grzelec K, Hebel K. Ensuring sustainable development of urban public transport: a case study of the trolleybus system in Gdynia and Sopot (Poland). J Clean Prod 2021;279:123807.

[10] Zhang H, Song X, Long Y, Xia T, Fang K, Zheng J, et al. Mobile phone GPS data in urban bicycle-sharing: layout optimization and emissions reduction analysis. Appl Energy 2019;242:138–47.

[11] Zhang H, Song X, Xia T, Yuan M, Fan Z, Shibasaki R, et al. Battery electric vehicles in Japan: human mobile behavior based adoption potential analysis and policy target response. Appl Energy 2018;220:527–35.

[12] Yu Q, Zhang H, Li W, Sui Y, Song X, Yang D, et al. Mobile phone data in urban bicycle-sharing: market-oriented sub-area division and spatial analysis on emission reduction potentials. J Clean Prod 2020;254:119974.

[13] Strömberg H, Karlsson ICM, Sochor J. Inviting travelers to the smorgasbord of sustainable urban transport: evidence from a MaaS field trial. Transportation 2018;45(6):1655–70.

[14] Barreto L, Amaral A, Baltazar S. Urban mobility digitalization: towards mobility as a service (MaaS). In: 2018 International conference on intelligent systems (IS). IEEE; 2018. p. 850–5.

[15] Jang S, Caiati V, Rasouli S, Timmermans H, Choi K. Does MaaS contribute to sustainable transportation? A mode choice perspective. Int J Sustain Transp 2021;15(5):351–63.

[16] Arias-Molinares D, García-Palomares JC. Shared mobility development as key for prompting mobility as a service (MaaS) in urban areas: the case of Madrid. Case Stud Transp Policy 2020;8(3):846–59.

[17] Lopez-Carreiro I, Monzon A, Lopez E, Lopez-Lambas ME. Urban mobility in the digital era: an exploration of travellers' expectations of MaaS mobile-technologies. Technol Soc 2020;63:101392.

[18] Esztergár-Kiss D, Kerényi T. Creation of mobility packages based on the MaaS concept. Travel Behav Soc 2020;21:307–17.

[19] Zhao ZD, Huang ZG, Huang L, Liu H, Lai YC. Scaling and correlation of human movements in cyberspace and physical space. Phys Rev E 2014;90(5):050802.

[20] Zhuge H. Semantic linking through spaces for cyber-physical-socio intelligence: a methodology. Artif Intell 2011;175(5–6):988–251019.

[21] Fenton P, Chimenti G, Kanda W. The role of local government in governance and diffusion of mobility-as-a-service: exploring the views of MaaS stakeholders in Stockholm. J Environ Plan Manag 2020;63(14):2554–76.

[22] Smith G, Sochor J, Karlsson ICM. Intermediary MaaS integrators: a case study on hopes and fears. Transp Res Part A Policy Pract 2020;131:163–77.

[23] Li Y, Cook S, May A. Understanding the exclusion issues of mobility-as-a-service (MaaS): the potential problems of older travellers' involvement. In: International conference on human-computer interaction. Springer; 2020. p. 269–87.

[24] Canale A, Tesoriere G, Campisi T. The MAAS development as a mobility solution based on the individual needs of transport users. In: AIP conference proceedings, vol. 2186. AIP Publishing LLC; 2019. p. 160005.

[25] Aapaoja A, Eckhardt J, Nykänen L, Sochor J. MaaS service combinations for different geographical areas. In: 24th World congress on intelligent transportation systems, vol. 29; 2017. p. 8.

[26] Dong Y, Jin G, Deng X. Dynamic interactive effects of urban land-use efficiency, industrial transformation, and carbon emissions. J Clean Prod 2020;270:122547.

[27] Wong YZ, Hensher DA, Mulley C. Mobility as a service (MaaS): charting a future context. Transp Res Part A Policy Pract 2020;131:5–119.

[28] Fang C, Yu D. Urban agglomeration: an evolving concept of an emerging phenomenon. Landsc Urban Plan 2017;162:126–36.

[29] Krugman P. Increasing returns and economic geography. J Polit Econ 1991;99(3):483–99.

[30] Ma X, Chen X, Li X, Ding C, Wang Y. Sustainable station-level planning: an integrated transport and land use design model for transit-oriented development. J Clean Prod 2018; 170:1052–63.

[31] Alonso-González MJ, Hoogendoorn-Lanser S, van Oort N, Cats O, Hoogendoorn S. Drivers and barriers in adopting mobility as a service (MaaS)—a latent class cluster analysis of attitudes. Transp Res Part A Policy Pract 2020;132:378–401.

[32] Sui Y, Zhang H, Song X, Shao F, Yu X, Shibasaki R, et al. GPS data in urban online ride-hailing: a comparative analysis on fuel consumption and emissions. J Clean Prod 2019;227:495–505.

Index

Note: Page numbers followed by *f* indicate figures and *t* indicate tables.

D

J

K

L

M

R

S

Kolibri 360 GmbH
Ferdinand-Jühlke-Straße 7
99095 Erfurt, Deutschland
produktsicherheit@kolibri360.de